SCIENTISTS IN POWER

Lew Kowarski, Hans Halban, and Frédéric Joliot
around a cloud chamber, 1939

SCIENTISTS IN POWER

Spencer R. Weart

HARVARD UNIVERSITY PRESS

Cambridge, Massachusetts, and London, England

1979

Library of Congress Cataloging in Publication Data

Weart, Spencer R., 1942-
　　Scientists in power.

　　Bibliography: p.
　　Includes index.
　　1. Science—History.　2. Physics—History.
　　3. Atomic energy—France—History.　4. Atomic energy—
　　Social aspects—History.　I. Title
　Q125.W34　　539.7'62'0944　　78-21670
　　ISBN 0-674-79515-6

Les cerveaux sont fâcheusement pourvus d'estomacs.

—Jean Perrin

Preface

Knowledge is power itself; but in our times the scientists who discover knowledge seldom control the power that results. In many cases they have little control over even the funding and organization of their own work. A prolonged struggle to reverse this situation—to seize the reins of scientific life—was waged a generation ago by French scientists. In their moment of triumph some of them suddenly found a greater possibility, that of directing the advance not only of scientific work but of the whole of society. It was their science itself which gave them this chance by revealing uranium fission.

Frédéric Joliot and his collaborators at the Collège de France in Paris were the first group in the world, and for a year almost the only group, to take fission dead seriously. In the early months of 1939 they led the way in discovering how a chain reaction could be ignited in a mass of uranium. By the middle of the year the team had written a secret patent on a crude uranium bomb, sketched out a workable nuclear reactor, and persuaded the French government and private industry to underwrite their research. A year later they were well on their way to building a reactor and had made arrangements to try to keep some measure of control over the future development of nuclear energy. The German invasion of France cut short this work but did not stop it altogether, for two Frenchmen, Hans Halban and Lew Kowarski, escaped to Britain. Throughout the war they worked with considerable success to influence the growth of the British and American fission projects. As a direct result, at war's end Joliot was able to set up a nuclear energy program that eventually turned France, despite her severely limited means, into one of the nuclear powers. These actions of the Collège de France team—before, during, and after the war, in France and outside—are the ostensible subject of this book.

But because science has close connections with everything that happens in the modern world, the book also tells the history of particular social circles, of ideas and ideals, of bureaucratic organizations, and of technologies. Whereas a conventional history of science would have drawn chiefly upon laboratory notes, papers in scientific publications, and letters, my research required me to look as well into legal contracts, political party platforms, newspaper articles, diplomatic memoranda, budgets, engineering accounts, and annual reports of corporations. I attempted, while keeping in touch with laboratory life, to spread into the region where pure science, joined by technology, mingles with every current of society.

In the development of nuclear energy in France this little-understood region proved to be at the fulcrum of history. French scientists helped bring into being the nuclear weapons now wielded by France (in the earliest case of nuclear "proliferation") as well as by Great Britain, India, perhaps Israel and other nations too; recent major attempts to overcome the world's growing deficit of energy resources; the present organization of science in the democracies; and related matters that will largely determine our future. We must be curious to learn how such a set of objects—hundreds of power plants, thousands of bombs, tens of thousands of people massed in national establishments—can be traced back to a few people sitting at laboratory benches discussing the peculiar behavior of one type of atom.

And we should not stop there, but should ask how people came to be sitting at such benches discussing such things. I have a personal interest in this subject, having once been a working physicist myself. I am curious to learn how scientists, like workers in any trade, are encouraged or thwarted by their surroundings, to what extent society directs their actions, and to what extent they personally control events.

The detailed investigation of a particular case is one way to study such issues. Although I came upon French nuclear physics by accident, I could not have devised a richer and more compact example. Scientific discoveries and political arrangements which developed confusedly in other countries seemed almost transparent in France. The French history also offered me, as an American, the slight advantage of distance.

Different sections of the book were written in different ways. The opening and closing chapters express only preliminary thoughts about the tangled movements of the time. They probe into scientists' political life in little-known areas where I was only occasionally able to touch ground in documents which came to light at random; the interest of this subject may attract other scholars to fill out and correct my impressions. The central sections of the book, dealing with the years 1939-1940, are more narrow and detailed, although still hardly complete.

The study of fission was so new in the 1940s that it was not yet overburdened with intricate experimental and theoretical techniques, so any reader with normal intelligence and sympathy for science should be able to understand the technical points. The transition from pure physics to engineering introduced further technicalities, but these too should offer no serious difficulties. In fact the physics, which is the backbone of the story, is also its simplest part. Scientists are not bare intellects but whole human beings, members of a community preoccupied with status, money, and war. Everything that happened in this history flowed from their interaction with leaders of government, industry, and the military. At first sight this interaction might appear to be a simple one, the scientists dependent on the backing of more powerful figures. Yet in the long run the pressure did not all go one way—for the movements of society, like the movements of planets, are not indifferent to the laws of physics.

The student of recent events is sometimes granted the privilege of meeting subjects in the flesh. It is a painful privilege, for the subjects do not always conform with the student's prejudices, and they do always produce a disquieting impression that human lives are too rich to pin down in a paragraph. For sharing their time and knowledge with me, and in some cases for reading and commenting on drafts, I am most grateful to participants in the events I have partly recorded, including among others Jacques Allier, Pierre Auger, Pierre Biquard, Maurice Dodé, Otto Frisch, Wolfgang Gentner, Bertrand Goldschmidt, Jules Guéron, Lew Kowarski, Rudolf Peierls, Francis Perrin, Boris Pregel, Richard B. Roberts, Pierre Savel, Maurice Surdin, George P. Thompson, Harold Urey, Pierre Villon, Victor Weisskopf, and Eugene Wigner.

The accumulated experience of historians and archivists is essential to a study such as this. Those who were generous with their help included above all Monique Bordry (Radium Institute) and Dr. Charles Weiner (American Institute of Physics), as well as Michel Antoine (Archives Nationales), Lorna Arnold (United Kingdom Atomic Energy Authority), Judy Fox (University of California, Berkeley), Professor Margaret Gowing (Oxford), Professor Roger Hahn (University of California, Berkeley), Dr. Richard Hewlett (United States Department of Energy), Professor Mary Jo Nye (University of Kansas), Joan Warnow (American Institute of Physics), and finally but in many ways foremost Professor John L. Heilbron, who gave me the use of the facilities of the Center (now Office) for the History of Science and Technology of the University of California, Berkeley. Dr. Kowarski, Joliot's daughter Dr. Hélène Langevin, and Dr. Gertrud Weiss Szilard not only gave me access to their invaluable family papers but went to no little trouble to help me use them. And Virginia LaPlante, as editor, suggested improvements for every page.

Thanks are also due Mlle. Angremy, Aage Bohr, John Cash, Robert Clark, Lady Cockcroft, Columbia University Libraries, Eugène Cotton,

Mme. Demanée, James Fisk, Michael Friedman, Mme. Gauja, Philippe Halban, Marie-Paul Lacoste, Mary Lawrence, Rolf Neuhaus, Arthur Norberg, Marguerite Oswald, Harry Paul, Agnes Peterson, Newman Pettit, Mrs. E. Placzek, Edward Rees, Richard Roberts, Jean St. Clair, Glenn Seaborg, Robert Seidel, Charles Susskind, Merle Tuve, and Gene Wilson. For permission to publish excerpts I am grateful to many of the above and also to the American Institute of Physics, for portions of my "Scientists with a Secret," *Physics Today*, February 1976, © 1976 by the American Institute of Physics; to Gertrude Weiss Szilard and the MIT Press, for portions of Leo Szilard, *His Version of the Facts*, ed. Spencer Weart and Gertrude Weiss Szilard, © 1978 by Gertrude Weiss Szilard; and to Bertrand Goldschmidt and Librairie Arthème Fayard, for portions of *Les Rivalités Atomiques*, © 1967 by Librairie Arthème Fayard.

The support of research being one of my subjects, I should say a word about the funding underlying this project. Most important was a Research Apprenticeship under Roger Hahn from the Institute for International Studies, Berkeley, which allowed me to do the bulk of my European research and my first draft. Through the Office for the History of Science and Technology at Berkeley, I was aided by a grant from the Program in History and Philosophy of Science of the National Science Foundation. Equally necessary to my work was the existence of a number of archives, of which the Radium Institute is particularly to be commended for taking better care of its historical records than most laboratories. Drs. Kowarski and G. W. Szilard are notable for the personal care and funds that they too have put into organizing the records in their possession. The editing of the Szilard papers, in which I participated, was also supported by the National Science Foundation. Essential to my work at every stage were the collections and facilities of the American Institute of Physics' Center for History of Physics in New York City. Finally, my work has been supported throughout by my family's savings and time.

Contents

THE FORT DE CHATILLON

ILLUSTRATIONS

The
Radium Institute

Pierre and Marie Curie in their garden, c. 1900

1

Professors and Politicians

To search out incalculable powers in the heart of the atom is one thing; to earn a living is something else. Each year Marie Curie gave some would-be physicist a chance to do both, hiring him as her laboratory assistant so he could have a chance at a career in science. Her choice for 1925, Frédéric Joliot, was an elegant young man who had little training in scientific research. With rough, handsome features under wavy dark hair, outgoing and persuasive, "Fred" soon formed a bond with Madame Curie's other assistant, her daughter Irène, heiress apparent to the family's talents and position. A plainspoken woman disdainful of appearances, two years older than Joliot, Irène Curie seemed his opposite in everything except a taste for science and outdoor sports. Within two years they married.

Many in the Paris scientific establishment slighted Joliot as an upstart. His parents had few intellectual credentials, and he had been educated as an engineer rather than a scientist. Yet now he was settled in one of the leading laboratories of Paris, Madame Curie's Radium Institute. He was called Irène Curie's "prince consort." Always one who wanted to be admired, Joliot had some bitter moments. He devoted himself to experimental studies of radioactivity, and soon Madame Curie could say proudly, "The boy is a skyrocket." Within ten years of his arrival at the Institute his work would raise him to the highest scientific circles; within fifteen years he would be one of a handful of scientists who would develop nuclear energy; and within twenty he would be among the most powerful men in France.[1]

Joliot's rise marked the crest of a rise in all French science. Both these advances were promoted by a close-knit group of professors and politicians who enthusiastically managed young scientists' careers and the career of science as a whole. Known to Irène Curie from her childhood, the group had taken form a generation earlier in the midst of radical scientific and political changes. The route Joliot and his colleagues followed with their nuclear physics was marked out in advance by this older generation, so it is important to see who they were, what their place in French society was, and what they thought their place ought to be.

A visitor who entered the garden behind Pierre and Marie Curie's house on a Sunday afternoon at the turn of the century might have discovered a number of young physicists and chemists sitting in the shade of a small cherry tree, eagerly discussing the latest problems of science, the arts, and politics. At the center would be the Curies themselves, two powerful intellects, he grave and penetrating, she intense and tenacious. Next to them one might find the chemist and artist Georges Urbain, the shy physicist Aimé Cotton, and the radiochemist André Debierne. Their earnest conversations would be punctuated by the cries of children, for the Curies' daughters, Irène and Eve, would be playing with the children of their parents' best friends, Paul Langevin and Jean Perrin.[2]

In 1900 Langevin was a serious young man whose profound black eyes revealed a sensitive mind in waiting behind his neat mustaches and goatee. Erect and reserved, he was sometimes mistaken for an officer in mufti. He had been a student of Pierre Curie's at the Paris Municipal School of Industrial Physics and Chemistry, where the chief product was engineers. But Curie had convinced him that science was a higher calling, so on graduation Langevin entered the elite Ecole Normale Supérieure, emerging three years later a brilliant physicist. In 1903 he began teaching at the prestigious Collège de France, supplementing this by succeeding Pierre Curie as professor of physics at the School of Industrial Physics and Chemistry. In 1906 Curie died in a tragic accident, and within a few years Langevin followed his honored teacher in another capacity: he became Marie Curie's confidant, sharing the passions she otherwise locked up within an increasingly somber demeanor. In both positions Langevin proved important to Joliot's career. Joliot too was educated at the School of Industrial Physics and Chemistry and, upon graduation, was ready to begin an engineering career until Langevin, recognizing his student's talents, turned him toward science by securing his appointment as Marie Curie's assistant.

Jean Perrin would eventually play an equally important role in guiding Joliot's progress. Around the turn of the century the Perrins had settled next door to the Curies, and Madame Perrin became Marie Curie's best friend. Perrin, a professor of physical chemistry, had already been Langevin's best friend since first meeting him as a student at the Ecole Normale. Their temperaments, however, were opposite, for Perrin was genial and exuberant. He attracted attention and affection with his strikingly beautiful face, framed by profuse chestnut curls, and his mischievous wit. Once Urbain carved a bust of him as Dionysus.

In the years after 1900 these intellectuals found many opportunities to see one another. Besides the Sunday afternoons at the Curies' house there were regular meetings every week which drew ten of fifteen people alternately to Perrin's house or to the apartment of Emile Borel, a rising mathematician. Tall and black bearded, as handsome as Perrin, Borel too attracted friendship with his looks and lively spirit.[3] Many of these friends were alumni of the Ecole Normale, where a student body of a few score, selected from the best young men of France by arduous competitive examinations, crowded together in little rooms for advanced study and schoolboy horseplay. The "Normaliens" worked furiously, argued intensely, and knew one another intimately. For generations the school had been famous as a nursery of intellectual brilliance and camaraderie.

Most of these scientists came from families in the middle and intellectual classes, having for example a doctor or a schoolteacher for a father, and few of them expected to rise much above this socially. If they performed particularly well in their studies, caught the eye of some senior professor, and showed unusual scientific talent, they might hope, perhaps around the age of forty, to win a major Paris professorship with a decent salary and recognized dignity. Meanwhile they devoted most of their hours to lectures and research. The public pictured such scientists as estimable toilers, useful as instructors of the young and as cultural adornments; scientists were expected to produce incomprehensible discoveries, but they were also expected to be somehow incomplete as people, absentminded and ignorant of real life. Although this public image accurately fitted some professors, there were others, and the Curies' friends above all, who were determined to be more than lecture-hall and laboratory drudges.

Discussions at the homes of the Curies, the Perrins, and the Borels often turned to politics. This was not idle conversation, for in France science and politics traditionally interlocked. Scientists needed politicians, politicians needed scientists, and at the turn of the century these needs were growing stronger. The government was in the hands of left-center republicans, who were occupied with an old and vicious struggle against backward-looking Catholics and monarchists. Needing an ideology of their own to combat these enemies, the anticlerical republicans welcomed support from anticlerical scientists. Society, these scientists declared, would advance in peace only as scientific truth displaced "medieval superstitions," for science alone consisted of objective facts on which all could agree.

Marcelin Berthelot, a chemist who became minister of public instruction and, briefly, of foreign affairs, was the epitome of the old republicans who controlled French science in these years. He was known not only for his discoveries in pure science but for his research in aid of the chemical industry and for his studies of explosives during the Franco-Prussian War of 1870. It was with emotion that Berthelot, who had thrown off the overpious Catholicism of his upbringing, regretted that "emancipating science" had been "bowed for centuries beneath the oppressive yoke of theocracy." By day the manager of a large laboratory devoted to the pursuit of fact, Berthelot at evening permitted himself to dream of the future, captivating his companions at fashionable soirées with visions of the year 2000 when science would have turned the earth into a garden where humanity would live amid abundance and joy. The source of both truth and prosperity, science had a clear right to moral and material power over the entire human domain. This was not only an ideology of science but a political ideal, as the minister made clear, for example, when he announced, "A society may live without official religion, without supernatural support, without prejudices, in a word drawing all its principles from the single authority of science and reason."

This was not abstract talk but a partisan program. Paul Bert, a physiologist and politician, stated the program more explicitly: "War on monks!" Catholic spokesmen struck back, maintaining that science was morally "bankrupt."[4]

A main battlefield in this struggle was the national educational system, under which nearly all French scientists held their jobs as professors. It was here that the republicans expected science to defeat outdated "mysticism"; moreover it would unite French society, since scientific truth was something everyone would accept. The government gave science the place of honor in its schools and universities, while Catholic professors, even Catholic scientists, who voiced any opposition found their careers blocked by the Ministry of Public Instruction. Scientists who upheld or at least did not attack the republican ideology were given excellent new laboratories, the equal of those anywhere in the world.

Not all French scientists approved of these policies, for in science, as in every sector of French society, there were conservative as well as liberal groups. The Curie circle descended from anticlerical republicans; other scientists kept to Catholicism and had reservations about mass democracy. The Curie circle looked back to the free-thinking Ecole Normale; others came from bastions of the military and managerial classes, such as the Ecole Polytechnique, an elite engineering school run under rigid discipline by the Army. The generation that produced Perrin and Langevin also produced, for example, Maurice de Broglie, a graduate of the Naval Academy and heir to a dukedom, who set up his own private laboratory where he contributed to the study of X-rays. Over the years relations between the de Broglie group and the Normaliens were always proper. But the conservatives, ill at ease in the climate of the Third Republic, kept to themselves, and the liberal professors rarely saw a need to court them.[5]

The Curie circle was caught up in political struggle when the Dreyfus Affair exploded shortly before 1900. Perrin, Langevin, Borel, and their friends joined the older republicans and anticlericals in the fight to win justice for Dreyfus and in the process to discredit the conservatives. In a few years the liberal forces won a complete victory, leading to a bitter and final separation of church and state. As religion withdrew in defeat from politics, it also separated from science. It was as if the priest and the physicist, dismayed at their argument, politely agreed that they no longer had anything important to say to one another. French scientists stopped proclaiming that science should displace religion in the moral leadership of society, not because they had begun to doubt this, but because they felt that religion had already lost its predominance.

With the nineteenth-century anticlerical battle behind them, French intellectuals faced twentieth-century social issues. A new industrial proletariat was struggling to advance in the teeth of entrenched and reactionary men of property. Because the growth of industry was a consequence of technology and therefore inseparable from science, the scientists could not avoid becoming entangled in the new issues and ideologies. The Curie circle, invigorated by their taste of political action and victory in the Dreyfus Affair, were prepared for the challenge.

That some of the Normaliens could go beyond traditional republicanism was largely the work of one man, a missionary for socialism, Lucien Herr. In 1887 Herr had taken the modest post of librarian at the Ecole Normale—his letter of application said that this job was his entire dream and ambition—for he had seen that from this library he could guide the reading and the thought of the next generation of French leaders. Among the students struck by Herr's social message were Langevin, whose father was a simple artisan, and Perrin, whose widowed mother made a meager living from a tobacco shop. Around the time of the Dreyfus Affair the two young physicists frequented a little bookstore in the Latin Quarter which Herr and others made a headquarters for propagating socialism. Here they often saw one of the most ardent of Herr's Normalien converts, Léon Blum, a rising journalist and politician.[6] From time to time Blum and others like him dropped in at the weekly gatherings in Perrin's or Borel's apartments.

Some of the liberal scientists took a direct part in politics. Thus Paul Painlevé, a Republican-Socialist and occasional visitor at Borel's, gave up a brilliant career in mathematics to win the Latin Quarter's seat in parliament. But most of the professors hesitated to play politics. Their real interest was science, and they may have agreed with the mathematical physicist Henri Poincaré that, since science needs money, it must never become possible for people who might someday be in power to call science their enemy. According to the novelist Emile Zola, even a scientist-politician like Berthelot was in the end above social issues, being "Republican under the Republic, but fully prepared to serve science under any master at all." Such scientists, Zola noted, held that one step forward in scientific discovery would do more to bring about the "City of Justice and Truth" than a century of politics and social revolution. Pierre Curie was another liberal who held this belief, and when his students discussed social problems, he used to tell them: "In the end it's not worth your while to fuss too much over all that. Physicists will resolve those difficulties simply by doing away with the problem, because they will eventually create enough wealth for everyone."[7]

By the end of the nineteenth century there was abundant evidence for the practical value of science. The French saw their stunning defeat by Prussia in 1870 and their increasing inferiority to the Germans in the new chemical and electrical industries as harsh lessons in the strength that a nation could draw from its scientists. Meanwhile science was making life better with improved agricultural methods and fertilizers, handsome new dyestuffs and plastics, and better means of preventing or curing disease. Paris was transformed by physics as brilliant strings of gas and electric lights spread down the boulevards where electric tramways bustled to and fro. Scientific knowledge was freeing mankind not only from ignorance and superstition but, perhaps more important, from many material burdens. It was little wonder that scientists felt their work deserved the highest respect and attention. For the most part Perrin, Langevin, the Curies, and their friends kept to their laboratories, but they did not believe that this meant they were turning their backs on human problems. They were convinced that their scientific work, beyond advancing their careers, beyond giving them personal joy in discovery, would lead to moral and material progress. It was this hope that, as Perrin later said, was "in the most elevated sense of the word, our religion."[8]

Nothing raised the physicists' hopes so much as the study of atoms and radioactivity. Like a fabulous philosophers' stone, radioactivity could transmute the elements; it had profound effects on living tissue; it drew on forces stronger than any known before. It fascinated researchers and the public with hints about the inner structure of matter and energy and with promises of practical applications.

In the late nineteenth century physicists had taken more and more interest in the various sorts of radiation that appeared in their electrical experiments. They had produced different kinds of light and radio waves, immaterial yet of great practical value; and they had produced sprays of material particles, for example the submicroscopic particles called electrons, whose electrical charge Perrin detected at the start of his scientific career in 1895. The next year a German physicist made a sensation by discovering still another kind of radiation, a sort of invisible light, the invaluable X-rays. With all this interest in different rays, it became more and more likely that someone would notice that certain minerals emit radiation spontaneously. It fell to the Paris professor Henri Becquerel to discover that some sort of penetrating radiation is emitted naturally by a scarce metallic element, uranium. This surprising activity, which the Curies named radioactivity, was soon detected in another rare metal, thorium. By 1900 the young Marie Curie, aided by her husband, had found two entirely new elements by laboriously tracking down the radioactivity of minerals: first she discovered polonium, naming it after her native Poland, and then came radium. Fiercely radioactive and incredibly scarce, radium caught the world's attention. Not only was it a scientific mystery, but its powers, people thought, were bound to have some sort of practical application.

Over the next few years scientists of several nations untangled the various radioactive substances and their radiations. The substances all proved to be among the heaviest elements, atom for atom more massive than lead. Each of these elements was characterized by the particular mixture of rays that it emitted. There could be just three kinds of these, which at first were denoted simply alpha, beta, and gamma. It turned out that gamma rays, the most penetrating, were much like X-rays; beta rays were streams of particles, in fact nothing but fast-moving electrons; alpha rays, the least penetrating, were streams of heavier particles. All of these had effects on living flesh, and if the effects were often destructive, that could be an advantage if the flesh was cancerous.

The emission of all these rays had an important implication which did not become clear until 1903 when a young student of Pierre Curie's attempted a delicate experiment. The student intended to find out whether the rays from radium exerted any pressure, but his apparatus acted strangely. Pierre Curie looked into the problem and noticed that the sensitive apparatus had been accidentally upset by a source of heat. It was the radium.[9] Curie and his student (and independently a pair of scientists working in Montreal, Ernest Rutherford and Frederick Soddy) soon found that radium was always a degree or so warmer than its surroundings. This simple fact was consequential, for it meant that radium was a natural source of power simply as it sat on the shelf. The substance gave forth energy of itself during its entire existence.

One possible explanation, which the Curies found attractive, was that radioactive atoms somehow tapped invisible sources of energy, perhaps an unknown radiation from the sun or beyond. The idea must have appealed to their hope of liberating mankind from material need. Around this time Pierre Curie, in a letter to Berthelot, told the old prophet of science about a dream of his that somehow chemical reactions could be devised to trap and utilize the limitless energy of sunlight which is lost on striking the earth.[10] The Curies may have wondered whether radioactivity would unlock such a possibility.

But over the next decade the work of many scientists proved that the truth was even stranger. All the energy of radioactivity came from within the atom. Moreover, when an atom emitted a ray, it might at that moment transmute from one element to another. For example, a radium atom, upon emitting an alpha particle, would transform itself into radon, a radioactive gas; this gas would transmute into polonium, which would eventually turn into ordinary lead. Unraveling the numerous interlocking modes of transmutation was a long task for the physicists and chemists who began to specialize in the field. The details were often confusing and tedious, but the workers were drawn on by a feeling that they were dealing with the secrets of matter. And there was no guessing how useful these processes might someday be.

Or how destructive. Already in 1905 Pierre Curie sensed this risk, as he explained in his Nobel Prize address: "One may suppose how radium could become very dangerous in criminal hands, and here we might ask ourselves if it is to mankind's advantage to know the secrets of nature, if we are mature enough to profit by them or if that knowledge will harm us." His answer, which was also the only public hint he ever gave of his political feelings, was optimistic: "The example of Nobel's discoveries is characteristic, for powerful explosives have enabled men to do admirable works. They are also a terrible means of destruction in the hands of the great criminals who are drawing the peoples towards war. I am one of those who believe with Nobel that Mankind will derive more good than harm from new discoveries."[11]

Pierre Curie was not the only one to sense the approach of war and the military power of science and technology. The public could read crude science fiction speculating about future world wars in which victory would depend on scientific novelties. "How awful!" exclaimed the heroine of a 1911 novel as guided missiles with exploding "radium" warheads destroyed a German fleet: "It really seems something more than human that you should be sitting in this little room and dealing out more death and destruction than might happen in a couple of hours' bombardment, by just pressing a button."[12]

The world war that broke forth in 1914 would do much to bring on that day, but at first it simply scattered the scientists from their laboratories. In each of the warring nations most young scientists became junior officers with traditional military duties; many descended into the lethal machinery of the Front. Although the French government organized an advisory commission of eminent older scientists, it neglected to give them authority or money, so at first science had little to do with the war. Scientists could only hope to contribute as soldiers, unless they acted on their own, as did some of the Curies' friends who began to figure out ways to locate enemy artillery by tracking the sound of its firing. Marie Curie herself worked indefatigably to bring medical X-ray apparatus to the Front, aided by her teenaged daughter Irène; the joint labor made a bond between them. But most scientists' special skills were wasted. As Perrin wrote in 1915 to Langevin, who was then a sergeant in an engineering battalion, "If you could only use your intelligence as a PHYSICIST you could be of more service than a thousand sergeants."[13] Once the war dragged on into a siege, this truth would become clear to all.

Late on an April afternoon in 1915, French Algerian troops in the trenches at Ypres beheld a yellow-green fog billow up from the German lines and drift languidly downwind toward them. When the weird clouds enveloped them the French soldiers began to cough and die. Many fled, trying to outrun the advancing fog, often in vain. Soon an entire division was gone, and German troops advanced past thousands of helplessly gasping or motionless young men.

The fog was chlorine gas, released in a field trial at the instigation of a German chemistry professor. The experiment produced a tactical victory for the Germans and, on the French side, consternation and immediate attention to science. Within a week of the disaster the French organized a Commission on Poison Gas on which scientists and military representatives met as equals.[14] The Commission's first task was to organize a defense against chlorine, for otherwise the war could soon be lost. They enlisted, for example, André Mayer, a physiologist from the Collège de France who had been called up in 1914 to serve as an ordinary doctor at the Front. Suddenly he found himself back in his laboratory, commanding far greater means than he had known in peacetime, scrambling to improvise gas masks. With his clipped mustache and clean features, Mayer was the image of a dashing army officer—but with a white laboratory coat thrown over his uniform. Meanwhile another group of scientists sought ways to strike back. Their chief was the chemist Charles Moureu, whose broad, balding forehead, beard, and mustaches gave him the traditional dignified appearance of the professor. In September Moureu's team struck at the enemy with shells containing a new gas suggested by the Curies' friend Urbain.[15]

For two years new gases and defenses against them were brought out by scientists on both sides of the lines, and a rough equilibrium was struck. Such close relations developed among scientists, industrialists, and the military that a new gas or mask could be rushed into large-scale production and then into combat within months. After a time the incessant improvements became routine, and gas came to seem only one more of the many harassments of trench warfare.

Then in July 1917, again at Ypres, another new gas fell upon French

troops. For an hour or two they experienced only minor discomfort behind their masks. Then their skin, even where it had been protected by
clothing or boots, began to break out into painful blisters, their eyes became inflamed, and they vomited. By the time they reached their casualty
stations, the victims were nearly blind and had to be led. Within four
days French scientists identified the new weapon as dichlorodiethyl sul-
fide, baptized *ypérite* or "mustard gas" by the troops. The French had
already thought of this chemical and tested it in Moureu's and Mayer's
laboratories, but had rejected it as not generally fatal. Now, as hospitals
began to fill up with young men who had fallen to the chemical, partly
blind, covered with persistent sores, coughing in agony from ruined
lungs, the Allies learned that an incapacitated soldier is more trouble to
his army than a dead one. Research again became frantic. Limited pro-
tective measures were devised, and by mid-1918 the French scientists,
working closely with industry, had their own mustard gas factories in
full operation; it was the German soldiers' turn to suffer.[16]

Gas warfare was not the only way science took a hand in battle. A few
months after the first Ypres attack the mathematician Painlevé, who had
risen swiftly in politics and was now a minister in the government, decided to set up a Directorate of Inventions Concerning the National Defense. Calling together Perrin, Borel, and many other scientist friends, he
set them to work sorting through ideas submitted by French inventors
and devising their own tools for warfare. The scientists in the Directorate
of Inventions and their counterparts in other countries reshaped artillery
operations, antisubmarine defense, and many other military activities,
reinforcing the lesson that scientists can add to national strength not only
in a vague long-term sense but in the most direct and deadly ways. It was
the start of a characteristic of twentieth century history, the warfare of
technologies.[17]

Yet at first none of this seemed to change the little society of French
scientists. The Directorate of Inventions and other bodies were largely
the scientists' own creation and simply reinforced and extended the
bonds that had formed before the war. This network of relationships
survived the conflict with all the tenacity of old school friendships. After
the victory, those young scientists who were left alive returned quietly to
their peacetime research.

By this time the group that had once gathered in the Curies' garden had
broadened its circle of acquaintances to interlock with many of the similar intellectual circles which flourished in Paris. And not only in Paris,
for during their leisurely summer vacations many of them could be found
in L'Arcouest. This Breton fishing village had been discovered at the turn
of the century by several professors, who founded a colony which came

to be called the *Sorbonne-plage*, "Sorbonne-by-the-sea." Far from the crowded streets of the Latin Quarter, L'Arcouest was made up of cottages of peasants and fisherfolk, scattered across windy slopes which led down to a sea studded with islands. In the 1920s Marie Curie built a summer house there, and near her settled Perrin, Borel, Debierne, and others. The professors and their children spent each August swimming, sailing, and talking together, expanding and reaffirming their friendships.[18]

Back in Paris as well as at L'Arcouest, the scientists found time outside their work not only for political and social discussions but for amateur painting and music and much more. Borel, for example, while pouring forth a stream of exciting lectures and mathematical discoveries, also paid close attention to literary affairs, maintained friendships with poets and politicians, and served as mayor of his home town in the south. Paris, which for centuries had sheltered perhaps the largest concentration of scientists assembled in any city as only one facet of its preeminence in all fields of civilization, was now in its last brilliant years as the unchallenged center of world culture. In cafés and walkup apartments its artistic and intellectual life burned and sparkled.

This brilliance was unsupported by any solid economic base. Behind the boulevards lived hundreds of thousands of foreign laborers not far from starvation, and among them would-be painters, poets, and physicists went hungry too. But the voters, ignoring the depressed condition of labor and a corresponding stagnation of enterprise, brought to power in the first postwar elections an unusually conservative coalition of parties, the National Bloc. Bitterly disappointed, the working class launched violent strikes and threatened revolution in the Russian fashion, but managed only to split their own ranks, Socialists against Communists. France's political life became more fragmented than ever, with factions heaping abuse on "selfish plutocrats" and "reactionaries" or "Bolsheviks" and "foreigners."

The scientists of Perrin and Borel's circle retained the political interests and socialist sympathies of their youth. Outstanding in his conviction was Langevin. When union agitation crested in an attempt at a general strike in May 1920 and the School of Industrial Physics and Chemistry was closed down so its students could work as strikebreakers, Langevin publicly protested the closing. And when a group of French sailors were brought to trial for having mutinied during a campaign against the Bolsheviks in the Black Sea, Langevin spoke out in their defense. His friends, while not going so far, tended to agree; Marie Curie wrote Irène that popular support for the Black Sea mutineers would "give the financiers and reactionaries something to think about."[19]

Marie, Irène, and Pierre Curie in their garden, 1904

Jean Perrin (right) using an apparatus to detect enemy airplanes by sound during First World War

Paul Langevin,
1922

Marie Curie,
1935

Another who admired Langevin's stand was his student at the School
of Industrial Physics and Chemistry, Joliot, just emerging from a comfortable bourgeois childhood. Joliot's father, who had concealed his proletarian background, had left the boy little beyond a permanent love for
hunting, fishing, and the outdoors. But Joliot's mother, forceful and
cultured, had taught her son a firm republicanism that left him open to
Langevin's influence. The teacher, who would fix his students with a
grave stare and puff on a cigar as he explicated a difficult point, moved
Joliot with the lucidity and passion of both his scientific and his political
thought.[20]

Although Langevin's occasional public statements put him to the left of
most scientists, many of them were concerned about the government's
policies. The National Bloc, far too conservative to pay for reconstruction by taxing the wealthy, and far too chauvinistic to be reconciled with
the national enemy, hoped to solve both problems at once by bleeding
Germany. Since the Germans were unable to pay their "debts," French
finances were crippled. This brought hardship to many groups but above
all to government employees, a group that included nearly all the scientists.

At the turn of the century the senior professors, if not wealthy, had at
least enjoyed the esteem of genteel society and a comfortable, secure live-
lihood. The new postwar society of masses and corporations did not
promise to honor professors in the old style nor even, given the financial
weakness of the government, to maintain their salaries and pensions
against a rapid inflation. By 1920 the Paris Faculty of Sciences could not
meet its debts. Membership in the French Physical Society, which had
doubled in the two decades preceding the war, dropped rapidly, and the
publication of scientific papers stagnated. The professors saw all this not
simply as a personal inconvenience but as a threat to French science itself.[21]

The war had undermined the French scientific community not only
financially but in a more direct way. Of the Sorbonne students who had
gone into uniform, some 40 percent were prevented by death or wounds
from ever returning to their studies. The Ecole Normale, whose tiny student body had supplied perhaps a third of all the physics professors of
France, suffered even more dreadful losses: of the 161 Normaliens in the
classes of 1911-1913, 61 were wounded and another 81 killed or missing
in action.[22] The carnage deprived France of a generation of scientists.

War losses were not the only thing depleting the ranks of the next generation of professors. The survivors, even those who had once been preparing for a career in science, were not always inclined after four years in
the trenches to bury themselves in study. As the president of the Paris
Academy of Sciences declared in 1921: "We are all agreed in crying out in

alarm over the diminution of genuine scientific vocations . . . Many young people on whom we might have based some hopes have rushed into industry." While a decline in the prestige of science was one problem, there was a more direct difficulty, as another professor explained: "When a ship is going to sink, the rats abandon it. When a career is badly paid, people flee it."[23] The scientists' salaries, fixed by the government, were worth less in purchasing power every day, but industry was not so restrained. The intellectuals sadly noted that a foreman was now as well paid as a professor.

Anxious that their position was slipping, the scientists were at the same time more convinced than ever that it must be raised, for the war had redoubled their belief that science was essential for the nation. Mayer, for example, reflecting on his experiences in gas warfare, wrote in 1920, "This war, as everyone saw, was as much a war of scientists as a war of soldiers." Borel had similar feelings based on similar experiences. He believed that his studies on the sound location of gunfire, done in collaboration with his friends in the Directorate of Inventions, had "enabled the French artillery to dominate the enemy artillery in 1918." Mayer, Borel, and others pointed out that if scientists, well funded and organized, could do so much to save the nation from conquest, surely they could do still better things in peacetime if they were given adequate means.[24]

Finding their science both more promising and more endangered in the 1920s than ever before, many professors joined forces in a long campaign that began by praising research and ended by greatly expanding it. This campaign was a main force in the lives of Joliot and Irène Curie. Whenever the professors won a gain, the young pair were among the first to enjoy the fruits. If Joliot eventually wielded great power, it was thanks to his seniors, who fought many battles to put this power into his hands.

2

A Campaign for Science

How was French science to be rescued from decay and made powerful and prosperous? The professors did not quite know. Through the 1920s they would push this way and that, trying out one scheme after another, sometimes running up a blind alley but sometimes making real progress.

Their first impulse was simply to appeal to public opinion, as French scientists had done at various times in the past. This past was a weakness, for the public had grown used to cries of poverty from scientists. During the nineteenth century French professors had enjoyed enough support to make notable discoveries and build international reputations, but they had let it be known that they could have done better still, given more money. Pierre Curie, for example, on being installed in a substantial new laboratory in 1904, had told a newspaper reporter that his facilities were poorly funded. Yet French science did not seem to have suffered, and the fact that by 1920 it was sliding toward inferiority was not yet obvious even to all the professors. The public had largely accepted an image of their scientists as self-sacrificing geniuses, like Louis Pasteur or Claude Bernard in their earlier years, producing marvelous discoveries in tiny ill-furnished laboratories. Some added Pierre Curie to the list of martyrs, helped by a biography of him that Marie Curie wrote in the early 1920s, which deprecated the barren shed in which the couple had made their chief discoveries. Nobody in France who picked up an article titled "The Material Situation of Scientists" would have been shocked by the first sentence: "It's miserable."[1] That this should be so was a respected tradition, and many saw no need for change.

Postwar conditions were different, and so was the vehemence of the scientists' publicity. In speeches, books, magazines, and newspapers they declared that the young people's "desertion of the laboratories is a peril for our country." They seized every opportunity to tell the public how science had helped save the nation from conquest, and they insisted that without a vigorous scientific life a nation could also be conquered in peacetime industrial competition, which was expected to become fiercer than ever.[2] According to the French professors, the Germans understood the situation and were funding their own science with disturbing generosity.

The scientists hastened to add that applied research was not the heart of the matter, for pure research needed support even more. Most scientists believed, as Jean Perrin later declared, that disinterested scientific discovery "has been the principal factor, and perhaps the unique factor, in human progress." And pure science deserved support, he said, not only because it was fundamental for national defense and material progress but also because it was needed to maintain national prestige abroad and the glory of French culture at home; further, it was a worthy end in itself, necessary for the dignity of the human spirit.[3]

Perhaps the most influential writer of "scientific propaganda," as Mayer called it, was Charles Moureu, the chemist and gas warfare expert. In works like *Chemistry and the War* he declared over and over that "knowledge is power." He held out the promise that science would level mountains, subjugate the seas and bend the forces of the atmosphere to its will: "If Man is chained to Matter by an infinity of servitudes, it will be the task of Science to free him bit by bit. Thus the conquests of Science must be an essential factor in social transformations. More and more the task of politicians and diplomats—and on occasion, alas! of armies—will consist of harmonizing the existence of Nations with the economic consequences of scientific discoveries."[4]

Shortly after the Armistice, Moureu decided to make the acquaintance of Maurice Barrès, a strong-willed novelist, propagandist, and deputy, who was well-known as a favorite of Catholics and monarchists. Over a long lunch, Moureu spoke to Barrès' ardent nationalism and enlisted him as a champion for French science. Campaigning in the press and the Chamber, Barrès simply repeated the scientists' arguments in his own lush prose: laboratories were a "school of Truth and model of social discipline;" scientists could give France strength and "propaganda" among the other nations; science could solve the problem of class warfare over the distribution of wealth by creating "limitless" riches.[5] He was demonstrating that the ideology of science could fit as easily with conservative as with liberal views. Nevertheless, the other National Bloc deputies remained indifferent.

Help in arousing public opinion arrived from a direction even more unexpected than Barrès: the women of America. In response to a magazine campaign organized in 1920 by a journalist, Americans poured in money to buy Marie Curie a gram of radium. This fiercely radioactive element was an essential tool for her work on radioactivity, but each gram had to be extracted at heavy expense from tons of uranium ore and then was bid for by doctors who intended to use it in cancer and other therapies. A single gram cost the Americans $100,000. Their gesture spurred the French to try to be as generous. In early 1921 a French periodical organized a gala in Curie's honor at the Paris Opera, presenting cultural pleasures and an appeal for funds for her Radium Institute. A Curie Foundation was organized, to which Lazard Frères, Dr. Henri de Rothschild, and others gave substantial sums.[6] These principal contributors were Jewish; that was common in French philanthropy, and the Curie circle, unlike some in France, had always welcomed Jewish scientists and intellectuals.

Encouraged by the Curie appeal's success, the Paris Academy of Sciences held a meeting to inform the press that the means for scientific work were inadequate. The Academy organized a "Pasteur Day" for May 27, 1923, to attract donations on behalf of research. The publicity raised millions of francs for laboratory construction and equipment, including a share of over two million for physics alone. Much of this share went into an electromagnet twice the height of a man, one of the world's first centralized facilities for probing atoms, developed by the Curies' old friend Aimé Cotton. He meanwhile raised funds directly from industrialists and others, engraving their names alongside the device on a marble plaque. Since appeals to the public appeared to be working, the scientists set up a National Committee to Aid Scientific Research in hopes of converting their initial success into a steady flow of donations.[7]

None of this support was entirely new. Since about 1900 wealthy Frenchmen, urged on by scientist acquaintances, had made gifts to aid research. Endowments had been set aside for such institutions as the Academy of Sciences and the University of Paris Faculty of Sciences. There had also existed since the turn of the century a Caisse des Recherches Scientifiques (Scientific Research Foundation), supported on a modest scale by state and private donations and devoted chiefly to biomedical research and public health measures. The Curies and their friends had benefited more than most scientists from all these little prizes and subventions. But the funds that philanthropy had given to French science before the war had not been large, and now inflation was attacking the endowments. Around 1920, for example, when Marie Curie received a typical grant from the Caisse des Recherches Scientifiques, it was only enough to build two instruments for measuring radioactivity.[8]

More significant was the pattern that had been set for dividing up the money. Usually researchers would submit proposals of a general nature, eminent scientists would make recommendations, and a committee would decide. The same names turned up in many of the funding committees. Rule by interlocking committees was a common feature of French administration, from the cabinet to the universities, and had an ancient tradition in the Academy of Sciences. The pattern would later be repeated in nearly every organization set up to support science, not only in France but abroad. The appeal for funds spread by the scientists and their friends in the early 1920s did not propose anything new in institutional arrangements. All that was new was the force and persistence of the plea.

There was disagreement over exactly which public the professors should appeal to. There were some scientists, among them Langevin and Urbain, who believed that the future belonged to the working classes. Urbain feared that the masses had been disillusioned by the "amoral side of science," which had revealed itself in the mechanization of war and the "abuse of industrialism." He and like-minded scientists spent part of their time publicizing pure science among "the people." But most leading professors followed more traditional lines, agreeing with Barrès in looking for support to the "leading and opulent classes."[9]

Among the most opulent to be drawn to the cause was Edmond de Rothschild, an aged connoisseur of art and music. Impressed by the important part that scientific technology had taken in the war, in 1921 he created a ten million franc foundation to aid research in subjects related to industry and defense. André Job, who had been a leading gas warfare chemist, persuaded him to divert the endowment toward pure research, with the funds administered by the usual committee of professors. This gift did not end the scientists' attentions to Rothschild. For example, in 1925 Perrin and some of his colleagues invited the baron to visit Perrin's laboratory on the top floor of the Sorbonne. The octogenarian was barely able to surmount the narrow, precipitous, and unlighted stairs which led to the laboratory, and on reaching it at last, he remarked that "science was decidedly ill-housed in France," which encouraged his hosts. He then asked to be shown Brownian motion. In this subtle phenomenon, whose study had made Perrin's fame, minute particles of resin are suspended in a liquid; the multiple invisible impacts of the liquid's atoms cause the particles to execute a jiggling dance, which can be made visible through a suitable microscope. The scientists scrambled to try to arrange the viewing for Rothschild, and to their relief the apparatus and the old man's eyes proved equal to the assignment. He felt as if he were witnessing the play of mechanical interactions which might lie at the basis of life itself. Such experiences reminded him of dinner conversations he had held long before with the physiologist Claude Bernard about the secret chemical forces of life, and Rothschild decided to endow an institution to pursue such research. In 1926 on the advice of Perrin and others he gave fifty million francs for an Insitute of Physicochemical Biology. The same year, urged on by Borel, he completed his coverage of the sciences by helping the Rockefeller Foundation create the Henri Poincaré Institute of Mathematics and Mathematical Physics. Borel became the director.[10]

Rothschild's new institutes went up on the Rue Pierre-Curie in the heart of the Latin Quarter, a few minutes' walk from the Sorbonne, the Ecole Normale, the Ecole Polytechnique, and the School of Industrial Physics and Chemistry (see figure). At the end of this short street stood Madame Curie's Radium Institute, completed just before the war with money from the government and the Pasteur Foundation, and behind the Radium Institute was Perrin's Institute of Physical Chemistry, built in the 1920s with university funds. Anyone looking down the quiet Rue Pierre-Curie, the south side lined with grave brick institutes, might feel confident that French science was well provided for.

But these buildings were as far as such philanthropy could go, and they were not far enough. Laboratories and their equipment were certainly important, yet as Borel put it, science is done in the first place not with apparatus but with brains. And as Perrin added, *les cerveaux sont fâcheusement pourvus d'estomacs*—"brains, annoyingly enough, are furnished with stomachs."[11] Wealthy friends of science might build all the laboratories they pleased, but there would still be the problem of filling them with researchers. It was not a simple problem.

Traditionally there had been little opportunity in France, as in most countries, to follow a life devoted to research alone. A young person who wished to be a scientist and who did not enjoy independent means of support would have to become a professor, devoting many hours to teaching and administrative duties within the university system. Not every good researcher could function as a teacher, and even those who could were often stuck for decades in inferior positions, out of reach of the research funds which were the prerogative of the full professor. It was no wonder that many of the brightest young people declined to attempt a career in science. This was a problem that money could not solve, so long as it was spent in the traditional way. But nothing could have been harder than changing the way a bureaucracy spent money.[12]

Perhaps the educational system could be induced, given more funds, to add positions that would be administered in the usual way but devoted exclusively to research rather than teaching. This idea was officially launched in 1921 at the Academy of Sciences when its president endorsed a plan to set up research positions equivalent to professorships, each with its own laboratory.[13] Even this modest plan, however, got no further than a draft bill in parliament. The only result of all this talk came with Rothschild's 1921 endowment, for the scientists persuaded him to put it entirely into scholarships to support young people doing research.

Joliot was among the students who were attracted by the promise of a career in industry. In 1922 he spent six agreeable weeks in an industrial laboratory, but during his subsequent term in military service he began to wonder whether a scientific career would suit him better. Hesitating to

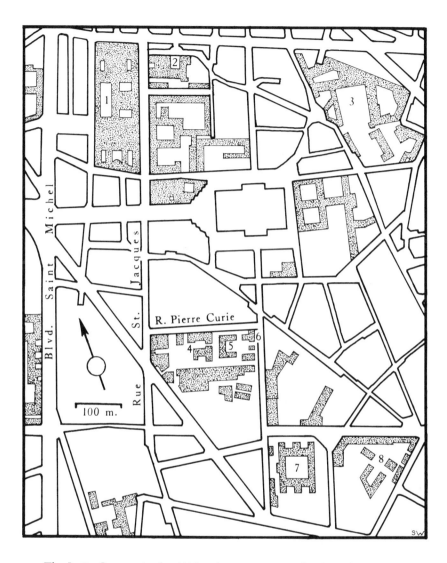

The Latin Quarter in the 1930s, showing major schools and research institutes: (1) Sorbonne, (2) Collège de France, (3) Ecole Polytechnique, (4) Institute of Physicochemical Biology and Henri Poincaré Institute, (5) Institute of Physical Chemistry, (6) Curie Laboratory of Radium Institute, (7) Ecole Normale Supérieure, (8) Municipal School of Industrial Physics and Chemistry.

approach his famous teacher Langevin, he allowed a school friend to make inquiries for both of them. Soon Langevin summoned the pair to his apartment and told them that, thanks to the Rothschild grants, they had a chance to enter scientific careers, although these would never pay them as well as industry. A few days later he told Joliot that Marie Curie would take him on as a junior assistant.[14]

Curie's other assistant, supported from the Institute's regular funds, was her daughter. Irène had been brought up so far as possible outside the rigid French school system and had been taught to respect nothing but the truth. She had spent her youth sharing her mother's work and concerns. By her twenties Irène Curie had already taken on some of the features that came to her mother only with age: awkwardness, shapeless dress, a remote and thoughtful gaze. Her fellow-workers at the Institute found her cold and strange. Working alongside her now was Fred Joliot, cheerful, careful of his looks, and popular with young men and women. Everyone was surprised when the two assistants found that they complemented one another. It took a long time for the gossip about the marriage to quiet. A decade later Joliot could still exclaim to an acquaintance: "Why are people so nasty? Why do they claim that I don't love my wife, and that I have married her just for the sake of my career? But I do love my wife. I love her very much." The feeling that the Parisian scientists would not entirely accept him was something Joliot never quite forgot.[15]

Joliot's job at the Institute paid little, so he also began to teach a course on electrical measurements at a private school, and his fee plus Irène's small salary allowed the couple to get by. Besides Joliot's substantial duties as teacher and research assistant, he had to work his way through a series of courses and examinations to qualify as a candidate for a doctorate. Despite all these tasks he published several scientific papers and in 1930 finished his doctor's thesis.

The philanthropy of Rothschild, though it helped some outstanding young people like Joliot, was not sufficient to support all the researchers that the French professors wanted, and few other wealthy Frenchmen followed Rothschild's example. To get the money they needed, French professors were expected to turn to the government. It was a discouraging prospect, for as late as 1920 the principle of research as separate from teaching was still not formally accepted. Aside from the little Caisse des Recherches Scientifiques there was no line in the government budget for anything like research, and the scientists often had to support their work, as Perrin admitted, by "illegal diversions" of appropriations intended for teaching.[16]

Alongside their appeals for public support, the scientists called on the government to accept the principle of a regular allocation for research within the budget of the Ministry of National Education. Spearheaded by

Barrès in the Chamber and supported by deputies on both the right and the left, the symbolic measure was pushed through. Meanwhile on March 30, 1921, the day after Marie Curie's day of honor at the Opera, a reorganization of the Caisse was completed. That year it could spend 48,000 francs for the physical sciences—more than twice as much as the total it had doled out during its entire previous existence, and including 13,000 francs for the Curie laboratory at the Radium Institute. Two years later the Caisse was giving the physical sciences ten times more, including some 22,000 francs for the Curie laboratory. Of this amount, 4000 francs (about 250 US dollars of the period) went to support the research of Irène Curie.[17] Such grants were still modest, but the principle of separate research funding, if not of fully independent research positions, had won a toehold.

This program was brought to a halt by the continuing collapse of the state's finances. By 1925 the value of the franc was again sliding steadily downward, and the scientists and their laboratories, dependent on inflexible government salaries and budgets, were badly hurt. As one deputy noted, the question now was not whether to support research, but where to find the money.[18]

With philanthropists unwilling and the government unable to help, some scientists began to look toward industry. After all, the great corporations had benefited enormously from science, and many were even based entirely on past discoveries. Could industrialists be persuaded that it was in their own interest to subsidize future research? This hope was a response to an international movement. Particularly in the United States and Germany, industrial corporations had supported some research both at the universities and within their own laboratories since the early years of the century, and by the 1920s these corporations were becoming generous patrons of research. The patronage was not caused simply by the expansion of capitalism, for the connection between scientific institutes and industrial organizations was also promoted in the Soviet Union.[19]

But when asked to dig into his pocket, the French industrialist turned away. Typically the paternalistic boss of a family business, he would rather preserve his share of a stable market than pay for research to develop something new; politically well to the right, he neither understood nor admired the scientists' plans for transforming industrial society. Many academics, for their part, still nourished a traditional disinterest in anything tainted by commercialism. The appeals of the 1920s to French industry brought little direct support.[20]

If industrialists would not willingly aid research, perhaps they could be forced to pay. Borel declared that considering how much wealth science ultimately created, it was "scandalous" that scientists were reduced to begging "for meager subsidies for scientific research, scarcely daring to

mention the personal poverty of researchers." Just as inventors of practical devices could grow wealthy through patents, so scientists should be
legally entitled to demand remuneration from any industry that made
money from their discoveries. The chance of getting such windfall pay-
ments would help attract young people back to science. Borel even dared
hope that all scientists as a body might claim a share of the profits of industries that gained from scientific work.[21]

This idea that scientists had "property rights" in their discoveries attracted many liberal scientists in the early 1920s. Marie Curie cam-
paigned for it tirelessly in the Commission on Intellectual Cooperation of
the League of Nations, winning international support. The problems
came with putting the idea into practice. Who would set the fees and dis-
tribute the proceeds? The suggestion that industrialists should take
charge was rejected outright by the intellectuals. Then would the scientists themselves control their "property" by licensing only certain applications of scientific discoveries? This idea offended many, and everyone conceded that it was unthinkable for scientists to exercise such power
over the uses of science. The longer the proposals circulated, the less they
were welcomed. By 1927, when a last attempt was made to establish scientific property rights, American scientists declared that "the protection
of the property rights of scientists in their discoveries is not feasible, and
is of doubtful desirability."[22]

All this effort was not wasted, however, for scientists were getting
used to the idea that the rich profits from their work should not flow only
into the pockets of businessmen. They would remember this the next
time they made a discovery with revolutionary practical applications—
such as nuclear fission. Meanwhile, if scientific property rights could not
be established, perhaps the government would tax industry more directly. This alternative had been out of the question under the National
Bloc, but the National Bloc had fallen.

In 1924 French voters sent a group of liberal parties, the Cartel of the
Left, to power. It was a high point for the "Republic of Professors." Presiding over the Chamber was the mathematician Painlevé; the new premier was a professor of literature, Edouard Herriot; and Léon Blum, the
old friend of Perrin and Borel, had risen to the head of a revitalized Socialist party. Borel himself was one of the newly elected deputies.

Though they had political disagreements, Painlevé, Herriot, Blum,
and Borel, old friends and Normaliens all, shared the scientists' vision of
a future, perhaps not too far off, made happy and free by the advance of
science. Such a belief in progress was the one fixed point in the confused
doctrines of the Radicals, the centrist party that held the balance of
power. A bare trust in progess, however, is no substitute for a coherent,
up-to-date vision of the future, the sort of vision a party needs to mobi-

lize enthusiasm and guide policy, and the Radicals never found such a vision. Their leader Herriot, when he spoke of science, simply followed the line laid down by Berthelot a generation earlier, adding only that the experience of the war showed that research should be centrally organized and coordinated with industry. If he had developed these views consistently, Herriot would have given strong attention to research. But he was not a consistent man: to friends he spoke like an erudite humanist; to voters he acted like a large, fleshy peasant in rumpled clothes; and he felt neither a personal nor a political urge to push an expensive science bill through the Chamber of Deputies.[23]

As in most parliamentary bodies, decisions were rarely made on the floor. The Chamber was a fine stage for speeches, shouted insults, and votes, but most of the difficult work took place in small committees and informal meetings, escaping the eyes of the public (and often of the historian). Borel thrived in these private encounters. A commanding figure, his massive head accentuated by a close-clipped beard once black but now grizzled, his eyes alight, Borel combined intellectual force with the warmth and peasant toughness of his forebears in the southern highlands. There were few he could not either befriend or sweep before him. He had many chances to do both in the spirited weekly lunches that he and his wife, a vivacious novelist, now held for politicians. Borel could continue his research in pure mathematics while maneuvering as a deputy without feeling divided, for in both careers he was moved by the same ideals. He believed that political combinations, even wars and revolutions, were passing accidents whose effects were small compared to the transformations which would be wrought by the applications of science. He therefore made it his first business as a politician to come to the aid of research.[24]

Shortly after the elections Borel offered an ingenious solution to the problem of funding science, taking advantage of a discussion in the Chamber on the "apprenticeship tax," a small levy on the payroll of French industry which raised some ten million francs a year to support technical training. Declaring that without scientific research much of modern industry would not exist, Borel proposed that the tax be raised by one-seventh and the increase be handed over to scientists. Since he had already won over Painlevé, Blum, and Herriot, the measure was quickly passed.[25] Borel's "Penny for the Laboratories" provided the long-sought mechanism to raise money for research. The professors' campaign in the press and parliament subsided.

Two years later conservatives returned to power and at last reorganized the budget and stabilized the franc. With the nation's finances in order and with Borel's addition to the apprenticeship tax in place, the scientists were better off than ever. Appropriations for laboratories set-

tled down at a satisfactory level. To pick one grant at random among many, in 1928 the Caisse des Recherches Scientifiques gave 40,000 francs to Marie Curie for her laboratory, more than it had given to all physicists together in 1921. The president of the Academy of Sciences in 1928 announced that the worries had been premature: students were returning to the study of pure science.[26]

But it was the president of the Academy who was premature. He could be optimistic only because he overlooked persistent weaknesses in the organization and support of French science, weaknesses that would be revealed once the high summer of the twenties was over. The professors would have to venture still deeper into politics before these problems could be overcome.

By 1930 the leading positions in French science were falling into the hands of a new generation: the members of the old Curie circle, now in their late fifties or early sixties. At last near the top of the educational bureaucracy, they were more determined than ever to change it, for they had kept the reformist spirit of their youth. Moreover, they were coming to realize that in fields like modern physics other countries had outstripped France. The new generation, and particularly Langevin, had long been fighting a frustrating campaign against elderly professors who looked on relativity theory and quantum mechanics with uncomprehending suspicion. An exciting new physics had been built around these theories, yet hardly any Frenchmen had helped in the building; the famous discoveries had been made largely by their deeply feared rivals, the Germans. In spite of her great traditions, Perrin noted, "France is now scarcely in the third rank in European scientific production, which is a humiliation, and also a danger."[27] Perrin and his friends realized that more money was not enough; institutional changes were also required.

French science was embedded in a flawed system of higher education. Salaries and promotions, strictly determined by the Ministry of National Education, were tangled in a web of inspectors' evaluations, letters of recommendation from friends and functionaries, committee reports, and gossip. Aside from minor funds controlled by the Academy of Sciences and similar bodies dominated by aged conservatives, the operating expenses of most laboratories were dictated along traditional lines by government bureaucrats. These officials worked closely with a small circle of established professors, some of whom had done little productive research for decades or were not even scientists at all. Favoritism and nepotism were common and old school loyalties taken for granted. When Langevin had found Joliot the position in the Radium Institute, he had warned his student that unless he made some notable discovery, he could never hope to go far, for he was not a Normalien. Perrin claimed that three-quarters of the potential scientific talent of France was wasted because young people of obscure origins could not win a position.[28]

It was true that there was little room for new blood. During the 1920s the number of science students rose to about double the prewar level, but the number of university science professors actually declined between 1914 and 1925, and by 1930 their number was still hardly above the prewar level. The secondary schools, which controlled the gates not only to science but to all the professions, had scarcely increased their enrollment since 1880, for the bourgeoisie and upper classes, fearful of competition and jealous of their hegemony, would admit no changes. The freeze on jobs for scientists was only one of many cases where France was locked in place, a "stalemate society."[29]

One of the students whose careers were blocked was Joliot. Having won his doctor's degree in 1930, he began to look for a full-time academic position. He found none. The Radium Institute could not support him as a student assistant indefinitely, and he and Irène now had a young daughter. One of his friends later recalled that at this time Joliot seriously considered leaving science for a career in industry.[30]

The group who had finally risen to the leadership of French science were determined to endure such weaknesses no longer. The movement for change found leaders among the directors of Rothschild's Institute of Physicochemical Biology, including Perrin, Mayer, and Urbain. In their weekly meetings the directors began to discuss the larger problems of French science and soon devised a plan to bypass the traditional system altogether. As Perrin explained, they decided that "the various categories of researchers should correspond to the hierarchy of the University system . . . each with the corresponding salary, so that a scientist could make a career entirely in Research with absolutely no other obligations."[31] This sweeping plan sought to create an entirely new *corps d'état* or division of government, directly under the minister of national education and in no way subject to the university bureaucracy. The principle of setting up research entirely outside education could be traced back through Rothschild's Institute to the Radium Institute and the Pasteur Institute; but this was the first time people had seriously proposed to create such a system wholesale.

One spur to this change was the scientists' wartime experience, for in the Directorate of Inventions they had already worked successfully as teams under a central administration of scientists. The French also saw that similar experiences had led to reorganization of scientific work in other countries, such as Germany, Belgium, and especially the Soviet Union, where scientific research had been granted new prestige and a new organization. One of the main bearers of the good news from Moscow was the increasingly leftist Langevin, who visited the Soviet Union several times around 1930, entertained visiting Soviet scientists in his Paris home, and presided over a Committee for Scientific Relations with the USSR. He also welcomed young British scientists like J. D. Bernal who, with an eye on the priorities that Soviet scientists enjoyed, were beginning a campaign in their own country to honor, finance, and organize scientific research.[32]

Perrin took charge of the French campaign. He began in June 1930 by going to the Academy of Sciences, winning the support of members of all political views. Since no plan could be enacted unless key personnel in the bureaucracy were won over, Perrin next set off on a round of visits. A Nobel Prize and a popular book on atoms had made him famous, and he looked more impressive than ever now that his abundant curls and beard had turned white. Borel's wife remarked that years before, Perrin had looked like a young archangel; now he resembled a prophet. He would need the force of his appearance, for in arguing that science had a moral right to a better position, that only so could it help create a millennial future, he would be speaking less as a practical man than as a prophet.[33]

First in line for persuasion was Perrin's administrative head in the university, the dean of the Faculty of Sciences, Charles Maurain. Maurain had known Perrin and Borel since the early 1890s when he had studied science with them at the Ecole Normale, and they had introduced him to Langevin, the Curies, and others of their circle. After the war the physicists had kept up their friendship in Paris and at L'Arcouest.[34] Maurain readily allied himself with Perrin.

Next in line above Maurain was the powerful head of the university bureaucracy, the director of higher education, Jacques Cavalier. Intelligent but unoriginal, laconic, punctual, dressed always in black, he was one of those reliable officials who governed behind the scenes while ministers came and went. Perrin and Maurain spent four hours in his office presenting their case. Before he turned administrator, Cavalier too had been a scientist, an expert on the chemistry of alloys, and they had all met and become friends as students at the Ecole Normale, so he was soon won over.

Perrin now had to go into governmental circles beyond the university community, and here he had less success. Then Herriot, an old friend and schoolfellow, invited him to give a speech to a teachers' group. Perrin repeated his usual arguments with unusual force: "Our existence *at this moment* is certainly poor and miserable compared with the existence we would be leading, this very day," if scientific genius had not been "senselessly" starved. After the speech, Perrin recalled, Herriot embraced him and praised him to the skies, "But I told him that my soul was base and that I wanted some money."[35] Cornering the bulky politician for a long talk, Perrin reached out to his faith in science and won his support.

Even with Herriot behind it, as well as Blum and Borel, the plan's success was in doubt, given the Chamber's conservative mood. Sometimes Perrin made use of Marie Curie, who obediently followed him from office to office, a frail pinched old woman draped in black, a legend of self-sacrifice. Her body, disintegrating after too much exposure to radioactive chemicals, meant little to her now, and she was more absorbed by the health of her Institute. Despite severe shyness, Madame Curie could grow fierce when the future of the Radium Institute was at stake, and on her visits to officials with Perrin, both her arguments and her mere presence strengthened his message. According to one observer, she had the appearance and moral force of a Buddhist monk from Tibet.[36]

Moral force can be a mighty lever in the material world, but it must have a material fulcrum to press upon. The question of what finally does or does not move may depend on whether such a fulcrum can be found—and where. What the scientists needed was a new source of money. In 1930 France was prosperous at last and enjoying a budget surplus, but it was one thing to see money available and quite another to divert some of it into a new organization. An approach was suggested by one of Perrin's aphorisms: "A corps of men used to considering things from the standpoint of scientific research can render as much service in time of war as a number of army corps." This sentiment appealed to every political party, for France was awake to the resurgence of Germany and the rise of fascism. Construction had just begun on the multibillion-franc Maginot fortifications. Herriot managed to secure five million francs of that allocation, a tiny fraction of the total National Defense budget, for the "research corps."[37]

To create a new institution would have required a protracted legislative struggle, so Herriot used an artifice: the money was turned over to the Caisse Nationale des Sciences, a minor state fund, independent of the bureaucracy, which till then had doled out honorific aid to aged and needy scientists (it should not be confused with the Caisse des Recherches Scientifiques, which continued to fund professors, among them Marie Curie, from a separate allocation). The Caisse Nationale was quietly reorganized along the lines of Perrin's original plan, with the details drawn up by Mayer. The Caisse Nationale would be responsible only to the minister of education. Following established practice, the money was allocated by committees, and here the old friends of the Curies predominated. Of the ten members of the Physical Sciences Section only three were conservatives, while the others were Marie Curie, Cotton, Langevin, Maurain, Perrin, and two more Normaliens who were old friends of the group.[38] This was the sort of control over the financing of research that the group had been fighting for.

In 1931 the Caisse Nationale opened its first salaried positions for research. These came just in time for Joliot, who was considering whether to give up his quest for a job in pure science. Perrin settled the matter by arranging a half-time research position for Joliot at the Radium Institute; later Irène Curie got a similar post. Joliot's position was at a low level, paying little more than the student post that now lapsed, and he needed something more. A chance came in the summer of 1932 at L'Arcouest where, as Irène Curie wrote to her mother, "Of his own accord M. Maurain has spoken to me about nominating Fred to an assistantship." The nomination was approved, and with this junior academic post added to his Caisse Nationale position and his wife's salary, Joliot could at last quit his time-consuming extracurricular job of teaching electrical measurement techniques. A combination of the new system and the old had succeeded in giving him and Irène Curie stability and time for research.[39]

This was opportune, for their chosen field of physics was becoming one of the most exciting areas of science, and Joliot and Irène Curie were moving to the forefront. Foreign physicists, who had once found little of interest in French scientific journals, now began to browse through them to see if there was anything new from the couple at the Radium Institute.[40]

Scientists in other advanced nations, after courting philanthropists and industrialists, had also begun to rely on government money, as if they were all pulled by some deep economic tide. The international growth of science was so rapid that the French, even with their increased funds, remained well behind. But unlike scientists elsewhere, many of whom were wary of government control even as they accepted government support, Perrin's circle were prepared to press on. Having been employees of the state all their lives, they did not fear state control of research but thought it desirable, even necessary, given the backwardness of French science.

Professors in France had traditionally dominated their personal laboratories behind a shield of prerogatives, jealously independent in their research or lack of it. The new system was supposed to restrain this independence, discouraging indolence, wasteful duplication, and the stubborn pursuit of trivial problems that had no possible connection to the practical needs of the nation. To be effective, Perrin felt, this control must be exercised not by politicians or bureaucrats, but by the most competent and progressive of the scientists themselves.

These goals were not assured by the reorganization of the Caisse Nationale. The Caisse could easily lapse into normal bureaucratic inertia; worse, it lacked the authority to defend itself against the established powers, who might seize it and direct its funds back into traditional channels so that the Caisse would end up supporting the system it was meant to supersede.[41] Therefore on Perrin's advice Mayer drafted yet another proposal, asking for an advisory council. Seemingly an innocuous administrative adjustment, this proposal was in fact among the boldest of the moves by which Perrin's group edged toward their objective, and it provoked a hidden struggle.

The old system for managing French science was symbolized by two institutions, the Academy of Sciences and the Superior Council of Public Instruction. Several of Perrin's circle were members of the Academy, but all their energy could not budge this little club of eminent and aged gentlemen, encrusted with centuries of tradition. Fortunately the Academy had no direct control over any funds but its own small endowments, so it could be circumvented. The Superior Council of Public Instruction was another matter. A formal assembly of senior figures of French learning, representing all the sciences and humanities, the Superior Council had long experience as lawful adviser to the education ministry. While ministers could overrule the Council, they rarely risked the uproar that this would provoke. One of the ministers said that there was nothing which awed him more than his ritual semiannual appearances before this body. He felt "a respectful fear" as he stood behind a little table with an ancient inkwell and green blotter, faced with row upon row of distinguished professors: "it wasn't at all the anxiety that your decrees and ordinances would be rejected, it was the obscure feeling of being judged yourself, while you were talking, by those augurs covered with diplomas. Once back among themselves, in their institutes and laboratories, they would make commentaries, emit lapidary evaluations, which would weigh more heavily on your reputation than all your citations by parliament."[42]

Representing every traditional pressure within the little world of Parisian professors, the Superior Council of Public Instruction could impel a minister to weaken the Caisse Nationale, and this threat could not be circumvented. It would have to be challenged head-on.

Mayer and Perrin's plan was to establish by law a new democratic assembly, a Superior Council of *Research*. Its members would not be drawn entirely from the old professional establishments but would in some cases come directly from the scientific disciplines themselves. Some representatives would be named by scientific societies and organizations, presumably on the basis of merit as scientists. Others would be elected only by researchers under forty years of age. To dilute further the influence of the elderly, when any member reached seventy, a new member would be chosen to join him on the Council. The Council, which would control the Caisse Nationale's money, would not only be a match for the conservative Academy of Sciences but would also be the juridical equal of the Superior Council of Public Instruction. In short, the scheme was aimed point-blank at the traditional domination of French science by cliques of old professors.[43]

As before, it was political developments that gave the reformers their opening. French voters moved leftward as the Great Depression deepened, and the elections of 1932 returned the Cartel of the Left and Herriot to power. "The elections are extremely satisfying," Irène Curie wrote her mother; she suspected the "banks and financiers" would try to sabotage the Cartel and was "rather inclined like Fred to think that it will be necessary to imprison a certain number of these people who engage in semi-legal operations." The Cartel proved to be less independent than she had hoped. Even on the question of the research Council, the new minister of national education, although persuaded by Perrin to endorse the plan, deferred the decision to the educational bureaucracy, whose leader was the sober Cavalier. Perrin and his friends bombarded Cavalier with notes, visits, and a petition signed by over a hundred leading savants and at last won their Council.[44]

The next big step was to make the Council the effective ruler of all French scientific research, not only outside but even within the universities. Perrin and his friends persuaded the government to effect this change in 1935, when the Caisse Nationale des Sciences and the Caisse de Recherche Scientifique were combined into a Caisse Nationale de la Recherche Scientifique (CNRS), which took over administration not only of these budgets but also of the universities' research budget and miscellaneous other funds formerly distributed by the educational bureaucracy. As a powerful and systematic organization for pure research, the CNRS had no precedent outside the Soviet Union. Its establishment was facilitated by the world economic collapse, which had weakened the traditional routines of government and led parliament to grant the administration almost dictatorial powers in matters of finance. "That was the era of decree-laws," an observer recalled of the birth of the CNRS, "and the juridical form of these creations was sometimes, it must be confessed, a bit surprising to confirmed republicans."[45]

But the reorganization of French science could mean little unless the CNRS had money to spend, and when the scientists asked for money, the government left them worse off than before. The Radical party and its allies lacked the power, the will, and the unity to stand against a financial oligarchy which insisted on a balanced budget as the answer to the Great Depression. Like an old-fashioned owner of a family business, the government simply cut the salaries and pensions of the people who depended upon it, and scientists, like other state functionaries, suffered. Decrees in 1934 and again in 1935 also eliminated so many professorships that by 1936 the system of higher education had lost over 250 positions. Other decrees slashed the research budget across the board. In 1933 the Caisse Nationale had a state subvention of 8.5 million francs; in 1934 it was told it would get only 7.6 million, and then a decree abruptly cut this back to 6.0 million. Digging into its reserves, the Caisse managed to keep its dependents, among them Joliot and Irène Curie, funded at the old level, but the reserves could not last long.[46] The tortuous advance of French science had come up against a wall. And the threat to science seemed to be only one part of an attack on all democratic society.

Every few months a new government would be formed, each as scandal-ridden and ineffectual as the last. Many people grew disgusted not only with politicians but with parliamentary democracy altogether. A new revolutionary right, inspired by the success of fascism in Italy and Germany, in 1934 brought rioting mobs almost to the doors of the Chamber of Deputies. Among the scientists Langevin now took the lead, explaining that he was no longer concerned with his personal contribution to science, for the first task was to defeat the fascists in order to maintain conditions where science would be possible at all.[47] Langevin, who had long been active in groups that supported human rights or opposed war, helped to found a Committee of Antifascist Intellectuals. Thousands of scientists and others, Joliot among them, rallied to the Committee.

Meanwhile the French Communist party, also recognizing that the battle against fascism took precedence over everything else, deferred its lonely dreams of revolution and formed a common front with the Socialists. When the Communists invited their old enemies the Radicals to join the common cause, the Committee of Antifascist Intellectuals encouraged the move, asking all three parties to attend a meeting in January 1935. The task of inviting the rival chiefs fell to Francis Perrin, Jean Perrin's son.

A member of the Socialist party since his schooldays at the Ecole Normale and an established mathematical physicist now in his thirties, Francis Perrin was nevertheless little-known; that made him a good emissary. He called on each party leader in turn, saying, "You must come, because the others will be there," even if he had not yet seen the others. The meeting succeeded, whereupon Francis Perrin went to a bar and, for the first time in his life, downed a big glass of beer. At a further meeting, with Langevin presiding, the parties began to organize a march for Bastille Day.[48] That July 14, 1935, saw one of the grandest demonstrations in the history of France as marchers poured down the boulevards by the hundreds of thousands. Langevin was at the front, linking arms with the leaders of the Socialist, Radical, and Communist parties. Pressed by a wave of popular emotion, all parties from the Radicals leftward united in a Popular Front, and the elections of 1936 brought the coalition to power. For the first time Blum was premier.

With his friendly air, walrus mustache, and round glasses, Blum looked more like a man of letters than a revolutionary leader. Yet he had the one thing most necessary to renew politics: a myth, a self-consistent vision of another world projected onto the future. He and his friends took the belief in science of Berthelot and mixed it with socialist thought, until images of laboratories and professors blended with images of modern factories and proletarians. Blum proclaimed that science would create a "paradisiacal" state where every worker might be a sort of Normalien, with the wealth and leisure for self-cultivation. Unlike many scientists, however, Blum never supposed that science was more important than politics, for he held that under capitalism, science would always be perverted to create wealth only for the privileged few. Therefore science and socialism were each "really the other's complement. Science develops humanity's riches; socialism assures their rational exploitation and equitable distribution." Jean Perrin, who had expressed similar views in a speech to the vast crowd at the Bastille Day demonstration, and his friends publicly supported the reform measures that the Popular Front enacted as soon as it took power. They particularly cherished the new laws mandating paid vacations and a forty-hour work week, for these laws reflected their demand that the benefits of technology be distributed to everyone.[49]

While liberal scientists supported the Popular Front's social programs, Blum gladly supported research. When at the start of the new government Perrin asked Blum to make official mention of the importance of science, Blum replied that he could do even better: he meant to raise the status of science by creating an undersecretary of state for scientific research. The job was offered to Irène Curie, who not only sympathized with the Popular Front's politics but in the last few years with Joliot had made a series of scientific discoveries that had won the public's attention and esteem. Though artless, devastatingly frank, and lacking a taste for political maneuvering, Curie accepted Blum's offer, because she was not sure that he would actually create the science post unless it were assured the Curie prestige, and also because it was one of the first high government posts to be offered to a woman. But she accepted it only on the condition that she could soon return to her scientific work, and after enduring her quota of ministerial meetings, she was replaced, by prearrangement, with Jean Perrin.[50]

The old dream of Perrin's circle was realized: one of their number was in power. It was not that they believed government should be run entirely by scientists and other technical experts, for they were too humanistic and politically sophisticated for such fantasies. Rather they felt that some scientists, like Perrin, should be in a strong position to guide and promote scientific research, and then to put the results of this research to work. Since they were convinced that science was the chief agent behind social change, such power seemed quite enough for their purposes.

Jean Zay, the minister of national education, recalled that while in government, Perrin "seemed naive and absent-minded, almost in the clouds, but in reality he was always attentive, precise, concentrated, if necessary crafty." With ardor, obstinacy, and patience Perrin set out to increase the funds for research. His closest ally was Zay, an energetic reformer, the youngest minister to serve in the Third Republic, and a warm admirer of scientists. Zay was deeply impressed by summer visits to L'Arcouest as Perrin's guest, by the mornings spent passing from boat to boat so nobody would feel left out and the evenings of sparkling talk.

Together, Zay and Perrin planned a science budget of twenty million francs. But then Zay came up against the Finance Ministry, which always had the last word on budgets and which now refused to allow science more than fourteen million. When Perrin heard of this, he decided that he and Zay would visit the finance minister. Zay recalled: "Jean Perrin took the floor as soon as he entered; he kept it for half an hour; our host couldn't get in a word. He was drowned in a flood of pathetic examples, of implacable reasonings, was seized by the arm, hustled, overwhelmed. When Jean Perrin left, he had obtained twenty-two million."[51] The total state allocation for research, including all CNRS and university research funds, leaped from eleven million francs in 1935 to twenty-six million in 1936 and thirty-two million in 1937. By 1939 the CNRS was providing partial or total support for about six hundred researchers, over half of the academic scientists in France.[52]

Whatever strains this rapid growth may have caused were hidden by the traditional privacy of the French bureaucracy. But there were quiet confrontations between the university "teachers" and the CNRS "researchers," for example in 1938, when the universities fought to keep control of the budget of their graduate student laboratories. Not only entrenched professors but some younger scientists grew wary of the power that Perrin and his circle had won over appointments and money. As one scientist recalled, "There were people who were not so much opposed to structures but who feared that the clan would get its hands on everything."[53]

Perrin took advantage of his position as undersecretary of state to strengthen the CNRS. His friends warned him that the Superior Council of Research, which met only from time to time, was not enough to protect the CNRS on the bureaucratic battlefield against other organs of state: without a permanent administrative corps of its own, the CNRS could disappear at any shift of government. Perrin determined to create such an administrative corps. Although he began modestly by asking for a dozen people, there was still a sharp struggle in the Senate. Perrin came up against Joseph Caillaux, the fierce old dictator of the Senate finance committee, who knew that in modern government a seemingly insignificant squabble over a new budget line can fix state policy for decades.

Worse, Caillaux believed that technological breakthroughs disrupted employment and lay at the root of the Great Depression, and he was notorious for his demands that science and invention be reined in.[54] But a number of politicians, such as Georges Cogniot—a leader among the many new Communist deputies, a Normalien, and a friend of Langevin's—were ready to fight for the scientists. Perrin's group pushed through their proposal in the closing days of 1936.[55]

The CNRS administrative corps was only one of Perrin's innovations. He created a technical corps of glassblowers, machinists, statisticians, and the like, who by 1939 numbered about a thousand. He founded a museum of science in Paris, the Palace of Discovery, which combined exhibits with lectures and demonstrations to spread scientific knowledge and ideals among the masses. He raised more and more money, for example by persuading the Finance Ministry to allocate millions of francs for science out of the military budget. And he created or took over a number of laboratories, thenceforth to be administered by the CNRS. The scientists had finally won a formidable and autonomous research organization.

Among the first of the new laboratories planned by the CNRS was one for Joliot; in this case the funds were diverted from the public works budget. This laboratory would prove to be one of the most significant fruits of the long campaign of Perrin and his allies. Yet they could only provide opportunities for Joliot; he and Irène Curie would not have profited so much from the growth of the CNRS if they had not meanwhile shown their excellence as scientists. Although they had supported their old friends' efforts, signing petitions and enrolling in organizations, the couple had spent most of their time away from the confusion of the streets and the Chamber, buried in their small rooms within the Radium Institute, pursuing seemingly unworldly puzzles of nuclear physics.

3

The Uranium Puzzle

The speeches of the professor-politicians were full of references to the discoveries of Pasteur and Faraday; they rarely referred to the scientific work of their own day. Political skills provided the money and organization that made research possible, but there the connection of politics with research appeared to end. This appearance was deceptive. For one thing, only those who could unravel scientific technicalities would rise to power. Madame Curie, Perrin, Langevin, Borel, and their companions were prestigious figures only because they had long since proved their excellence as scientists; Joliot and Irène Curie could win a like influence only if they too could make notable discoveries. More important, the French scientists, as they would have been the first to insist, were not paid simply to solve abstruse puzzles with no conceivable use for society. When Madame Curie and her disciples buried themselves in the study of radioactivity, they wished for themselves only to advance toward an ideal knowledge of the universe, but they never doubted that this knowledge would someday produce important benefits for society.

The practical potential of radioactivity was not hard to see. While the medical uses of radium were expanding, much greater applications were suggested. Perrin and others, for example, pointed out how atomic energy might be the source of the almost limitless power of the sun. Equally exciting was the prewar discovery by Ernest Rutherford and Frederick Soddy that, when radioactive elements emitted energy, they could transmute to different elements, as in the old dream of the alchemists. Soddy declared that since transmutation was linked to radioactivity, transmutation was the key to the vast energies within the atom. "A race which could transform matter," he announced, "could transform a desert continent, thaw the frozen poles, and make the whole world one smiling Garden of Eden." Concerned that the depletion of coal reserves must sooner or later bring industry to a halt, Soddy thought a solution might lie in the energy revealed by radioactivity. During the First World War he also noted the awful possibilities for weapons.[1]

If atomic energy sounded like a fantasy from a science fiction novel, the novelists were learning the notion from scientists, and many scientists did not disavow it. Perrin, for one, was a fan of H. G. Wells, who drew on Soddy's writings. By the 1920s everyone had heard such fantastic speculations about atomic energy. Charles Moureu made use of the notion in his propaganda on behalf of science. Choosing the heaviest known radioactive element for an example, he wrote, "The energy which sleeps in a kilogram of uranium may be equivalent to what five hundred tonnes of coal could supply . . . Probably gold will someday be a metal as common as iron . . . entirely new atoms will come to be . . . Strange surprises surely await us, more astonishing than those attributed to the philosophers' stone and the elixir of life of the perennial alchemists." This vision of material power was directly linked with demands that scientific careers be given funds and prestige. The flow of power *from* science and the flow of power *to* science were inseparable.[2]

Nevertheless in the 1920s most scientists still thought that the promise of energy from radioactivity was remote. They had worked for decades without learning how to make atoms obey their commands. If they kept on working, it was largely because, as Perrin insisted, science was an aesthetic activity, and the scientists found pleasure simply in working through their puzzles and gaining an occasional hint about the nature of things.

They had already learned some surprising facts. Even before the First World War, those who studied radioactivity had recognized that all atoms have an essential double structure. Although priorities in such slow realizations are hard to assign, it seems that Langevin, Perrin, and their circle were among the first to perceive this structure. Every atom has a husk, a swarm of electrons which brush against the world outside and whose configurations determine the external characteristics of substances, such as the hardness of iron or the blackness of coal. Every atom also has an essential kernel, the nucleus, identified by Rutherford in 1911. This is the part that governs the nature of the atom, surrounding itself with the proper configuration of electrons so that an iron atom shall always behave like iron rather than, say, copper. The nucleus was not affected by anything that the scientists of the period could do to it in their laboratories. But it sometimes transmuted all by itself, a change signaled by radioactive emissions.

In 1919 Rutherford announced that he had discovered a way to transmute a nucleus at will. He had used radioactivity itself, the only available source of energy powerful enough to overcome the forces within the cores of atoms. By playing a stream of alpha particles onto nitrogen gas, he had "disintegrated" some of the nitrogen atoms, knocking chips off them. The chips turned out to be nuclei of hydrogen gas, the simplest of all elements.

Soon after, Perrin pointed out that Rutherford had not so much disintegrated a nitrogen nucleus as plastered an alpha particle onto it, thus forming a new and heavier nucleus, which immediately spat out the hydrogen nucleus by a sort of instant radioactivity. Perrin hoped that important practical results would follow. If ways were discovered to create new radioactive elements at will, one might be able to release nuclear energy in bulk more or less as one burned coal. "Next to that discovery," he wrote, "the discovery of Fire would be a small matter in the history of Humanity."[3]

But over the following years more detailed study of nuclei, far from clarifying their structure, turned up baffling puzzles, and the study of radioactivity came to appear less promising than other fields of physics. Once the two parts of the atom had been clearly distinguished, many physicists undertook to study the exterior electron swarm and thereby to understand all the external, everyday characteristics of matter. This enterprise was succeeding remarkably in the 1920s with the advance of quantum theory. For the relatively small number of scientists who chose to work instead on the inner nucleus, understanding came more slowly, and is not complete even today. Thus the study of radioactivity went on a bit outside the triumphant mainstream of physics during the twenties, and was pursued intensively at only a few centers. A leader among these was the Radium Institute in Paris.

Although the Radium Institute included a large medical section, it was better known to physicists for its other section, the Curie Laboratory, built for Marie Curie and entirely hers. These laboratories and workshops were housed in a long brick building of narrow corridors and many small rooms crowded with chemical glassware and homemade electrical instruments. The Institute taught university students courses on radioactivity and was the world center for certifying the strength of industrial and medical radium supplies; nevertheless its heart was devoted to pure science. Packed with three or four dozen researchers from a number of countries, it offered daily opportunities for young scientists to learn ideas and techniques from their seniors. Curie herself insisted on discussing everyone's work in detail. The moment she arrived in the morning she was surrounded by researchers, and she often stopped to sit on the stairs in the narrow entrance hall, talking briskly with her "Soviet."[4]

In its organization the Radium Institute resembled certain earlier laboratories, each ruled as a fief by a single professor, but it groped toward a new and more efficient structure.[5] Curie recognized that the increase of knowledge and the climbing costs of research would force laboratories to become more and more specialized. She therefore devoted the Radium Institute to the study of radioactivity. A scientific institute, she felt, should be constituted of research teams tending in the same general direction, but each with its own independent objectives and its own apparatus. From year to year some groups would expand and others diminish, but only rarely was a new group formed. When inexperienced workers joined the Institute, they were attached to a seasoned leader, as Joliot and Irène Curie served Madame Curie herself. Once the new workers learned their way around and proved their worth, they could be given independent work. Only the exceptionally qualified would eventually get their own apparatus and their own group.[5] Joliot and Irène Curie, as Madame Curie realized, were just such rare people, so after a few years she allowed them to strike out together in their own direction.

Toward the end of the 1920s, when Joliot and Irène Curie began working together, nuclear physics was less a research field than a pack of riddles. Scientists had been unable to form a sensible picture of nuclei, for they had few means to investigate them. The only radioactive elements known were the handful like radium and polonium that could be extracted with great pains from naturally occurring minerals. Not yet invented were machines that could hurl a particle into the core of an atom; nuclear physicists had to use as best they could the alpha particles, electrons, and gamma rays emitted by the natural radioactive elements. Theory was in such confusion that some scientists began to doubt that energy was conserved in nuclear reactions.[6] All this would change within the next decade, with Joliot and Irène Curie hurrying the change along.

By 1929 each of them had published separately some useful pieces of research, but neither had made a mark as an exceptional scientist. They had grown closer, swimming and sailing at L'Arcouest in the summers, skiing in the winters, raising their little daughter. But the chief bond between them was their scientific vocation, and they now began to follow it in collaboration.

The first large task they undertook together was neither particularly original nor striking, for it was precisely the sort of work that had been a specialty of the Radium Institute since its foundation: the extraction of a large quantity of a radioactive element. The couple decided to isolate a supply of polonium, one of the rarer elements. Minute amounts of this element were present in the nearly two grams of radium that Madame Curie had acquired through her own work and by the gift of the Americans. More was to be found as traces in thousands of ampoules filled with radon gas, their radioactivity largely spent, which she had laboriously gathered from doctors all over the world who had used them to treat cancer. Joliot and Irène Curie devised several chemical techniques to extract the polonium from these raw materials. The work was laborious, highly technical, and—since radium, radon, and polonium are all intensely radioactive—dangerous. They managed to create little plates covered with powder, the purest and most radioactive samples of polonium anywhere. This material could have been divided into bits and distributed for various small researches by different students in the laboratory, but the pair decided to concentrate it in one powerful mass in an attempt to force important discoveries.[7]

Radium emits both alpha particles and quantities of gamma rays, whose effects were difficult to disentangle from one another, whereas polonium has the useful property of emitting abundant high-energy alpha particles while producing almost no other radiation at all. At the beginning of 1932 Joliot and Irène Curie used this property of polonium to attack an unsolved problem. In 1930 two German scientists, Walter Bothe and Herbert Becker, had taken a bit of beryllium, a lightweight steely metal, and bombarded it with alpha particles. This sort of experiment was common in that stage of nuclear physics: the scientist took two substances, one radioactive and the other not, put them next to each other, and studied what came out. In principle the scientist could then deduce what happened as individual particles hit individual atoms within the substances, but in practice it was not always possible even to tell what was coming out. When alpha particles bombarded beryllium, Bothe and Becker found that what came out was some sort of radiation which, to their surprise, could penetrate two centimeters of lead. The two Germans, followed by Joliot, Irène Curie, and many others, assumed that this peculiar radiation consisted entirely of some new sort of gamma ray.[8]

Curious about these rays, the French couple first confirmed that when they exposed beryllium metal to their powerful polonium source of alpha particles, the beryllium did emit penetrating radiation. Then, in an exploratory mood, they tried putting bits of various elements in the way of the new radiation. One of the elements they wanted to try was the lightest of all, hydrogen, so they exposed some paraffin wax, which consists of hydrogen atoms stuck together with carbon atoms, to the new radiation. They were astonished to see some of the hydrogen nuclei come flying out. They could not understand how a gamma ray could strike a hydrogen nucleus such a blow as to knock it altogether out of the paraffin. Energetic electrons also came out, so the couple began to study these in hopes of getting some clue. Meanwhile they published their preliminary results.[9]

Their paper was read with amazement by physicists at the Cavendish Laboratory in Cambridge, England. Among the readers was James Chadwick, who immediately recognized that the peculiar new radiation included not only gamma rays but also a subatomic particle never before seen, one he had been looking for without success for nearly a decade, the "neutron." Lacking an electrical charge, the neutron could penetrate lead or any other material with ease, and since it was as massive as a hydrogen nucleus, it could easily knock hydrogen out of the paraffin. As it happened, Chadwick had just made a powerful polonium source of his own from used radon ampoules, so within a few days he could confirm his guess with a few well-turned experiments.[10]

A word about the Cavendish Laboratory, the Radium Institute's great rival for leadership in nuclear physics. Like many of the French institutes, the Cavendish held aloft the banner of disinterested research while standing on a foundation of potential uses of science to industry and the state. Founded in 1870, the laboratory had been devoted in its early years not only to teaching Cambridge University students but also to establishing standards needed by the fast-rising electrical industry. During the First World War the Cavendish personnel, like their French colleagues, had been brought back from the Front to carry out important military research. Immediately after the war Rutherford, the new Cavendish Professor, set out to increase his laboratory's funding by emphasizing the value of physics for warfare, industry, and national prestige.[11]

Some therefore found it strange, even reprehensible, that Rutherford dedicated the Cavendish to work on radioactivity, a subject relatively remote from industry. Although he had been among the earliest to point out the immense potential of nuclear energy for peaceful and destructive uses, by the 1920s he had grown reluctant to claim that research on radioactivity could lead to spectacular applications. The possibility nevertheless remained in the back of the Cavendish physicists' minds. In 1930, for example, Rutherford and two others, noting that an atom of any heavy element, like uranium, has a mass greater than the sum of the masses of its constituent parts, pointed out that the surplus mass was a sign of strong energies locked within the nuclei. "If the uranium nucleus were taken to pieces," they wrote, "it is clear that a considerable amount of free energy would be liberated." Yet Rutherford's crew were less attracted by practical possibilities than by the unsolved puzzle of the nucleus itself. As Chadwick recalled, doing nuclear experiments was simply "a kind of sport. It was contending with nature."[12]

It was also contending in a friendly way with other scientists. Joliot and Irène Curie were at first incredulous, then chagrined, to learn how close they had come to the solution of the beryllium radiation puzzle without grasping it. They had handed a decisive clue about the neutron to Chadwick, yet the fame of discovering the new particle justly descended on him. Racing to keep up with a crowd of physicists attracted to the problem, Joliot and Curie set out to study Chadwick's new particle. They used an instrument which Joliot loved to construct and improve, a Wilson cloud chamber, which revealed the paths of charged particles as spidery tracks of mist. In the course of their work the couple happened to notice an odd phenomenon: on a few occasions the tracks of energetic electrons curved the wrong way in a magnetic field. They speculated that these electrons traveled not away from but toward the source of radioactivity.[13] Before Joliot and Curie could investigate the phenomenon in detail, physicists elsewhere found the true answer: the tracks were caused by yet another new type of particle, the positively-charged electron or "positron." Again Joliot and Curie had come close to a great discovery without quite grasping it. Disappointed but refusing to be discouraged, they set out to investigate positrons. Their labors produced important results but still nothing famous.

Meanwhile the discovery of the neutron had thrown open new doors into the nucleus. Theoreticians were finding they could make plausible models of a nucleus if they put neutrons inside. Once stuck in almost hopeless confusion, theory was on its way to making sense. Experimentalists were even happier, for they could study the neutron and its effects on nuclei with relative ease. All they needed was a little radium or radon, available on loan from any large hospital, and some beryllium; the combination gave out abundant neutrons. Meanwhile in 1932 John Cockcroft at the Cavendish Laboratory and Ernest O. Lawrence at the University of California at Berkeley and their collaborators devised machines that could accelerate particles to high velocity, making them powerful tools for poking at the nucleus. The field of nuclear physics became more chaotic than ever, but it was an exciting chaos in which solutions could at last be seen rising toward the surface. These discoveries came at an opportune time, for other physicists had recently completed the most successful of all theories, quantum mechanics, which could explain nearly every ordinary property of matter outside the nucleus. Many physicists had therefore lost interest in studying the electron husk of atoms and were moving into nuclear physics.

This progress did not show the way to any practical applications, and some leading physicists continued to deny that nuclear energy would ever be directly exploited. The research was therefore less like an ordinary enterprise with a hard goal than like a puzzle-solving contest, in which part of the excitement came from working through an intriguing problem and part came from scrambling after the substantial rewards that would go to the first person to come up with a solution. The work itself was rewarding, the French scientists stressed, largely because it brought the researcher closer to nature. As Irène Curie explained in one of the radio talks that she and her colleagues gave during the 1930s as part of their program of spreading scientific culture to the masses:

If an explorer tries to satisfy a taste for adventure in Research, it would seem that the tranquil life of the laboratory has little to offer in this regard. Yet we may find ourselves in the presence of singular facts, sometimes of interest only to specialists, sometimes striking enough to be popularized. A task once begun develops in an unexpected fashion, opening new paths for future work. And thus we satisfy our spirit of adventure. Another attraction of scientific life is the almost childish joy one feels while watching natural phenomena, even well-known ones. I feel the same pleasure every time I see a brightly colored precipitate form during a chemical procedure, or watch a radioactive substance glowing in the shadows.[14]

Curie did not mention the pleasures of working toward a theory that could reveal the inner nature of the universe, for in fact she and Joliot were not much interested in theory. Their satisfaction came from skillful experiment, from their delicate instruments of brass and glass and the curious phenomena these revealed.

The next year, 1933, gave them full opportunity for such work. In rapid succession they turned out important papers on a variety of subjects, always basing their work on their polonium source. Although a crowning discovery continued to elude them, their reputations grew. In October they were honored by an invitation to the Solvay Congress, to present their latest results before a few dozen of the world's premier physicists.

The young couple were particularly proud of one recent experiment. Again the scheme had been to place a radioactive substance next to something else, study what came out, and try to deduce what this meant for the individual atoms within the substances. In this case they had put polonium next to some aluminum foil and had found that sometimes what came out was not the expected hydrogen nuclei, but neutrons and positrons together. This encouraged them to suppose that a hydrogen nucleus is a compound of one neutron with one positron—a hypothesis that quite a few people were trying to prove or disprove. Joliot and Curie suggested that their experiment was the much-sought proof.[15]

As soon as they described their work, they received a sharp setback. Lise Meitner, a Berlin radiochemist with a reputation for exacting work, announced that she and a colleague had recently done exactly the same experiment but, unlike the French, had "been unable to uncover *a single* neutron." Although this was one occasion when Meitner was wrong, her flat statement put Joliot and Curie's work in doubt. "In the end," Joliot recalled, "the great majority of the physicists present did not believe in the exactitude of our experiments. After the session we were quite downhearted, but at that moment professor Niels Bohr took us aside, my wife and I, to tell us that he found our results very important. A little later Pauli gave us the same encouragement."[16] The combination of general skepticism with private encouragement from two famous physicists was exactly what Joliot and Curie needed. While other scientists did not trouble to follow up the putative results of irradiating aluminum with alpha particles, the French couple, convinced that they were on the track of something important, pursued the problem vigorously once they got back to the Radium Institute.

They felt that if they were creating a neutron and a positron simultaneously, the process should depend on the energy of the alpha particles that provoked it. Joliot tried an experiment, moving the polonium source of alpha particles a little back so the particles would be slowed down by passing through some centimeters of air. He was surprised to see that positrons kept coming out. Even when he took away the polonium altogether, he still saw thin, sudden lines of mist appear in his cloud chamber, the signature of radioactive emissions. He immediately called in his wife and moved to a geiger counter to pin down the phenomenon with numerical measurement.[17] When they irradiated the aluminum and then removed the alpha particle source, the chatter of the geiger counter died down over several minutes. This was a familiar sound to the pair, for gradually decreasing emission was a characteristic of natural radioactive atoms. A mass of such atoms loses half its radioactivity in a characteristic time, the "half-life," then loses half its remaining radioactivity during a second half-life period, and so on. Was that what was happening here? If so, then the ordinary, stable aluminum had developed radioactivity. It was as if handling a stick of wood could induce it to burst into flower.

Exactly the same slowly dying chatter could also be heard from a defective geiger counter. In the 1930s operation of a geiger counter was a black art requiring luck, trial and error, and infinite patience, so that one never quite knew when a given counter would misbehave. Joliot called in the laboratory's geiger counter expert, Wolfgang Gentner, a visiting student from Germany. After some hours Gentner reported that the counter was working perfectly.

Joliot and Irène Curie immediately guessed that they had stumbled upon an entirely new reaction, of a type physicists had been seeking for years. A decade earlier Perrin had mentioned his hopes for "transmutations arranged at our will, where we would fashion heavy atoms by bombarding light elements" with alpha particles or the like. Perhaps this was what Joliot and Curie were now doing. If so, the aluminum nuclei were swallowing up alpha particles to become heavier nuclei. Adding the weights of an alpha particle and an aluminum nucleus and subtracting the weight of the neutron that flew out showed that the new element should be a new sort of phosphorus never found in nature. They calculated that this artificially constructed phosphorus should be unstable: it should rapidly transmute to ordinary silicon, in the process emitting the positron which had led everyone astray.[18]

The couple set out to verify their guess by determining whether phosphorus was indeed created in the aluminum foil. The task was peculiarly hard, for the new element ceased to exist within a few minutes. The couple recalled that one of them happened to meet a chemist who worked in another laboratory off the Rue Pierre-Curie and asked him "to suggest a procedure which would permit the separation of phosphorus from aluminum within three minutes. He threw up his hands, never having envisaged chemistry from that point of view. Nevertheless we did find a way." When the phosphorus was found as predicted, Joliot began to run and jump about the basement room with childlike joy. The discovery was beautiful, significant, and entirely their own. "With the neutron we were too late. With the positron we were too late," he told Gentner. "Now we are in time."[19]

They immediately called in Madame Curie to witness the experiment, and she arrived quickly, accompanied by Langevin. By this time she was chronically ill, for at first she and other scientists had not understood the dangers of radioactivity, and even when these became clear, she had scorned to let precautions interfere with her work. To verify the discovery for herself, she took the tube of newly created phosphorus in her fingers and held it in front of a geiger counter, which duly began to crackle. "I will never forget the expression of intense joy that seized her," said Joliot long after. "It was without doubt the last great satisfaction of her life."[20] A few months later, in 1934, she died of a leukemia which was probably brought on by her long exposure to radioactivity. The next year her daughter and son-in-law received a Nobel Prize for their discovery of "artificial radioactivity."

The French scientists and public joyfully welcomed this Nobel Prize, particularly since there had been a certain lack of French scientific heroes recently. The glory of the couple's discovery was redoubled by the fact that they had made it jointly in the tradition of Pierre and Marie Curie, peerless romantic figures of French science. Artificial radioactive elements were also expected to be as beneficial to humanity as radium itself.[21] Perrin was especially pleased by the development, for he could henceforth use Joliot as a stock example of how CNRS support could help a young person of obscure origins rise to the peak of scientific accomplishment.

During this period of increasingly violent politics it was also helpful to Perrin and his politician friends that Joliot and Irène Curie agreed with their views. In an interview Joliot told a journalist that he was not a revolutionary; artificial radioactivity represented only an improvement within the existing framework of knowledge, and in society, too, the heyday of those who would overthrow things was past. He had joined the Socialist party after the riots of 1934, reflecting that "such things happen because too many people remain unorganized." With other scientists he was also shocked by the turmoil in German universities, where the new Nazi regime was dismantling the finest engine for research that the world had yet known, assaulting intellectuals and driving Jews into exile. Joliot became an officer of Langevin's Committee of Antifascist Intellectuals, explaining that it was "impossible for a disinterested scientist to accept passively the ruin of science and the general retreat of civilization which the advent of fascism would involve in France."[22] And Irène Curie lent the Popular Front government her prestige in 1936 by becoming undersecretary of state for scientific research. Meanwhile the state expanded science funding, and Joliot got the laboratories in which he and his colleagues were to develop the consequences of nuclear fission.

Shortly after Joliot's Nobel Prize was announced, Langevin proposed him for a vacancy at the Collège de France. The Collège's laboratories of inorganic chemistry had been without a professor since the death of the former chair-holder, and Langevin hoped that the chair would be renamed "nuclear chemistry" and given to Joliot. A few science professors opposed the move, feeling that one of the Collège's two chemistry chairs should not fall into the hands of a physicist. There were also problems of precedence with the Collège's humanities professors, for the scientists were trying to give Joliot a chair with unaccustomed haste. The vote was close, but the fame of his Nobel Prize in chemistry carried the day.[23]

Stately and austere, the buildings where Joliot now took his place had housed leaders of science and the humanities for centuries. The Collège de France was the center and symbol of French learning, where bourgeois who came to a public lecture could lose themselves, as a journalist wrote, among colonnades and passageways, "dim rooms with a patina of noble dust, illustrious professors whose knowledge is so specialized that they are almost the only ones to understand it." Tradition was strong at the Collège, and Joliot would often have been reminded of an earlier chemistry chair-holder, Berthelot, if only because one entered the Collège by way of the Place Marcelin-Berthelot.[24]

Joliot took on a number of advanced students in the chemical department, which had been thriving under the subdirectorship of Henri Moureu, Charles Moureu's son. Joliot grafted onto this establishment an unprecedented second section for nuclear physics under another subdirector, Pierre Savel, a loyal helper from the Radium Institute. While the chemistry section remained large and flourishing through the thirties, the nuclear physics section rapidly surpassed it.[25]

Even before Joliot had been formally voted into his chair, Perrin had promised him a generous share of the Popular Front's largesse, some four million francs, to build new laboratories. Joliot could not have won this money through family friendship or scientific accomplishment alone. Like Perrin, he was now moving on his own in high social and even aristocratic circles, lionized for his Nobel Prize and appreciated for his personal charm. And like Perrin, Joliot was beginning to show an extraordinary ability to deal with the various men who controlled the funds. One government official recalled that when he went to visit laboratories, most scientists were irritated or reminded him that they had been his teachers at the university; Joliot alone welcomed the stranger enthusiastically, explained lucidly what was being done in his laboratory, and revealed an affable but obstinate will to get the means necessary for his research.[26]

Over a decade earlier Barrès had chosen the old laboratories of inorganic chemistry as a pitiful example of scientific penury: "rooms without light, damp, with decrepit walls, where the instruments rust." With his new funds Joliot renovated them from the ground up and for a year they were filled with dust and workmen. He also moved into a more modern, multistory building crammed into a little lot near the main Collège buildings, and there in the deepest cellar, ten meters below ground level, he began to build a cyclotron. This device, invented by Ernest Lawrence, was a flat cylinder in which particles whirled in circles, kicked to higher speeds each time around. Such particle accelerators had recently become the most exciting tools of nuclear physics, for they were stronger and more controllable sources of projectiles than were elements like polonium.

Joliot decided he would also build a different type of machine, invented elsewhere, an electrostatic accelerator. Such devices were towering tubes down which high voltages accelerated particles to great velocity in a single shove. With one or two technical assistants Joliot tried various inexpensive schemes for building an electrostatic accelerator, working first in a government testing laboratory and then, having won permission from the Compagnie Générale d'Electrocéramique, in their industrial facility for testing high-voltage insulators, located at Ivry on the outskirts of Paris. Once he got the money from Perrin, he bought the Ivry facility outright, renaming it the Laboratory of Nuclear Synthesis. It was "an immense hangar like a shelter for dirigibles," a journalist reported, "where gigantic apparatus, tall as towers and silent, stand ready for a potential of 2,600,000 volts." Joliot converted these testing apparatus to electrostatic machines.[27]

Meanwhile Joliot helped to direct the construction of a similar device, ten times his height, which looked like a particle accelerator but was not fully equipped as one, as the centerpiece of Perrin's Palace of Discovery. Visitors to this science museum could see demonstrations of artificial radioactivity and witness a spectacular display of lightning from Joliot's machine.[28] In the public image of nuclear physics during the thirties, "atom smashers" with their millions of volts loomed as emblems of prodigious energy. Nor was the image false, for these monumental devices were needed for research precisely because the forces within the nucleus, which physicists using the accelerators were beginning to map, themselves ran into millions of volts.

However, the accelerators required a great amount of power to transform only a few atoms, and therefore scientists seemed as far as ever from releasing nuclear energy wholesale. Most would probably have agreed with the astronomer Arthur Eddington: "The cupboard is locked, but we are irresistibly drawn to peep through the keyhole like boys who know where the jam is kept." Physicists differed on the question of whether anyone could ever find a key.[29]

Joliot, as well as many other scientists, had a different sort of use in mind for radioactivity. Alongside the hangar housing his accelerators he built a separate four-story biological laboratory. The accelerators were to produce intense neutron beams that would be sprayed on various elements to create artificially radioactive substances. These would then be carried safely on a conveyor belt into the biology building, where the substances would be ingested by living creatures in the same way as ordinary elements; then their radioactivity would allow the atoms' paths to be traced with geiger counters through the intricacies of biological processes. Such a use of radioactive tracers was as exciting a field for research as nuclear physics itself, and seemed likely to be of more immediate use to humanity.

To secure the salaries needed by the chemists, physicists, and others who depended on him, Joliot had to work within the CNRS. He also extracted money from, for example, the Rockefeller Foundation, a longtime supporter of Marie Curie; the Air Ministry, for work on alloys; and the Ministry of Agriculture, for research on the production of fuels from vegetable matter. Since the Ivry laboratory's neutron beams and radioactive tracers were largely for biomedical research, they had the potential to unlock many other sources of money.[30]

In this private fund-raising Joliot was following in the footsteps of his older colleagues, and after his Nobel Prize he joined them in addressing the public as well. For example, in 1935 he gave an open lecture at the Sorbonne on artificial radioactivity, at which, according to a newspaper report, he was "elegant, discreet, fluent, dominating with sovereign ease." Joliot told his audience that compared with the neutron, the philosophers' stone of the alchemists was a childish fantasy. Doubtless people would soon be able to change lead into gold, but this would be an insignificant by-product compared with the enormous quantity of energy that would be released. Chemistry would produce an infinite variety of unheard-of products, while medicine would make marvelous advances.[31] In short, Joliot held forth the same scientific utopia that his seniors preached.

Frédéric Joliot, 1930

Paul Langevin, Paul Rivet, and Pierre Cot, giving the Communist salute atop a taxicab during July 14, 1935, demonstration

Irène Curie
in the late 1930s

Jean Perrin in the 1930s

Joliot was also following older professors like Perrin and Marie Curie in organizing his laboratories as a complex of little independent research groups, funded from a variety of sources and run informally. But he was getting near the maximum workable size for this sort of organization. In addition to numerous graduate students, he had a laboratory staff of more than twenty secretaries, technical assistants and so forth, and, by 1939, was host to some fifteen post-graduate researchers. A colleague recalled that Joliot had mused on the possibility of a laboratory organized still more effectively, "after the fashion of an industrial enterprise. There should be one man for each type of machine, one man to run a particle accelerator, one who runs a Wilson cloud chamber, one who runs photographic counters, and so on—and one man in the middle who tells the others to perform various experiments."[32]

Even if sheer size did not change the nature of the laboratory, the particle accelerators would. For the big machines demanded a concentration of money, personnel, and administration unknown in the earlier years when each scientist built and in effect possessed his own apparatus. A reporter visiting Joliot's laboratories at the Collège de France in 1938 got "the impression of having strayed into a factory. I survey a long glassed-in corridor full of the murmur of machines, I climb an iron staircase hung with electrical switchboards where ammeters quiver."[33] The reporter would have found the Laboratory of Nuclear Synthesis at Ivry even more like a factory, since the building's original use had been industrial.

Probably Joliot's first protracted experience in negotiating with heavy industry had come over the transfer of the Compagnie Générale d'Electrocéramique's facility at Ivry. He was subjected to bureaucratic criticism because the deal he struck with the company gave them a good sum for obsolescent equipment, and the purchase was delayed until May 1937. Even so, the transfer was a case of unusually close cooperation between industry and science, which in France had traditionally stayed farther apart than in other countries.[34]

Perrin and his circle, while never forgetting pure research, had always insisted that science should be applied to human needs as rapidly as possible, and they had assumed without question that this application must take place by way of the industrial companies. For decades they had deplored the suspicions between industrialists and scientists and had urged businessmen to aid science. Now that they were in a position of power, they could be more direct, and they characteristically set up a centralized government agency, an applied science counterpart to the CNRS. The new Centre National de la Recherche Scientifique Appliquée (CNRSA), starting up in 1938, had a large budget which it parceled out in research contracts to both academic and commercial laboratories. The allocations were made by several dozen committees of specialists, each including representatives of the academic world, industry, and government. The first meetings proved difficult, as one participant recalled:

The academic people asserted their lack of interest in the problems of application; the Army men excused their partial silences on the ground of a necessity for secrecy in regard to all problems involving National Defense; and the industrialists were extremely reserved, due to an anxiety to keep the control of their secrets of manufacture in their own hands—and also to hide, under a veil of silence, the poverty of the research work undertaken by their companies! It took several rather gloomy sessions before contact could be established and the spark kindled into flame. But always, after a period more or less long, these committees became effective.[35]

The CNRSA was one of the last ambitious moves of the government before it suffered the usual fragmentation of French governments, its difficulties increased by the decline of all France. When George Orwell passed through in late 1936 on his way to the Spanish wars he found Paris "decayed and gloomy, very different from the Paris I had known eight years earlier . . . everyone was obsessed by the high cost of living and the fear of war." Unable to solve the problems of inflation or of Spain, the Popular Front fell from power in June 1937, leaving the government once again in the feeble hands of the Radicals. Perrin departed with Blum. These were painful events for Perrin's circle. Like many on the left, Perrin, Langevin, Joliot, and Irène Curie had lost faith in the Popular Front, largely because it failed to support the Spanish republicans.[36] To Blum, the threat of fascism imposed caution, while to many others it commanded action. On this issue Joliot broke with the Socialist party.

Now that the Socialists as well as the Radicals had lost all coherence, many people saw no alternative but the Communists. In lonely struggles through the 1920s the Communist party had refined itself into a tightly disciplined, Stalinist organization, but the party's belligerence had repelled reformist intellectuals, including most scientists. For example, the liberal belief that science would progressively unify society was opposed by the Marxist doctrine that science, as the force behind the expansion of the industrial system, would be a direct cause of the coming revolution. However, when the Communists put off their plans for revolution in 1934 and joined the Popular Front, they set to work to attract intellectuals. By 1937 Langevin's Communist friend Cogniot could boast that thousands of teachers and other intellectuals, once estranged from the party, had recently joined. The marriage would not be easy, he warned: "For our Party, intellectuals can be the best or the worst of things," depending on whether or not they had "the spirit of discipline."[37]

When in June 1938 the leader of the Communist parliamentary bloc,

Jacques Duclos, addressed a large assembly of intellectuals at the Centre Marcelin-Berthelot, Joliot and Irène Curie occupied the foremost place among the audience, and science occupied the foremost place in the speech. The principal goal of Communist policy, announced Duclos, was that man's energy "should no longer be turned against himself, but should be employed to master the forces of nature . . . It is because we are conscious of the greatness of this mission which falls to the lot of French science that we protest against the miserliness of the funds placed at the disposal of our scientists . . . We think that steps should be taken to shelter scientists from material cares, to permit them to devote themselves without restriction to their work."[38] In short, under capitalism science must languish, while under communism it would be first in line for prestige and funds. The scientists were at last hearing a major political party adopt precisely their own goals.

The Communists, even more aggressively than the Socialists, also matched the scientists' belief in internationalism. The liberal group of scientists had always scorned chauvinism, claiming as one of the chief virtues of science that it could transcend national boundaries. Marie Curie, herself a Polish immigrant, had welcomed researchers from many countries into the Radium Institute, and Joliot was proud to carry on the tradition in his own laboratories. According to an acquaintance, he was "deeply interested in the internationalizing influence of science. He had noticed that men of all nationalities, when they have worked together in the laboratory, tend to preserve contact when they return to their own countries, whereas the study of literature seemed to encourage nationalism." That science had a moral authority above the archaic struggles of peoples was a belief shared by scientists in many countries, who even in the 1930s hoped science could help promote world understanding.[39] The Communists, with their appeals for Franco-Soviet amity or aid to Spain, could satisfy both the scientists' antifascism and their internationalism.

While Joliot joined a few liberal organizations and made some public statements, little of his time went into such affairs. Between his manifold functions of sitting on government committees, administering his laboratories, raising funds, teaching, and tending to his family (he now had a son as well as a daughter), he could not give much attention to politics or even to his own research.[40] Meanwhile Irène Curie was following a more traditional path. After leaving her post as undersecretary, she had taken on the duties of an associate professor at the Sorbonne, while continuing her research in the narrow laboratories of the Radium Institute. Working now without her mother's advice or her husband's help, she pursued the consequences of the discovery of artificial radioactivity.

As soon as artificial radioactivity had been announced, Enrico Fermi in Rome had pounced upon it. Like Joliot, who was almost the same age,

Fermi was a brilliant new star in a country where science was in a twilight period. A genial, ardently competitive physicist, he led a rising team of young Italian scientists, who had not yet made their mark but who over the past few years had been building up their laboratories and going abroad, sometimes to the Radium Institute, to train themselves. Two things held the group together: Fermi himself, matchless at experiment, theory, and teaching, and the political patronage of Orso Mario Corbino. Corbino, a physicist himself before turning to a career as senator and minister, a government specialist in electrical industry affairs, loved science as much as politics. Like his French counterparts, to win support for scientists he emphasized that research could enhance the national prestige and also lead to practical benefits. Nuclear physics, Corbino noted, was particularly likely to help the nation.[41]

In 1934, when artificial radioactivity gave the Rome team an opportunity to enter the nuclear physics field, Fermi quickly discovered that he could create new atoms most easily by irradiating substances not with alpha particles but with neutrons. Lacking an electrical charge, a neutron could easily slip into an atom and be absorbed by the nucleus to make a slightly heavier atom. The newly constructed atom usually turned out to be unstable, emitting radioactivity as it transmuted to an ordinary atom.

The Rome team marched through many of the elements, irradiating each in turn with neutrons. At length they came to the heaviest of known elements, uranium. As expected, neutrons changed some of the uranium atoms into new radioactive substances. Fermi's team, finding that these were not any known heavy element, thought it likely that they had created a new element heavier than uranium. In fact they had not; they had split the uranium nucleus in two; but more than four years would pass before scientists realized this.[42]

The key to the discovery of nuclear fission was a radioactive substance with a half-life of about 3½ hours, first seen in 1935 by Irène Curie and her students. Following the lead of Fermi's team, they irradiated thorium, a heavy metallic element like uranium, and studied the resulting substances. It was not a simple task, for all atoms of a given element are not exactly the same. If a neutron is added to the nucleus, the new nucleus is an isotope of the element, slightly heavier, with a markedly different type of radioactivity. Adding or subtracting more neutrons makes other isotopes. But none of this changes the atom's external swarm of electrons, and therefore all the isotopes of a given element behave the same way in chemical treatments.

Curie's task was to unmix a broth of various elements and their isotopes, all transmuting into one another minute by minute. Moreover, the radioactive substances were produced in invisibly small quantities, detectable only by their radioactivity. It was therefore a common trick of radiochemists to add some chemically related element as a "carrier" of the radioactive atoms. This gave the scientist a mixture big enough to see and handle, and the radioactive atoms would "follow" the carrier through the chemical treatments. When Curie added lanthanum, a rare metallic element, as a carrier to her solution, then precipitated the lanthanum back out, she found the 3½-hour radioactivity mixed with the lanthanum rather than in the solution. In short, this radioactivity came from a substance chemically similar to lanthanum.

Nobody imagined the substance could actually *be* lanthanum or any element close to lanthanum in weight, since lanthanum atoms are only a little over half the weight of thorium atoms. At that time no nuclear physicist could imagine atoms gaining or losing more than one or two particles at a time, which would change the weight of a heavy element only a little. The 3½-hour substance would therefore be an element with nearly the same weight as thorium. Curie and her students guessed it was a new isotope of actinium, a heavy, intensely radioactive element chemically similar to lanthanum. The supposed actinium isotope was only one of several new substances they had under study, and they did not think it particularly important.[43]

This sort of work was traditional in the Radium Institute—painstaking, highly technical, and hazardous. Irène Curie, who had learned all the techniques of radiochemistry from her mother and her mother's friends, taught it in turn to the students who filled the Institute's laboratories. In radiochemical research it helped to have many hands. A typical experiment began with an irradiation; next came elaborate chemical separations; there followed tedious days and nights watching the gradual change of a substance's radioactivity hour by hour. There were few places where much work of this kind was done.

Life in the Radium Institute was not confined to radiochemistry. As one student recalled, "the laboratory's atmosphere was extremely friendly. Little groups were often found on the entrance hall stairs, sometimes including Irène Joliot-Curie standing against the radiator, animatedly discussing not only scientific problems, but also the burning political questions of those years of mounting danger in Europe." Curie and her students kept in close touch with other scientists in Paris. Between the sturdy brick structures that made up the complex of institutes on the south side of the Rue Pierre-Curie were walkways and alleys and tiny plots of grass and shrubs, separate but intercommunicating, forming a sort of scientists' cloister. On special occasions, if the weather was good, tables would be set out in the little garden behind the Curie Laboratory and the company would eat cakes from darkroom trays and drink tea from laboratory glassware. These laboratories were only a short walk down the narrow Rue Saint-Jacques from the Collège de France, and Irène Curie's and Joliot's groups were in daily contact. Considerable communication took place at seminars that Joliot held every week in his laboratory so that physicists and chemists from Paris and elsewhere could get together; Langevin too held seminars. But most important were the meetings presided over by Perrin in his physical chemistry laboratory adjacent to the Radium Institute. Here numbers of scientists (and women friends, for Perrin particularly enjoyed the company of women) crowded together every Monday afternoon in the lecture hall to drink tea brewed in a laboratory flask over a Bunsen burner, eat biscuits, mingle with famous visitors from abroad and occasional notables from the world of letters, talk over the latest problems, and perhaps listen to a speaker. Perrin's teas were the soul of Parisian physics. In all these gatherings Curie's work was vigorously discussed.[44]

Meanwhile the problem of heavy elements irradiated with neutrons was worked over by Otto Hahn and Lise Meitner at the Kaiser-Wilhelm Institute for Chemistry in Berlin-Dahlem. They had gotten onto the same track as Curie, and they were as expert and diligent as the radiochemists in Paris. They too had published a paper in 1935 on the results of irradiating thorium and, as Hahn wrote Rutherford, had been "a little angry with Mme. Irène Curie for not having cited us properly . . . We regret this very much, for a scientific echo has never been as necessary to us as just now."[45]

A word about the Kaiser-Wilhelm Institute for Chemistry. Like many laboratories elsewhere, it was dedicated to the disinterested pursuit of knowledge, but on the assumption that this work served vital interests of industry and the state. The various Kaiser-Wilhelm institutes had been established with national prestige and practical needs clearly in mind, and most of them were connected with particular industries such as coal, leather, or medicine. During the First World War the Kaiser-Wilhelm Institute for Physical Chemistry, transformed into a militarized installation with 150 academics and 2000 other employees, became the center of German gas warfare. Hahn himself was one of the first chemists recalled from the front lines to work on poison gas.[46] After the war, scientists established close ties with the new government, while the chemical industry continued to provide most of the funds for the Institute for Chemistry itself. Hahn was awake to the possibility of industrial applications of the Institute's work, having earned a good sum before the war by helping a chemical firm to manufacture radioactive substances.

Yet Hahn and Meitner—the one unpretentious, the other shy—made few claims for the value of their own work. While the Nazi government stifled university life in Germany, they continued to concentrate on research. Meitner studied basic problems of nuclei, while Hahn devoted himself to applying radioactive tracers to chemical problems, work that bore not only on pure science but also on technological concerns. Meanwhile they worked together on irradiating uranium and thorium with neutrons. Like Irène Curie, the pair could not have expected either immediate practical results or fundamental discoveries from the uranium puzzle. They probably wished simply to tidy up the odd variety of heavy radioactive isotopes and perhaps with luck to discover altogether new elements heavier than uranium.[47]

During the late 1930s Hahn and Meitner, later joined by Fritz Strassmann, built up an elaborate table of supposed heavy isotopes all produced by irradiating uranium. As the table grew larger, it had to hold more and more isotopes, unlike the similar but simpler tables that had been made for better-known elements. A number of nuclear physicists found this excess of isotopes odd, but only at the Radium Institute did anyone begin experiments.[48] Curie looked into the problem in 1937, along with Pavle Savitch, a visiting Yugoslavian scientist. Pondering how to untangle the many isotopes in the radioactive broth made from irradiated uranium, Curie hit on the idea of focusing on whichever isotope had the most penetrating beta rays. She and Savitch covered their broth with a sheet of copper and studied the one substance whose radiation got through. They found it was a substance that their colleagues in Berlin had overlooked, one with a 3½-hour half-life. Curie failed to identify this with the substance she had produced from thorium and called actinium two years earlier, for actinium is substantially lighter than uranium and known processes could not produce so great a change in an atom. She and Savitch supposed that they had found a new isotope of thorium.[49]

The Berlin group could not believe that they had missed an important substance. They had been irradiating uranium for nearly three years and had identified four supposed elements heavier than uranium, each with a number of isotopes. There was no place in their elaborate scheme for a new thorium isotope. After reading Curie and Savitch's paper, they tried their experiments again, still without finding the isotope. Meitner and Hahn wrote Curie a letter, giving her the details of their new experiments and politely suggesting that she had committed a gross error. Perhaps she had found one of the substances they had already discovered, with a 2½-hour half-life, and imagined it to be a new substance with a much longer half-life. With characteristic courtesy they suggested that if she would make a public retraction, they would not publish their criticism.[50]

Curie and Savitch replied by publishing a retraction which was even more disturbing than their original claim. While admitting that the 3½-hour substance was not thorium, they reaffirmed its existence and reported that it followed lanthanum in chemical separations. From the chemical point of view, they said, it might be either actinium, which was chemically extremely similar to lanthanum, or some new and peculiar element. But "from the physical point of view, both hypotheses run up against considerable difficulties."[51] If they were right, there was something badly wrong with Hahn and Meitner's work.

In May 1938 Hahn met Joliot for the first time at an international congress of chemistry in Rome. While expressing respect and friendship for Joliot and his wife, Hahn did not hide his skepticism about Irène Curie's work. He told Joliot that he was going to repeat her experiments and prove she was mistaken.[52]

Curie and Savitch meanwhile decided to check their hypothesis that the substance with 3½-hour activity was actinium. Years before, Marie Curie had devised a laborious procedure to separate actinium from lanthanum, and when Irène Curie and Savitch went through the steps, they found that the 3½-hour substance followed along with the lanthanum, not the actinium. They presumed that they had some peculiar new element heavier than uranium, failing to imagine that the substance could be lanthanum itself. Not only was it unthinkable that a uranium atom should change into an atom half its weight, but also there were signs of subtle chemical differences between the 3½-hour substance and lanthanum. Probably they were testing not pure lanthanum but a mixture of lanthanum with other rare elements that have properties deceptively like lanthanum's.[53] Curie and Savitch, mystified by their results, could only suggest hypotheses which they admitted to be thoroughly improbable.[54] They talked over their problem at length with their friends at Joliot's and Perrin's seminars, but nobody could solve the puzzle.

Further experiments were delayed, first by the traditional August vacation at L'Arcouest and then by larger events. It had been an uneasy summer. Austria had become a province of Hitler's Reich in March 1938, and by September the dictator was laying claim to the German-speaking parts of Czechoslovakia, a country France was sworn to defend. In the streets of Paris white posters went up announcing mobilization for war. Gas masks were distributed, and ministries made contingency plans to evacuate Paris.

Hurrying back from the seacoast, the professors were dismayed to find that there were no plans for military security or air-raid protection at their laboratories and that their assistants, their students, and they themselves were being called to the colors. At a large meeting they held in the midst of the confusion, André Mayer took charge, proposing to the scientists that the laboratories be put under the protection of the armed forces, complete with military guards. A man liable to be called up for service who could be better used for war research would be granted a "special assignment," with orders to stay in his own laboratory.[55]

These plans had scarcely been laid when the crisis ended with the surrender to Hitler in Munich at the end of September. In the Chamber of Deputies the Communists alone were united in opposing the Munich agreement, and the Socialist party began to break apart over the issue of pacifism. Their disintegration left the Communist party more than ever the last resort for progressive intellectuals, and many French scientists sided with it against the Munich agreement. Around this time Langevin began openly to identify his goals with the Communist party's. While not going so far, Perrin, Joliot, and Irène Curie also denounced the policy of appeasing Hitler. In an open letter to the Radical premier Edouard Daladier a few months after the Munich conference, all three scientists declared: "We are afraid . . . that our external interests are being en-trusted in very weak men . . . We demand that no concession be made to the Italian and German demands."[56] Where physicists of an earlier generation might not have presumed to lecture their government on questions of international policy, Perrin's group felt duty-bond to join the debate.

During the gathering European crisis a scientific dispute was being played out, as polite as the public conflict was vicious. Esoteric and all but invisible, it would turn out to be as consequential as any of the events in the headlines. In Berlin during the Munich crisis Hahn received the paper of Curie and Savitch which reaffirmed the existence of a product of irradiated uranium with a half-life of 3½-hours, a strange product resembling nothing so much as lanthanum. Hahn's reaction was that this simply could not be, that Curie and Savitch had gotten hope-lessly muddled. But he could not ignore the French challenge. By now his system of supposed isotopes heavier than uranium had become so com-plex that he must have been almost as disturbed as Curie herself. Such an impasse can sometimes mean not defeat but progress, for when a scientific system becomes more and more elaborate and strained, it may eventually crack so that an entirely different idea can be born. Un-fortunately Meitner, Hahn's collaborator of thirty years, had been forced by Nazi anti-Semitism to flee Germany on a few hours' notice. From Sweden she urged on her colleagues by letter, but it was left to Hahn and Strassmann to go back to irradiated uranium and take a still closer look at the products.[57] They went systematically through a number of ex-periments. At first they thought that they had found three more isotopes of radium. But this conclusion stretched their system near the breaking-point, and they returned to the laboratory. Just as Curie and Savitch had used lanthanum for a carrier, so Hahn and Strassmann had used barium, an element about half the weight of uranium, as a carrier for their supposed radium. Now, for the first time, they went through the steps to separate radium from barium. Against their will, they were compelled to conclude that they had no radium at all, but radioactive barium instead.[58] Next they checked Curie and Savitch's work and found lanthanum in their broth.

"We publish these peculiar results rather hesitantly," wrote Hahn and Strassmann. As chemists, they had to conclude that barium and lanthanum were somehow produced out of uranium, but with their knowledge of physics, "we cannot bring ourselves to take such a drastic step, which goes against all previous laws of nuclear physics." After they sent the report in for publication, the thought came to Hahn that a barium atom plus an atom of another medium-weight element which could be in the broth would add up to give the mass of a uranium atom. Had the uranium nucleus somehow split in two? He revised the report in proof to hint at this idea; he also altered his statement that the results were "against all previous laws" to read, "against all previous experience." The laws had fallen.[59]

So far the study of uranium had been followed by a very few masters of the intricate and recondite techniques of radiochemistry; it was a pursuit beset with wrong turnings, gross errors, and confusion. After Hahn and Strassmann's paper was published, scientists suddenly stumbled out of a nearly impenetrable jungle into the open. Simple, straightforward experiments would follow one another swiftly. The difficulty now would be the very breadth of the prospect: few physicists saw the direction, among many which lay open, that would lead to the exploitation of nuclear energy.

The Collège
de France

Frédéric Joliot and Irène Curie, 1935

4

Formation of a Team

Hahn and Strassmann's paper, published January 6, 1939, arrived in Joliot's office by mail about ten days later. Noticing the authors' names, he immediately looked into the paper and recognized that it indicated uranium atoms might be split in two. Hans Halban, a researcher in the laboratory, recalled that Joliot "in a rather moving meeting made a report on this result to Madame Joliot and myself after having locked himself in for a few days and not talked to anybody." The news struck the Paris physicists like a thunderclap. Irène Curie was exasperated to find how close she had come to the truth without actually seizing it; she and Savitch had previously toyed with and rejected the notion that uranium might break in two. Another researcher, Lew Kowarski, recalled that she exclaimed, "What fools we have been!" and bitterly regretted that she and her husband had not worked together on the problem, for their combined brains might well have solved it.[1]

Dropping his other activities, Joliot turned all his attention to the new phenomenon. Within a few days he deduced two simple but momentous consequences. First, each splitting, or "fission," should release a large amount of energy; atom for atom, fission would be the most powerful process ever produced in a laboratory. Second, the fission fragments, such as barium or lanthanum, when added together would have fewer neutrons than the original uranium atom, and therefore some extra neutrons might be emitted after each fission.

Before anything else Joliot had to find out whether Hahn and Strassmann's tentative guess was true. Focusing first on the neutrons that might result from fission, he set up a quick experiment to find them. But this turned out to be difficult, and within a few days he thought of a swifter, surer approach. Fission should release so much energy that the fragments would fly apart and be easy to recognize.[2]

Joliot already had a feel for how far a heavy charged nucleus can plow through the air, for in 1931 he had briefly studied this phenomenon. The experience helped him estimate that the fission fragments would be thrown apart with enough energy to escape from the sample of uranium altogether and go a few centimeters through the air. They could then be collected and observed. Once he had the idea, it took him only a few days to set up the experiment. It was the sort of experiment that Joliot, like many French scientists, loved best, in which the apparatus seems to be simply a crystallization of logic into the natural world. Joliot needed only a small sample of uranium, a standard source of neutrons to irradiate the uranium, and a clean surface on which to catch the fission fragments. Nothing was found on the surface if either the uranium or the source of neutrons was removed, but if both neutrons and uranium were in place, the surface collected a wealth of radioactive fragments. Joliot could tell that these fragments were identical to the substances known to Curie's and Hahn's groups, for their radioactivity fell off over time in the same way. Joliot was now convinced that nuclear fission was a fact.[3]

Joliot evidently discussed these matters in detail with his wife, for they now turned at the same time to study thorium, starting separate experiments around January 27. Curie and Savitch, having finally recalled the 3½-hour substance produced from thorium in 1935, realized that it might not be an isotope of actinium, as everyone had believed, but rather the same 3½-hour lanthanum that resulted from fission. Within a few days they found that this was so, which implied that thorium, like uranium, could be split. Meanwhile Joliot confirmed the fission of thorium in his own way by collecting and identifying the fragments.[4]

Kowarski, who was in the room when Joliot first confirmed uranium fission on January 26, recalled that upon being congratulated, Joliot simply remarked that no doubt the same sort of experiments were being done elsewhere at that moment. In fact the first confirmation of fission was already a fortnight old. Otto Frisch, an Austrian physicist working in Copenhagen, had paid a visit over the Christmas holidays to his aunt, Meitner, who had just received a letter from Hahn describing the startling existence of barium in his broth. The first scientists outside Berlin to hear the news, Frisch and Meitner were also the first to understand that it was possible for a uranium nucleus to break in two. Frisch soon returned to the laboratory in Copenhagen and thought of a way to verify the idea. Beginning on January 12 and working long after midnight for several nights in a row, he detected the fission fragments.[5]

Even before this confirmation, Frisch had discussed fission with Niels Bohr, the creator and head of the Copenhagen institute. Bohr was just leaving for the United States, and he and his news reached scientists at Columbia University in New York on January 16, exactly the same day that Hahn and Strassmann's paper reached Joliot. At the moment that Joliot was finishing his experiment, physicists at Columbia were also confirming that fission occurs, and many other experiments were getting underway around the United States.[6] Although Joliot missed being the first to confirm fission, his experiment was essential to him and other French physicists. By convincing them that fission was real, weeks before the news would arrive through foreign publications, it put them at the head of the race. This was a position they never gave up.

Joliot had an exceptional opportunity to study fission. For several years he had been engulfed in administrative duties and had written only occasional scientific papers of slight value, but his laboratories were finally near completion and he was ready for a major new task. At hand were not only all the facilities he had gathered or constructed but also a trained technical staff and, above all, excellent postdoctoral students of nuclear physics. Among them were Halban and Kowarski.

Lew Kowarski was a huge man, heavy-footed, with a deceptively innocent face and an astonishing appetite. He had been born in St. Petersburg in 1907, son of a businessman and a singer. His father was Jewish and his mother Christian, a difference that nobody in Russia could overlook, so their two sons were born out of wedlock. After a few years the parents drifted apart. "It all contributed to create a fundamental feeling of borderline life, not quite belonging to the core of things," Kowarski recalled, "which characterized my whole career later on." In the thick of the Revolution his father extricated the boys from Russia and brought them to Wilno, a town which in the next few years often changed hands. Kowarski grew up an insecure outsider in a threatened community. He showed an early talent for music, but because his family had little money left, he needed a more secure vocation. In 1928 he took a degree in chemical engineering and moved to Paris, where he found a part-time job with a company that manufactured steel gas mains. In his remaining time he prepared himself for higher degrees, for he was determined to free himself from his low station by trying his talents as a scientist. He entered Jean Perrin's laboratory of physical chemistry, prepared a doctor's thesis, and then, like many Eastern Europeans, found his way into the Radium Institute. He served as a part-time research assistant to Joliot, all the while continuing to work for the steel pipe company. Meanwhile Kowarski married and applied for French citizenship. When in 1936 a daughter was born, he had to supplement his small wages by working part-time as Joliot's personal secretary, a job which opened his eyes to a larger world of scientific and political affairs. On moving to the Collège de France, Joliot got Kowarski a small CNRS fellowship to augment his pay as a secretary, enabling him to leave his job at the steel pipe company at last. Again the CNRS had freed a young outsider to do research.

Despite his efforts, Kowarski was still marginal to the French scientific community, a foreigner with few publications and little experience in nuclear physics. Since he was so large, he was jokingly referred to at the laboratory as Joliot's "little typist." He was well aware that his position was insecure. In the summer of 1938 he approached Halban, who decided to initiate him into the techniques of nuclear physics.[7]

Hans von Halban, Jr., had followed a career as smooth as Kowarski's was rugged. One year younger, he was born of Austrian parents residing in Leipzig, where his father was an assistant in a physical chemistry laboratory. In 1930 his father became professor at the University of Zurich, where he divided his time among solid scientific work, cultural life, skiing, and good company. Halban Junior, whose mother had died early, took after his father. In 1935 he won his doctorate by measuring the absorption of light in vapors, a topic related to his father's research.

Such work was less attractive than the sparkling new discoveries in nuclear physics. The young Halban, drawn by the reputation of Joliot and the Curies, went to the Radium Institute. He learned radiochemistry as an apprentice to Irène Curie and collaborated in several important studies, including the one where the $3\frac{1}{2}$-hour substance was first seen. Next he went off for a year's work at Niels Bohr's Institute for Theoretical Physics in Copenhagen where, working with Frisch and others, he sharpened his knowledge of the experimental techniques of nuclear physics. Back in Paris around the end of 1937, he followed Joliot to the Collège de France and was awarded a CNRS position. Meanwhile he found an attractive and well-to-do wife. With characteristic energy he began independent studies of deuterium, which is hydrogen with an extra neutron stuck on. Then he took on Kowarski as apprentice and collaborator and began studies of neutrons. When the news of fission arrived, he and Kowarski were planning a search for neutrons in cosmic rays.

Halban had come to Paris simply because the Radium Institute seemed a good place for an ambitious young scientist to work, but the Nazi takeover in Germany had also barred him from the normal scientific career there, for he was part Jewish. In 1937, on Joliot's advice, he applied for French naturalization. After Hitler annnexed Austria in 1938, Halban renounced his Austrian citizenship to avoid carrying a German passport. By 1939 Halban was keeping his laboratory notes in passable French, although his best friends were still German-speaking scientists with whom he kept up a cheerful correspondence. Like his father, he had cultivated tastes, a love for high society, and some financial independence, as well as a nose for important scientific problems. With over two dozen substantial publications behind him he was already an expert on the measurement of neutrons and their absorption in matter, a subject which in early 1939 had unexpectedly become of the highest importance.[8]

The scientists at the Collège de France suspected that fission was more than a laboratory trick. It might be the mechanism, long dreamed of, which would unlock the enormous energies Pierre Curie and others had discovered within the nucleus. The possibility that there might be some way to propagate a release of energy from one atom to another in a chain reaction, perhaps as an unprecedented explosion, had been suggested long since by respected scientists like Frederick Soddy. Probably the first to bring the idea into public view had been H. G. Wells. In *The World Set Free*, a science fiction novel which deeply impressed Kowarski when he read it as a boy in 1917, Wells fantasized a future in which infernal "atomic bombs" crushed hundreds of cities with fire, blast, and radiation. Catastrophic war was followed by an ideal world state which used nuclear energy to transform society and the earth, building garden cities in deserts and arctic wastes. Joliot, in his 1935 Nobel Prize address, predicted that nuclear energy might somehow be liberated in a chain reaction, one transmutation provoking the next. Such liberation of energy could serve humanity, but Joliot also warned scientists to be careful lest by mistake they blow up the world.[9] Halban, during his stay in Copenhagen, had heard talk about possibilities for a chain reaction. All that was needed was a process in which one neutron struck a nucleus and liberated more than one neutron. These neutrons would then swiftly strike other nuclei, releasing energy and still more neutrons, until the entire mass was transformed. The scientists at Copenhagen decided such a process was unlikely. The discovery of fission reopened the question. Although nobody had mentioned chain reactions just yet, Joliot recognized that he should find out whether or not neutrons were indeed liberated in uranium fission. Theoretically, as many scientists independently realized, there was a good chance that they were.[10]

It would be difficult to detect the occasional neutrons produced by fission amid the flood of neutrons that were needed to induce the fission in the first place. Joliot brought the problem to Halban, the laboratory's expert in neutron measurements, in mid-January, and Halban agreed to help work out a solution. He also told Joliot that Kowarski knew enough to assist, with the result, as Halban recalled, that "we took him as a collaborator on equal terms. Joliot had an uncanny knowledge of the importance of what was done. When I told him, 'I will work with you, but I would like to take Kowarski with me,' he said, 'This is important. This might make him a name. Do you intend to do that?' I said, 'He is fully qualified for it.' It is useful to mention because it shows that Joliot had a great foresight of the importance of what we did then."[11]

Up to this point Joliot had never formed a large research team; indeed, few physicists anywhere had done so. He had directed a few engineers working on the construction of particle accelerators; he had organized a laboratory in which individuals or groups of two or three pursued independent studies, as in the Radium Institute; and he had done close collaborative work with his wife. The new field which was opening up would force him to bring all these activities together and form a larger research team. But at first all that Joliot noticed was that it might be useful to work with two of the senior students.

Joliot privately discussed the scientific problem with Kowarski, and within a few days Kowarski came up with an idea for an experiment. This clinched his membership in the team. Meanwhile Halban independently suggested a different type of experiment. Around the start of February the three decided to pool their resources and do both experiments together. If only one experiment worked, Halban recalled, they would all share the credit.

After that came a period of hectic collaboration. When I say hectic, I mean we were twelve to fourteen hours a day in the laboratory and spent many hours together discussing frantically. It is now and was then impossible to say who contributed what. This question arose from time to time of somebody saying you had a good idea yesterday, I thought about it, and soon we had a kind of conspiracy agreement, we said really we must stop attributing ideas to one of us, because it might cause one day jealousy and there is a cross-fertilization of ideas if three people work together, and you must attribute whatever we did to the group.[12]

The team obeyed this ethic throughout their joint work. The historian may try to look deeper into the workings of a group and, to show how the group functioned, note that a certain action was taken by a particular person, but any credit belongs unquestionably to all the members of the team jointly. They collaborated very closely. In the surviving papers one may find that the first draft of an article was written by one member of the team and the final draft by a second, with significant corrections by the third.

The first of the team's joint experiments to be finished was the one suggested by Halban. It used a technique he had mastered under Frisch's tutelage at Copenhagen and which he himself had used when he came to the Collège de France. This method, first developed by Fermi and his colleague Edoardo Amaldi in Rome and extended by Frisch, served as the foundation for most of the French work on fission and much of the work done outside France.[13] It involved plunging a source of neutrons into the middle of a tank of water.

For the neutron source one usually used either radium or radon combined with beryllium. As in Joliot and Curie's 1931 experiments, rays from the radioactive substance struck the beryllium and ejected neutrons. Joliot's team never lacked powerful sources since they could borrow all the radium or radon they needed from the Radium Institute. The source was then sunk in a tank of water. The neutrons diffused through the water, bouncing around off the hydrogen and oxygen atoms, slowing down as they spread farther from the source. To detect the neutrons at a given point, one put in a "detector," a small strip of some material such as a metal. The neutrons induced artificial radioactivity in the strip, the intensity of radioactivity depending on the number and energy of the neutrons which bombarded it. The radioactivity was measured by setting the detector before a geiger counter or the like. Repeating the process for different points in the tank, one could map out a curve showing where the neutrons were dense and where scarce. By custom, the intensity of the detector's radioactivity, called I, was multiplied by the square of its distance from the source, r^2, to indicate the number of neutron captures, Ir^2, in a spherical shell surrounding the source.

This "density-distribution curve" would rise to a maximum at some distance from the source, partly because the neutrons would not induce radioactivity in the detector until they had slowed down somewhat; as Fermi had discovered, fast neutrons tended to shoot right through an atom without being absorbed. Still farther from the source the number of neutrons fell off and the curve dropped. In a water tank of practical size the curve would not drop to near zero. But if the tank were large enough, one could make a good approximation of the remaining "tail" of the curve by assuming that it fell off smoothly, like the exponential curves of mathematics (see figure).

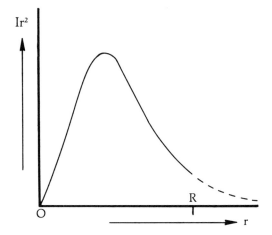

Schematic neutron density-distribution curve, used to test whether a chain reaction is possible. O=center of tank; r=distance from center; R=distance to edge of tank; I=intensity of radioactivity; —— =neutron density (I times r squared); - - - =extrapolation of curve outside tank.

This procedure demanded care and patience, but it was worth the trouble, for a density-distribution curve spoke volumes about how neutrons interacted with matter. Amaldi and Fermi, for example, who had used different elements as detectors, found that the measured curves had different shapes. The Italians deduced that a nucleus captured neutrons most avidly when the neutrons were of a particular "resonant" speed, which was different for each element. This resonance capture of neutrons of certain speeds later played a key role in work on chain reactions. Another use of the density-distribution curve was made by Frisch, Halban and Koch in Copenhagen in 1937. Pouring one or another chemical into the liquid in their tank, they could determine from the curves how easily carbon, oxygen, and other elements absorbed neutrons. Halban now turned this method on its head to find not the absorption but the production of neutrons in uranium.

It was known that the total area under the curve was proportional to the total number of neutrons in the tank at a given moment. To find the area under the curve, the team could simply measure the neutron densities and plot them on graph paper, then count the little squares under the curve. If they put some uranium in the tank, there would be two possible effects: The uranium would certainly capture some neutrons and so tend to decrease the area under the curve; but if fission produced neutrons, the uranium would tend to increase the area. If the density-distribution curve covered a larger area when uranium was in the tank than when it was not—in other words, if adding uranium raised the total radioactivity that neutrons induced in detectors scattered throughout the tank—they could be sure that fission produced neutrons.

First the team needed a source of low-energy neutrons. If high-energy neutrons were used, they might produce extra, unwanted neutrons through a familiar type of reaction in which one fast neutron struck a nucleus and two slow neutrons came out; these would mask any neutrons produced by fission. The relatively common and cheap radon-beryllium neutron sources could not be used, for they produced neutrons whose energy was too high, but the well-stocked Collège de France laboratory already had a radium-beryllium source of suitably low energy.

The team plunged their slow-neutron source into the middle of a tank a half-meter across, filled the tank with water, and dissolved a uranium compound in the water. Then they laboriously measured the neutron distribution with detectors of dysprosium, a rare silvery metal. For comparison, they also measured the distribution when the tank was filled with a uranium-free solution. The measurements, begun on February 19 and lasting for two weeks, were accompanied by checks of the chemical composition and concentration of the solutions, standardization of the radiation counters, and all the other tedious details required by a quantitative experiment. The team did not find any increase in the area under the density-distribution curve. The uranium absorbed too many neutrons.[14]

Nevertheless by the end of February Kowarski wrote to Joliot, "Up to now all the indications concerning uranium in water are *positive* (the explosion gives neutrons! not many, however)." The proof was unexpectedly given not by the area but by the shape of the curves (see figure). Absorption of neutrons made the peak lower in the curve for the uranium solution, but this curve did not fall off so steeply as the other. Far from the central source, there were more neutrons in the uranium solution. There was no way this could happen unless neutrons were being emitted by the uranium itself. Either energetic neutrons were being produced by uranium fission near the source and traveling farther than the original source neutrons, or else neutrons were being produced far from the source by uranium fission.[15]

The team at the Collège de France were eager to publish this important result. But they paused to make the fact completely certain. To check the crucial part of the curve far from the source, where neutron intensity was low, they decided to use a much stronger source of slow neutrons, so

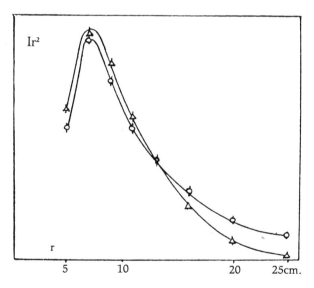

Experimental neutron density-distribution curves, published by the French team in *Nature*, March 18, 1939. Δ = ammonium nitrate; o = uranyl nitrate (uranium). The curve with uranium is higher far from the center of the tank, which demonstrates that neutrons are produced in fission.

Halban requested a large quantity of radium from the Radium Institute. The functionary there defended the radium "like a dragon," as Halban reported, but one phone call from Joliot secured it. The team quickly completed their measurements.[16]

They agreed to publish their paper in the British journal *Nature*, not only because this was one of the main outlets for fission research but also because they wanted to publish an illustration (see figure), which would delay publication in the French *Comptes Rendus*. The point of science is to discover things, and it means little to be the second to make a discovery; by universal agreement among scientists the first to publish is the first and often the only one to get credit, so in a fast-moving field much can depend on speed of publication. On March 8 Kowarski went out to Le Bourget airport to post the letter in order to make sure it would reach the publisher swiftly. "We knew that the whole scientific world was boiling over these things," Kowarski recalled. "We did not realize how much we were in advance of the others . . . We thought there were far more numerous attempts in progress, so why not secure priority?"[17]

Meanwhile the team worked on their other experiment, the one proposed by Kowarski. The idea here was simply to detect fission neutrons by their energy. If the team again took a source of low-energy neutrons, surrounded it with uranium, and observed high-energy neutrons coming out, these fast neutrons could come only from uranium fission. The team took the large neutron source that they had just used in the other experiment, covered it with a layer of uranium compound, and submerged it in a solution containing sulphur. If fast neutrons were produced in the uranium, they would react with the sulphur to create radioactive phosphorus.[18]

In their chemical work the team were aided by Maurice Dodé, a chemist working in Moureu's section of the laboratories. Dodé heard the idea of the experiment at one of Joliot's laboratory seminars and agreed to help with the delicate separations of phosphorus from the solution. This was only one of the many times that the team used their laboratory's unusual resources of equipment and personnel. In their density-distribution experiment they had already been helped by Pierre Delattre, a young laboratory technician who carried out some of the routine measurements.[19]

Dodé waited a week to let substances accumulate in the solution, then analyzed it and found the characteristic radioactivity of phosphorus. The French now had confirmation that neutrons were produced by fission, and also proof that at least some of these neutrons were emitted with high velocities. "With this positive information," Halban recalled, "we were thoroughly convinced that the conditions for establishing a divergent chain reaction with neutrons could be realized."[20]

Halban, Joliot, and Kowarski were making such rapid progress because they constituted a true team, each member bringing unique abilities. Halban, as the expert on neutron measurements, could guide the details of the experiments and interpret the results, meanwhile offering new ideas. Equally important, he brought a powerful drive to make the work succeed. Kowarski's particular contributions were conceptual. The field was so new that mathematical expertise, which he lacked, was less important than the originality and meticulous simplicity which characterized his thought. His big hands were not adapted to the construction of apparatus, but he contributed both hard labor and ingenious ideas to the experimental side of the work. Joliot was the unquestioned chief and, as a colleague said, "the cement and the *animateur* of that team."[21] He contributed in many ways, not least by providing a well-equipped laboratory and all the help necessary from governmental and private sources. Within a few years both Halban and Kowarski were to show that they too could promote large scientific organizations, but in 1939-1940 this task was Joliot's. Besides his vital political skills, Joliot had a penetrating logic which helped clarify and organize the team's ideas. Moreover, he possessed a broad scientific background which took him beyond the limits of his younger colleagues' knowledge, as when he helped work out radiochemical procedures or the detailed design of a mechanical device. It is impossible to imagine the team succeeding so well without all three of its members.

Their personalities differed as widely as their skills. Halban was a dynamo of aggressive enthusiasm whirling within the urbane exterior of the cultured European. To the subordinates in the laboratory, used to the genteel traditions of French science, he seemed the model of a brilliant but offensively authoritarian German professor. Kowarski had the big man's tenacity and unexpected scrupulosity. He was notoriously independent and blunt-spoken, and if the French laboratory workers sometimes saw Halban as a Junker, they sometimes saw Kowarski as a Russian barbarian. Although he owed Halban a great deal, Kowarski would not suffer himself to be subordinated by him, particularly when he seemed finally near his heart's desire of becoming a scientist in his own right. There was a potential for conflict here, but for the moment Halban and Kowarski pulled together in close and usually friendly collaboration.

Joliot himself was always surrounded, in the laboratory and outside, not only by friends but by devoted followers. He was not as handsome as in earlier years, for his face had grown thinner, his cheekbones and nose more prominent, but few people found him unattractive. It has been suggested that some hidden insecurity drove him constantly to court both companions and success. Whatever their origin, his charm and drive served to keep the team and their helpers moving smoothly and vigorously. An onlooker called him "less a laboratory director than a coryphaeus, a moving spirit."[22]

The three men had only a few things in common. They were all in their thirties, Joliot the oldest by a few years. They all had intelligent wives and small children, with Halban's first child born that August. And they were all, each for his own reasons, more ambitious than most men; this was the ultimate precondition for their success.

5

The Secret of the Chain Reaction

In early February 1939, when the team at the Collège de France were working on their first experiments together, a letter came from Leo Szilard in New York, addressed to "Professor M. Joliot" and dated February 2. It read as follows:

Dear Professor Joliot:

The only reason for my writing to you this letter to-day is the remote possibility that I shall have to send you a cable in some weeks, and if that happens this letter will help you to understand what the cable is about. This letter is therefore merely a precaution, and we hope an unnecessary precaution.

When Hahn's paper reached this country about a fortnight ago, a few of us got at once interested in the question whether neutrons are liberated in the disintegration of uranium. Obviously, if more than one neutron were liberated, a sort of chain reaction would be possible. In certain circumstances this might then lead to the construction of bombs which would be extremely dangerous in general and particularly in the hands of certain governments.

It is of course not possible to prevent physicists from discussing these things among themselves, and, as a matter of fact, the subject is fairly widely discussed here. However, so far, every individual exercised sufficient discretion to prevent a leakage of these ideas into the newspapers.

In the last few days there was some discussion here among physicists whether or not we should take action to prevent anything along this line from being published in scientific periodicals in this country, and also ask colleagues in England and France to consider taking similar action. No definite conclusions have so far been reached in these discussions, but if and when definite steps are being taken I shall send you a cable to tell you what is being done.

We all hope that there will be no, or at least not sufficient, neutron emission and therefore nothing to worry about. Still, in order to be on the safe side, efforts are made to clear up this point as quickly as possible. Experiments at Columbia University are in charge of Fermi and will perhaps be the first to give reliable results.

Perhaps you have also thought of the same things and have contemplated or started such experiments. Maybe you are able to get definite results at an earlier date, which, of course, would be very valuable help towards ending the present disquieting uncertainty. Whatever information on the subject you might care to transmit by letter or cable at some later date will, I am sure, be greatly appreciated. Also, should you come to the conclusion that publication of certain matters should be prevented, your opinion will certainly be given very serious consideration in this country.

(signed) Leo Szilard[1]

Szilard was a Hungarian refugee physicist with neither a job nor a home, a short, stout, impetuous man bursting with ideas. Like Kowarski, though much later, he had read H. G. Wells' *The World Set Free* and was struck by its vision of nuclear energy devastating and transfiguring civilization. Szilard believed with Wells that the world was headed for social catastrophe and could be saved only by the combined action of the best people, including scientists like himself. In 1933, having just emigrated to London to escape the Nazi persecution of Jews, Szilard was jobless and had time to think about both society and the booming field of nuclear physics. Nuclei were not his specialty, but he now chanced to realize before anyone else that there was a serious possibility of creating a nuclear chain reaction. He predicted that if an atom of some element could be bombarded with a particle, such as the newly discovered neutron, then the atom could be made artificially radioactive. He further predicted that if this radioactivity was such that the atom immediately emitted two particles, then a chain of reactions could swiftly spread through a collection of such atoms. This possibility seemed to come closer in 1933 when Joliot and Curie confirmed Szilard's first prediction by discovering artificial radioactivity. Wells' novel of 1913 had imagined just such a discovery to occur in 1933, as the forerunner to the unleashing of nuclear energy. At this point Szilard, possessed by his dream of chain reactions, decided to devote himself to nuclear physics.[2]

Thinking at first that he might be able to provoke neutron emission in beryllium, an element that some physicists thought had a peculiar instability, he sank part of his savings into a block of the rare metal and planned some experiments. But beryllium turned out to be more stable than scientists had supposed.[3] Szilard's next candidate was indium, another rare metal with an instability that was not understood, and by 1935 he thought it as likely as not that he could start a nuclear chain reaction in this element.

From the start he recognized that his work might be consequential. He warned the British Admiralty of the small but real possibility of constructing "explosive bodies . . . very many thousand times more powerful than ordinary bombs." And he brashly tried to alert British scientists. Because the study of chain reactions, he told one professor in 1935, could bring about an industrial revolution but might also cause a disaster, an attempt would have to be made "gradually to bring about something like a conspiracy of those scientists who work in this field." To the head of physics at Oxford University, Szilard suggested a secret agreement to make scientific results in the "dangerous zone" available only to selected groups of nuclear physicists in Britain, America, and perhaps one or two other countries.[4]

Szilard also foresaw only too well the likely reaction to his efforts: "Unfortunately it will appear to many people premature to take some action until it will be too late to take any action." The senior physicists in Britain were cool to Szilard's obstreperous advice. They thought his proposed chain reaction mechanism thoroughly unworkable, as was in fact the case for indium; they suspected him of seeking personal cash profit, a motive incompatible with the tradition of disinterested science; and they found the idea of secrecy in pure science altogether alien. Even British scientists like Bernal who spoke strongly about the responsibility of scientists for the use of their discoveries nevertheless maintained that secrecy was abhorrent. To interfere with the normal international exchange of ideas would be to impede scientific progress, to pervert it and all human progress.[5]

Szilard seemed all the more suspect because he was looking for money, a few thousand pounds to finance his research. As early as 1934 he had thought up two sorts of experiments. He might assemble a mass of a given element, expose it to a source of neutrons, and study the resulting density of neutrons within the mass, using small pieces of another element as neutron detectors. Or he might search for the production of neutrons by using an element sensitive only to fast neutrons. These two methods were the ones that several years later the French would choose to find the neutrons produced in uranium fission, and in thinking up such techniques when neutron physics was still scarcely begun, Szilard showed the full powers of his active imagination.

Even if the tests should prove negative for indium, Szilard was prepared to press on and try all the other elements. For one thing, he was already growing suspicious of Hahn and Meitner's increasingly elaborate table of heavy isotopes supposed to be produced from uranium. Bombarding indium produced one or two isotopes that puzzled physicists, but bombarding uranium produced a dozen. "The case of uranium," Szilard wrote Bohr in 1936, "seems to me somewhat analogous to the case of indium."[6] But Szilard never raised the money to do all the experiments he planned. If he had, the Second World War might have had a different character.

Supporting his work out of his savings, Szilard was able to study only indium, and it took him until 1938 to unravel its peculiarities and find that they could not lead to neutron multiplication. Although his private discussions had made a number of leading physicists familiar with the notion of neutron chain reactions, there seemed to have been no further effect.[7] Discouraged, homeless, and jobless, Szilard was ready to give up on chain reactions when in January 1939 the news of the discovery of fission revived his hopes and fears. He was now in New York, for after the Munich crisis he had predicted that there would be war in Europe and saw no point in staying. He began to discuss his worries with the scientists around Columbia University.

The leading nuclear physicist at Columbia was Fermi. Having fled Mussolini's Italy when his Jewish wife was no longer welcome there, Fermi had arrived with his family in New York to take up a professorship at the start of 1939. He was half bald now, a small man with a long nose and a humorous smile, who drove himself so intensely that nobody resented his intellectual superiority. Like Joliot at this time, Fermi was at loose ends and on the watch for a new problem, and as soon as he arrived in New York, he began to read the scientific literature and take up theoretical work. When the news of fission broke, Fermi took it seriously. By one account he made a rough calculation of the size of the hole in Manhattan Island that a kilogram of uranium would make if it all exploded: a very big hole.[8] But Fermi concluded that this outcome was unlikely, for fission probably would not release enough neutrons to sustain a chain reaction.

Shortly after hearing the news of fission, Szilard asked I. I. Rabi, a physics professor at Columbia, to warn Fermi that the matter should be kept secret since it might lead to the construction of bombs. A few days later Rabi told Szilard that Fermi's response had been: "Nuts!" According to Szilard, he and Rabi went to Fermi's office to ask for an explanation:

Fermi said, "Well . . . there is the remote possibility that neutrons may be emitted in the fission of uranium and then of course perhaps a chain reaction can be made." Rabi said, "What do you mean by 'remote possibility'?" and Fermi said, "Well, ten percent." Rabi said, "Ten percent is not a remote possibility if it seems that we may die of it. If I have pneumonia and the doctor tells me that there is a remote possibility that I might die, and it's ten percent, I get excited about it."

From the very beginning the line was drawn; the difference between Fermi's position throughout this and mine was marked on the first day we talked about it. We both wanted to be conservative, but Fermi thought that the conservative thing is to play down the possibility that this may happen, and I thought the conservative thing is to assume that it will happen and take all the necessary precautions.[9]

Rebuffed by Fermi, Szilard remained alert for a way to control events. Sometime in late January a telegram arrived at Columbia from Halban to his friend George Placzek, another Central European refugee physicist. As Szilard recalled, the telegram was opened by a secretary by mistake and Szilard learned the contents: "JOLIOT'S EXPERIMENTS SECRET." It is not clear what Halban and Placzek were actually corresponding about, but because Placzek had just come from Paris, Szilard assumed that he had learned of an experiment that Joliot was doing, which Joliot had now decided to keep quiet. Szilard had little doubt what experiment could be so important as to require secrecy. But Joliot might have been invoking only the sort of secrecy that had been traditional in science for centuries, secrecy stemming from the caution of the scientist who holds back results until ready to publish them, so that they will not be broadcast in a distorted form and so that others will not take advantage of a hint to beat him to the next result. This was quite different from the sort of secrecy that Szilard desired.

Convinced that Joliot's group was working on fission, Szilard sent him the letter of February 2 suggesting a secrecy pact. This message was a purely personal venture. Szilard, who had wandered for years in Germany, Britain, and America without a permanent job, had no formal affiliation with Columbia or any other institution. In his letter he admitted that nobody had yet agreed on secrecy, a fact that may have been reinforced by a letter Fermi sent Joliot two days after Szilard's. "I am presently engaged," Fermi wrote, "like, I think, in every nuclear physics laboratory, in trying to understand what goes on in the castastrophal disintegration of Uranium." Having thus informed Joliot that he had competition, Fermi went on to other matters without saying a word about keeping secrets.[10]

Even as a personal request, Szilard's letter made little impression on the French. When weeks passed and the cable promised in the letter did not appear, the French, as Kowarski recalled, "considered that probably the whole idea was abandoned." That the Columbia group was hot on their heels they knew not only through Szilard's and Fermi's letters, but also through news reports of the work and through a letter from Hugh Paxton, a cyclotron worker in New York, saying that the "first job that the cyclotrons here find themselves involved in is the Uranium split business with which half the world seems to be occupied."[11] There was every reason to believe that Fermi would publish first if the French held back their own results, and moreover that many other scientists were feverishly working on the problem.

In fact, despite an initial flurry of interest in fission, hardly any physicists were attacking the problem in a way that would lead to the exploitation of nuclear energy. For example, the French might have expected competition from physicists at Berkeley, California, where Ernest O. Lawrence had put together one of the world's strongest teams in nuclear physics and where a radiochemistry group also flourished. The Berkeley scientists had recognized at once, as Lawrence wrote a colleague, that if neutrons were emitted in fission, "it will mean very real prospects for useful nuclear energy." But while a spate of papers on fission flowed from Berkeley in the early months of 1939, none was about the point that concerned the French team. The Berkeley scientists confirmed the existence of fission in various ways, measured the ranges of the fragments flying through air, and studied the chemistry of the fragments, but did nothing to learn whether a chain reaction could be started. Only Luis Alvarez, one of Lawrence's brightest young physicists, came close. He had recently built a device that produced a pure beam of slow neutrons from the Berkeley cyclotron, the only beam of its kind then existing. He signed out a bottle of uranium oxide from the chemical storeroom, put it in his beam, and hooked up a radiation counter to see whether fast fission neutrons would come out. "When I didn't see any effect in a couple of minutes," Alvarez recalled, "I merely said, 'Too bad,' and went back to what I was interested in at the time. There is no doubt that had I taken an hour off to move the counter closer to the cyclotron, and to collect some more uranium, the counts would have been there." Alvarez and the others in Berkeley were struggling with more fundamental problems in nuclear physics and with the construction of a new and bigger cyclotron; they had little time to spare on fission. [12]

The French could also have expected competition from Hahn and Strassmann and from Meitner and Frisch, who had a few weeks' head start on everybody else. But these workers only produced more studies on the chemistry, energies, and radioactivity of the fission fragments. These were interesting questions, but they did not bear directly on chain reactions. In Berlin, Hahn and Strassmann made one attempt to find the neutrons resulting from fission, but with inadequate equipment; when they had no success, they returned to their radiochemistry. In Copenhagen, Frisch thought of looking for the fission neutrons but could not figure out a way to find them with the equipment at hand. A colleague pointed out that uranium fission might start a chain reaction. "The idea clearly opened fascinating vistas," Frisch recalled, "but I did nothing about it. Threat of war had become palpable in the spring of 1939 and I did not feel like starting any new work that might soon be stopped." The German threat was serious for Frisch, his father having just been released from a concentration camp, so he gave up on fission work and began to look for a way to move to Britain. [13]

The only serious rivals to the Collège de France team were in New York. Columbia University was unusually well set up for fission work. Besides Fermi, there was a group under John Dunning and George Pegram, who had been working there on neutrons since 1932 and had just finished building a cyclotron. Although Fermi was reluctant to jump into experiments so soon after his arrival, Herbert Anderson, a strong-willed graduate student who was excited about fission, persuaded him to start some experiments. Anderson, Fermi, and several others collaborated on a quick test that confirmed the existence of fission; their experiment produced results on January 25, 1939, a few hours before Joliot's confirmation of fission. Over the next two months Fermi and his group prepared further experiments remarkably similar to the independent French ones. Placzek, who made frequent visits to New York, wrote Halban in early March that "Fermi has made a remarkable reorganization of the whole Columbia operation and has everything well in hand, and some really good physics should come out of there now." Although Fermi refused to believe that a chain reaction was feasible until he saw undeniable proof, he never doubted that the problem was worth a close look. As a joke, he said that someday Anderson would be president of a Uranium Corporation of America.[14]

A word about the Pupin Laboratories, home of the Columbia University Physics Department. Like most of the places where nuclear physics was developed, these laboratories held aloft the banner of disinterested research but stood on a foundation of potential uses of science to industry and the nation. Michael Pupin, an immigrant physicist who had been deeply impressed by the search for "eternal truth," was nevertheless best known for his inventions. His patents became the basis of a fortune which he donated in part to the service of physics at Columbia. During the First World War Pupin and his colleagues threw themselves into the research on defense against submarines, in parallel with Langevin and Rutherford. After the war, interest in practical applications strengthened not only at Columbia but throughout American physics until Pupin could say that "the universities are beginning to be led by the industries, instead of vice versa," a development he considered a triumph of scientific idealism.[15]

Shortly after the Physics Department opened their new, thirteen-story laboratories in 1927, they defined the goals of their research into matters like the nature of the atom:

The solution or the partial solution of such problems . . . while they may start with very abstruse results, will inevitably lead to consequences of the most practical kind . . . One question of the very greatest importance to the human race is that of our future supply of energy. Whatever it may be, whether we shall, as is quite conceivable, be able to tap sources of energy as yet untouched, or whether it will be secured by a better utilization of the sun's radiation to us, or however else it may come, it is probably the largest single problem for applied Physics.

Of course, it is the prime function of the Physicist to extend the borders of knowledge, and while he is engaged in doing this he need not concern himself with the question of the immediate utilization of the knowledge that he obtains. His discoveries will in these days be quickly taken up by engineers and inventors.

The physicists in the Pupin Laboratories made good use of their ideology as they developed the connections that would help make their predictions come true. For example, to build up their cyclotron, they were persuasive enough to get free equipment from the United States Navy, the Bell Telephone Laboratories, and the United Electric Light and Power Company.[16] It was thus in the local tradition for Fermi's group to have pure science and potential applications in mind at the same time.

Szilard, who was not in the Columbia group, was strongly interested in their work. He noticed that the scientists would not get reliable results with the radon-beryllium neutron source that they were using, since the fast neutrons from this source might cause a multiplication of neutrons by reactions that had nothing to do with fission. A slow-neutron source, such as the radium-beryllium one the French used, was not available at Columbia, but the resourceful Szilard wrote to Oxford to ask for a large cylinder of beryllium that he had bought in 1934 when he first began to search for nuclear chain reactions. Meanwhile he rented a gram of radium at $120 a month out of his own pocket. He had to borrow $2000 from a friend so that he could keep the radium for several months and could hire assistance. Szilard told people that the money came from a non-profit "Association for Scientific Collaboration," a hopeful fabrication. As a private person not attached to a university, he had trouble insuring the valuable radium. Negotiations with the radium company dragged on for weeks, ending successfully at the beginning of March.[17]

When Szilard now appeared at the Pupin Laboratories with his neutron source, it aided him in injecting himself into their activity. He was granted permission to work there as a guest for three months. In collaboration with a younger physicist, Walter Zinn, he immediately found the energetic neutrons resulting from fission. The two scientists used the same concept as the French had used in their second experiment (Kowarski's), detecting fast fission neutrons produced by bombarding the uranium with slow neutrons. Then Szilard pressed his neutron source on Fermi, who with Anderson and others used it for a density-distribution experiment that was nearly identical with the French team's first experiment (Halban's). All these results were ready by about March 15, independent of the French and scarcely a week after Kowarski had gone to the airport to rush off to *Nature* the Collège de France team's report on the production of neutrons in fission.[18] If Szilard had been able to get radium with one phone call, as the French had done, the two groups would have finished at about the same time.

The similarity between the experiments in New York and in Paris stemmed from other similarities: these were the only two places where physicists saw the immediate possibility of nuclear energy, had the means to pursue that possibility, and did so with ingenuity and urgency. Like the French, the scientists in New York wanted a quick answer to the question of whether uranium chain reactions were feasible. Now that they had a partial answer, they paused to decide what to do with it. Meanwhile, unknown to the New York group, the team at the Collège de France were pressing forward.

The French, who were getting increasingly excited about the possibilities for chain reactions, still had no idea whether or not these possibilities could ever become realities. They needed not new experiments but a new way of looking at the experiments they had already done. Fermi's group had confirmed that fission, like artificial radioactivity, was provoked much more easily by slow neutrons than by fast ones; the fast neutrons would shoot through a nucleus without having time to react with it.[19] This fact suggested that a complex process had taken place in the tank of water. Whenever a uranium atom split apart, the fast neutrons that came out would bounce around in the liquid until they slowed down, and only then would be absorbed by uranium atoms to start new fissions. A chain reaction thus depended on the water in the tank, for it was collisions with the water's hydrogen atoms that slowed down the neutrons. A mass of pure uranium would not do; there had to be a "slow-downer" or "moderator."

This deduction suggested other important conclusions, which the team, unused to the logic of chain reactions, had difficulty grasping. Hydrogen could sometimes absorb a neutron, and for that matter, uranium too could sometimes absorb a neutron without breaking apart. The team groped toward the implications of this neutron absorption, inventing obscure arguments about "cross-sections," a term that called up a picture physicists had used in their thinking for nearly half a century. When a beam of particles, in this case neutrons, was thrown upon a thin layer of a substance, in this case uranium, either the particles were absorbed or they passed through, and the cross-section of uranium for neutron absorption was defined as the probability that a neutron in the beam would hit a uranium nucleus and be absorbed. It was like shooting bullets at a row of bottles and reckoning the cross-sectional area of the bottles, which would tell the probability of hitting a bottle rather than missing them all. However, this traditional image would not help, for the group of neutrons diffusing about among the atoms in the tank at the Collège de France was nothing like a stream of bullets. It was more like a batch of balls knocking about in a three-dimensional pinball machine, where at each impact a ball might slow down, or speed up, or be captured, or drop into something that would spit out more balls. The team did not know how to analyze such chaos.

But they recalled that there were more than enough neutrons in the tank to account for all the neutrons produced by the central source, and from this fact alone, ignoring all the puzzling details of absorption and fission, they could reach a key conclusion. For they saw that the set of neutrons was like a human population in which many individuals die without producing children, analogous to the neutrons that were absorbed without producing more neutrons through fission. Such a population will maintain itself only if those who become parents average more than one child apiece. Similarly, since the neutron population in the tank did maintain itself, each fission had to be giving birth to more than one neutron. Joliot jotted down the conclusion and underlined it: *"several neutrons are emitted per explosion, a necessary condition for the production of chain reactions."*[20]

At first the team were not clear about this, but through March they moved haltingly toward the new "population" way of thinking. It was a sort of argument hitherto used chiefly by chemists concerned with atoms reacting in solutions or gases. The team may have found this approach easier because they all had some training in industrial chemistry. Joliot, while a young student, had briefly studied hot gases for an industrial firm; Halban's doctoral research in Zurich had concerned vapor pressures and chemical reaction kinetics; Kowarski's outside job at the gas pipe company had forced him to understand diffusion thoroughly, and he had just published a book on commercial gas distribution.[21] The movement of neutrons among the atoms in a liquid, resembling the movement of molecules diffusing through a gas, was something they could learn to visualize more quickly than could most physicists.

The first thing they wanted to know was the "birth rate" in their tank, that is, the number of neutrons produced per fission. They named this number with the Greek letter v ("nu"). There was no way to predict how big the birth rate would be, for there was no fundamental reason for it to be small or large. From the standpoint of pure theoretical physics, predicting the size of v would be as difficult as predicting how many pieces a glass will break into if it is dropped, and about as uninteresting. But the French, unlike many physicists, saw that this number was of major interest: the bigger it was, the better the chances that nuclear energy could be released on a large scale.

To find the birth rate, they would have to construct an equation, plug in some measured numbers, and crank out the result. Even before building the equation, they could see what sort of parts had to go into it. They would need to know the total population of their little nation of neutrons, and they would need to know the "death rate." They already had a measure of the total population, for the area under their density-distribution curve increased or decreased as the number of neutrons in the tank rose or fell. Neutron deaths occurred through various processes. A neutron might be absorbed so as to cause fission, and the cross-section for such absorption had already been measured and published in February by the Columbia group. Or a neutron might be absorbed without causing fission, and there were two ways this could occur, either through absorption of slow neutrons or through "resonance" capture of faster neutrons. The team would have to measure the cross-sections for both processes.

If a uranium nucleus absorbed a slow neutron but failed to fission, the process would be simply the familiar one that caused artificial radio-activity. With the neutron stuck on, the nucleus would be a slightly heavier, radioactive isotope of uranium. The team could find the probability of this sort of absorption by irradiating uranium with neutrons and measuring how much of the new isotope was produced. Once again the team called for outside help, in this case Savitch from the Radium In-stitute, to help with the chemical separations, for the new isotope quickly went through a set of transmutations and therefore the separations had to be done within half an hour after the irradiation was finished. These experiments were completed at the end of March 1939.[22]

Meanwhile the team had been measuring the resonance capture of faster neutrons, and this measurement was trickier. The difficulty was that only neutrons within a narrow range of speeds would be lost through resonance capture. Thus each particular mixture of uranium in water, with its own proportion of uranium and its own distribution of neutron speeds, would capture a different percentage of the fast neutrons while they were slowing down. When the French team set out to measure the probability of capture, they stuck as close as they could to the traditional way of experimenting in terms of cross-sections. Instead of measuring the captures directly in the same tank used in their density-dis-tribution experiments, they sent a beam of fast neutrons into a two-centimeter thickness of the uranium-water solution and measured the proportion of neutrons that got through. The measurement was not conceptually ideal and it seriously overestimated the resonance capture probability, but it was quick and gave at least an approximate number.[23]

Now the team could construct an equation, feed in their measured numbers, and turn out a value for v, the number of neutrons produced per fission. Since their equation would have to include not only the various cross-sections but also the area under the density-distribution curve, the team spent a few days rechecking and extending their measurements of the curve.[24] When they wrote down their equation, however, they made a subtle error that illustrates how chain reaction theory was hampered by traditional ways of thinking. Because their mathematics resembled the sort that physicists had always used for particles shooting through a thin layer, they overlooked the fact that in their tank neutrons did not just go through the uranium once and for all but were continually bouncing around and reproducing among the uranium atoms, which is the essence of a chain reaction. The simple point was so unexpected that no physicist saw through it until the following year. Their errors caused the French team to overestimate v. Having neglected the proliferation of neutrons in chains, they needed a high birth rate if single fissions were to produce all the neutrons observed. They came up with 3.5 neutrons produced per fission with an uncertainty of ±0.7, whereas the true value should have been close to 2.5.[25] This error made the team's overall conclusion seem much stronger: a number of neutrons were emitted in each uranium fission, apparently three or four, which was almost certainly enough to sustain a chain reaction. They reached this conclusion around the end of March. From this time forward, Halban recalled, "We were absolutely bent on creating a nuclear chain reaction which could be used for industrial power."[26]

Fission might become a source of virtually free energy which would transform the world. Moreover, where the Laboratory of Nuclear Synthesis had been producing new elements by transmutation only in microscopic quantities, the nuclear industry of the future might produce them by the kilogram for both practical and purely scientific uses. The vision of scientists since Pierre Curie, the promises made by Perrin and his friends when seeking support for science, were becoming facts. The team saw that a great opportunity lay in their hands. As Kowarski recalled: "To be the first to achieve the chain reaction was like achieving the philosophers' stone. It's far more than a Nobel Prize. We were perfectly aware of that. Whether it was disinterested glow of scientific curiosity or sordid interest of self-exaltation, I don't know—mixed motives, human nature."[27]

As the team at the Collège de France were finishing their experiments and calculations on v and planning to write up the results, the rival group at Columbia University were also thinking about how important fission might be. The scientists in New York had not yet received the French team's first report, rushed off to *Nature* and published there on March 18, but from their own similar work they already knew the chief conclusion, that fission produces some neutrons. Unlike the French, the Columbia group had not gone on to measure exactly how many neutrons were produced per fission, and from rough estimates Fermi concluded that there probably would not be enough to sustain a chain reaction. Nevertheless Szilard stressed the danger. His old friend and fellow-Hungarian Eugene Wigner, a theoretical physicist at Princeton, on a visit to Columbia insisted that the matter was so serious that the government should be informed.

On March 15, as the Columbia group finished writing up their experiments for publication, German troops invaded the remnant of free Czechoslovakia and immediately subdued it. The next day Dean George Pegram, administrative patron of the Columbia group and a respected physicist, telephoned the Navy Department to arrange a conference for Fermi, who happened to be on his way to Washington. That same day, March 16, the Columbia physicists sent reports on their neutron production experiments to the *Physical Review*. On March 17 Fermi met with a group of naval officers and told them that bombs might someday be made which could blast out craters a few miles in diameter. Of more immediate interest to the Navy, uranium power might be used to propel submarines. The Navy people promised to keep in touch with Fermi's work and await further results of his experiments; neither they nor he were interested in more direct government participation.[28]

Two days later Fermi, still in Washington, met with Szilard and another refugee Hungarian physicist, Edward Teller. They discussed whether the Columbia reports just submitted to the *Physical Review* should in fact be published. Szilard and Teller offered a proposal, disturbing to Fermi, that the results should be restricted to a limited group—"I don't know whether it would be called a secret society or what it would be called," Fermi later remarked. But after a long discussion, as Szilard recalled, "Fermi took the position that after all this is a democracy; if the majority is against publication, he will abide by the wish of the majority." Outvoted by Szilard and Teller, Fermi returned to New York, selflessly argued the case for self-censorship with Dean Pegram, and told the *Physical Review* to hold back publication.[29]

Szilard was now on the point of cabling Joliot, but before he did so, he heard of the French team's paper, published in the March 18 issue of *Nature*, which announced that fission releases at least some neutrons. Beaten to the discovery, Fermi felt that there was now no secret to keep and thus no longer any sense in refusing to publish. Szilard replied that, "If we persist in not publishing, Joliot will have to come around; otherwise he will be at a disadvantage, because we will know his results and he will not know of our results." Fermi, unconvinced but determined to be fair, agreed to put the matter before Pegram. Pegram delayed his decision. Szilard's determined arguments were countered by others at Columbia who felt that any attempt to restrict publication would be both futile and a breach of scientific custom.[30]

While Pegram deliberated, Szilard pushed ahead on his own. He convinced Victor Weisskopf, an émigré Austrian physicist at Princeton, and other friends to try to impose secrecy by themselves. "We knew this was a hopeless thing, but we thought we had to try," Weisskopf recalled. "We were very much afraid of the Nazis." On March 31 Weisskopf cabled P. M. S. Blackett, a leading British physicist, to ask whether it would be possible to obtain the cooperation of *Nature* and the Royal Society to circulate papers on fission only in private, "IN VIEW OF REMOTE BUT NOT NEGLIGIBLE CHANCE OF GRAVE MISUSE IN EUROPE." Meanwhile Wigner had written P. A. M. Dirac, another noted British physicist, urging him to cooperate: "It is my impression that there is some urgency in the matter. Although there exists apparently a great willingness for cooperation here, it is realized that the interests of the scientific workers in the U.S. might be prejudiced to some extent if America abeyed alone by the proposed procedure." A scientist's reputation is built upon his published work; an American physicist would make discoveries in vain if the same discoveries were published elsewhere while he kept silent. About a week later, the British replied: They would support the scheme.[31]

Szilard and his companions also talked about secrecy with Niels Bohr, who was still in the United States. Bohr strongly doubted that fission could be used to cause a devastating explosion. And he thought it would be very difficult if not impossible to keep truly important results secret from military experts; the matter was already public. Nevertheless he agreed to go along with the attempt and drafted a letter to his institute in Copenhagen, although he apparently did not mail it immediately: "The question which has attracted chief interest is the neutron emission which accompanies or follows fission . . . The Columbia group is busy organizing cooperation among all the physics laboratories outside the dictatorship countries, to keep possible results from being used in a catastrophic way in a war situation, and I must therefore ask you, if work along these lines is going on in Copenhagen, to wait before you publish anything."[32]

Another group that had to be brought into the pact was in Washington, D.C., where Richard B. Roberts and others worked under Merle Tuve at the Carnegie Institution. They had been among the most eager students of fission since the discovery was announced, and Roberts felt that the discovery "brings back the possibility of atomic power." In February the group found some neutrons that appeared after uranium fission. But these neutrons dribbled out over a period of some seconds after the fission; they were not true fission neutrons but occasional neutrons emitted as a side-effect of the radioactivity of the fission fragments, and they could not sustain a chain reaction. The group published their result on March 1. Teller, who often visited them and knew about their work, wrote Szilard that the publication was not his fault and at any rate could cause no harm.[33]

The Washington group's work had already been announced in a Science Service news release dated February 24, written by Robert D. Potter, a conscientious journalist who had once worked in the Carnegie Institution laboratory simply to get a feel for research and who kept in close touch with the New York scientists. Infected with the excitement that ran through both groups, Potter headlined the possibility of an explosive chain reaction propagated by neutrons. He carefully noted that the Washington group's delayed neutrons might not be enough to sustain a chain reaction, but he quoted Fermi as saying that the possibility of a chain reaction was certainly present.[34]

In early April Szilard and his friends approached the Washington scientists, who promised to cooperate in withholding future publications. The self-censorship proposal spread further by letter and word of mouth. The scheme to restrict information to a circle of physicists outside the dictatorships was thus well underway.[35] It lacked chiefly the acquiescence of the French.

When Szilard and his friends set their scheme in motion, Joliot was uppermost in their minds. Szilard did not know Joliot but Weisskopf did, and Weisskopf knew Halban even better, for they had met when Halban was a student in Zurich, and the two had become close friends. Szilard and Weisskopf therefore drafted a telegram to Halban, which Weisskopf signed and sent off on March 31. The telegram informed the French that other physicists too had written papers about the production of neutrons. "AUTHORS AGREED HOWEVER TO DELAY PUBLICATION FOR REASONS INDICATED IN SZILARDS LETTER TO JOLIOT FEBRUARY SECOND AND THESE PAPERS ARE STILL HELD UP . . . IT IS SUGGESTED THAT PAPERS BE SENT TO PERIODICALS AS USUAL BUT PRINTING BE DELAYED UNTIL IT IS CERTAIN THAT NO HARMFUL CONSEQUENCES TO BE FEARED. RESULTS WOULD BE COMMUNICATED IN MANUSCRIPTS TO COOPERATING LABORATORIES IN AMERICA ENGLAND FRANCE AND DENMARK." The telegram also said that the British were being asked to cooperate, and concluded, "PLEASE CABLE."[36]

At first Halban thought that the telegram was an April Fool's joke. The message, he later wrote Weisskopf, had an air of Jules Verne or H. G. Wells. But during the day the evident cost of the 139-word telegram, and its contents, convinced the people at the Collège de France that the matter was serious.[37] They knew what Weisskopf's request for censorship implied. For the past half-dozen years displaced scientists and other refugees from Central Europe and Italy had been fleeing through Paris, some using the Collège de France laboratories as a resting-place before continuing westward. The French physicists despised the anti-Semitic governments that had driven these people from their homes, and Joliot had spoken out repeatedly to condemn French appeasement of the fascists. As for Halban and Kowarski, each was half-Jewish. All of them would hesitate to allow the dictators to get a new weapon. In France, much more than across the Atlantic, the German seizure of Czechoslovakia was recognized as a fatal sign, and nearly everyone expected war within a year or two. Halban, for example, promised to pay Weisskopf a visit the following year, "if it happens that there is still no war."[38]

The French scientists, however, believed like Bohr and Fermi that a nuclear bomb was not likely to be built for many years, if ever. With the knowledge available at that early point, they could see some possibility of releasing nuclear energy but no way of releasing it all at once as a powerful explosive; their water tank was nothing like a bomb. They therefore wondered whether secrecy in scientific research might not do more harm than good.

It was up to Joliot, as head of the team, to answer the telegram, but he discussed it at length with his colleagues. There were a number of points that may have affected their decision. For one thing, Joliot continued to believe firmly in an international community of scientists all sharing their discoveries. He was "naturally opposed to secrets," he wrote Bohr a few years later. "It is clear that a temporary advantage cannot be maintained in the face of the certainty that secrecy measures would be extended to every area of science and would thereby slow down or even halt the entire progress of civilization (return to the Middle Ages)."[39] Further, Joliot knew that if his team failed to publish, they might well be eclipsed by others who did publish, for it was hard to believe that everyone would adhere to an unprecedented pact, pushed forward, so far as the French knew, merely by two Central European refugees on the outskirts of the New York scientific community. Had Fermi, Bohr, or a leading American scientist written, the French might have found the scheme more plausible. If someone published discoveries ahead of them, the French could lose not only glory but also the money they would need to pursue their work. Moreover, even if all the laboratories joined the agreement and stuck by it, there would remain a powerful objection, the same one noted by Fermi and Bohr: it was scarcely likely that copies of papers circulated privately around America, France, Britain, and Denmark could be kept out of Germany and the Soviet Union. Even if they could, German and Soviet scientists were surely aware already of the importance of fission chain reactions.

A more immediate influence on the French team's deliberations came from a Science Service news release of February 24, summarizing their own report, published in *Nature* on March 18, that some neutrons resulted from fission. The news release said that they had been beaten to the discovery by Roberts' group in Washington, which "had been able to observe the same important reaction in atomic trasmutation."[40] This error made it seem that the Americans were busily publishing important facts.

For all these reasons, Joliot, Halban, and Kowarski cabled Weisskopf a discouraging reply on April 5: "SZILARD LETTER RECEIVED BUT NOT PROMISED CABLE. PROPOSITION OF MARCH 31 VERY REASONABLE BUT COMES TOO LATE. LEARNED LAST WEEK THAT SCIENCE SERVICE HAD INFORMED AMERICAN PRESS FEBRUARY 24 ABOUT ROBERTS WORK."[41]

Szilard, who was well informed on the work of the Washington group through their publications and through letters from Teller, answered the French on the next day, Weisskopf having left New York: "ROBERTS PAPERS CONCERNING DELAYED NEUTRON EMISSION WHICH IS MUCH WEAKER THAN HE THINKS AND HARMLESS. HOWEVER TUVES GROUP WAS RECENTLY APPROACHED AND PROMISED COOPERATION. WE HAVE SO FAR DELAYED PAPERS IN VIEW OF POSSIBLE MISUSE IN EUROPE. KINDLY CABLE AS SOON AS POSSIBLE WHETHER INCLINED SIMILARLY TO DELAY YOUR PAPERS OR WHETHER YOU THINK THAT WE SHOULD PUBLISH EVERYTHING." Joliot received this on April 7 and on that same day sent the French team's final answer: "QUESTION STUDIED MY OPINION NOW IS TO PUBLISH REGARDS."[42]

This reply, along with the earlier French announcement in *Nature* that fission produces some neutrons, doomed the attempt to restrict publication. Pegram, who was not aware of how much progress Szilard and his friends had made before the negative French reply, after some deliberation decided that any attempt to impose secrecy was hopeless. Szilard was forced to give in, and the Columbia scientists asked the *Physical Review* to print the papers on neutron production that they had submitted in March.[43]

There were many reasons for the failure of secrecy, of which two are likely to arise again if the need for secrecy in basic science should recur. First was a failure to understand how serious the problem was and how much secrecy might accomplish—a natural failure, since secrecy could work only if it was imposed before anyone could be sure it was needed. Second was a failure of communication, in which nobody quite understood what was happening or how good the chances of getting agreement really were, and this failure too was natural, since any attempt to impose censorship in itself implies restricted communications.

The scheme's breakdown had immediate and heavy consequences. On April 7, the day of their final reply to Szilard, the French sent *Nature* the results of their latest experiments and calculations, giving the number of neutrons emitted per fission as 3.5. The report, duly published on April 22, convinced many physicists that uranium chain reactions were a real possibility. George P. Thomson, a professor at the Imperial College of Science and Technology in London with important discoveries to his credit, was struck by the French results. Feeling, as he recalled, like a character in a third-rate thriller, he decided to warn his government of the dangerous prospects and meanwhile to begin experimenting with uranium. A German scientist, hearing a report on the French experiments at the Göttingen Physics Colloquium, wrote a letter to the Reich Ministry of Education, while independently but simultaneously two other German scientists wrote a joint letter to the War Office. News of the French work may also have played a role in the start-up of Soviet nuclear energy research, perhaps provoking the letters on uranium that physicists sent about this time to the Soviet Academy of Sciences.[44] Thus in Britain, Germany, and possibly the Soviet Union, publication of the French results precipitated state-supported programs of research into nuclear energy.

Nobody took the paper more seriously than the French scientists themselves. While Fermi estimated that less than two neutrons were released per fission, and others thought the matter unsettled, the Collège de France team put considerable faith in their conclusion that v was between three and four neutrons per fission and that therefore a chain reaction could be started easily. Excited to see the door open, they were exhilarated to find themselves in the lead among the scientists crowding to enter it. But if they were to keep their lead, they would have to leave the familiar confines of small-scale laboratory experiments and push blindly into an arena where great sums of money and vital national interests were at stake.

6

A Contract for Nuclear Energy

From the beginning Halban, Joliot, and Kowarski discussed their research with other scientists, especially over the cups of tea at Jean Perrin's Monday seminars. Perrin's son Francis, a leading theoretical physicist in his own right who worked just down the Rue Pierre-Curie in the Henri Poincaré Institute, regularly attended the seminar and took a passionate interest in the fission discussions. Francis Perrin never shared his father's glamour; he was smaller and slighter, more modest and reserved; but he had his father's skills in science and, as appeared during the formation of the Popular Front, in politics. In March 1939 Joliot asked him to help the Collège de France team with advice on the theory of chain reactions.[1]

Francis Perrin had already started to think about these problems, for he was in close touch with the team. He had known Kowarski since 1931 when Kowarski had begun work on his doctoral thesis in the elder Perrin's laboratory. Francis Perrin had guided this work and, once the thesis was done, had approached Joliot to secure Kowarski's entry into the Curie Laboratory. Perrin also knew Halban well, although the relationship was more professional than personal. He knew Joliot best of all. Francis Perrin and Irène Curie had grown up together as companions in Paris and L'Arcouest, and he had met Joliot as soon as Joliot had entered the Curie Laboratory. The two men, almost the same age, were good friends.[2]

Perrin was unusually well prepared for work on chain reaction theory. He was one of the few theoretical physicists in France, and on Borel's advice he had earned doctor's degrees in both physics and mathematics. His previous work had centered on two subjects which turned out to be relevant to chain reactions: Brownian motion, which made him expert in the mathematics of statistical processes such as diffusion, and nuclear physics, including the resonance capture of neutrons. He now had to apply statistical and diffusion calculations to neutrons and their capture.[3]

Perrin began by considering a theoretical picture resembling his colleagues' actual experimental setup, except that where they used a uranium compound dissolved in water, Perrin imagined a sphere of solid uranium oxide (U_3O_8) surrounding the neutron source. He wanted to estimate how much material was needed to start a chain reaction that would diverge, or grow without limit. He wrote a simple equation for the diffusion of neutrons in the uranium oxide and added on a term representing the neutrons produced in fission. From this equation he figured out mathematically that even if the central source were negligibly small, if the uranium oxide sphere were larger than a certain size, the flux of neutrons leaving it would be infinite. In short, above a "critical value" size, as Perrin wrote, "the chain reaction will be explosive." Putting into his formula the best available experimental values, including the over-optimistic $\nu \approx 3$ neutrons per fission, Perrin estimated the critical value of the mass to be forty tonnes of uranium oxide (a tonne being 1000 kilograms or 1.10 United States or short tons). If the ball were surrounded by a layer of heavy metal to reflect neutrons back in, the critical mass could be reduced to about twelve tonnes.[4]

This calculation was only a rough approach to the problem, for Perrin had deliberately made some gross oversimplifications. He assumed that only fast neutrons were involved, that these neutrons did not slow down, and that whenever a fast neutron struck a uranium atom, it would cause a fission. It is now known that this is not the case, that in fact no mass of natural uranium oxide, however large, can have a divergent chain reaction. But in 1939 it seemed that some sort of uranium bomb might be made quite simply. By the end of April Perrin and the others had worked out several tentative ways to make such a bomb. Different pieces of uranium, preferably the dense metal, would have to be assembled very rapidly and perhaps be compressed to still greater density with the aid of ordinary chemical explosives. A small source of neutrons might be needed to initiate the reaction.[5]

But Perrin's calculations showed that at the very least a bomb would need tonnes of uranium. Any explosion of so large a mass could not be efficient. "We did not discuss very thoroughly the force which the explosion could have reached," Perrin recalled, "because we saw clearly that when the chain reaction began to develop, before any extremely violent explosion could occur the rise in temperature would blow up the system, would disperse it, and make it fall below the critical size." Therefore, several tonnes of uranium would be no better than several tonnes of ordinary explosive. This reasoning was entirely correct for natural uranium. In 1939 no scientist, in France or elsewhere, realized that one could use natural uranium as a raw material to produce other substances which would be far more explosive. The French scientists set aside the bomb, never forgetting it, but feeling that their first job was to investigate uranium chain reactions methodically.[6]

They hoped to do so by building a nuclear reactor, in which a large mass of uranium would be mixed with a moderator to slow down the neutrons. This mixture should support a chain reaction that progressed only to a certain point, proceeding slowly and under control. Once they had built such a reactor, they would have a powerful tool for fission research. Perrin, returning to his calculations, considered the effect of adding hydrogen as a moderator. Again using the crudest and most optimistic assumptions, he determined that it might be possible to build a reactor of critical size with only five tonnes of uranium oxide.[7]

Perrin and the others found the practical prospects of nuclear reactors more and more exciting. While the French physicists were too skeptical on bombs, they were almost too optimistic on the future of nuclear power. Their hopes reminded Kowarski of H. G. Wells' fantasies: "We were talking about our changing the face of the world map, diverting the Mediterranean into the Sahara, things like that . . . I was always aware of science fiction; Francis Perrin too." Perrin, who liked to read such books, recalled that the group forecast great centralized atomic power stations; "It was altogether the atmosphere of a Jules Verne novel."[8]

Once Perrin's calculation of the critical mass and his other practical ideas had helped turn the group's attention away from the abstract phenomena of fission, they concentrated on the problem of building a reactor. Confident that a device could be made to work, they worried that it might work too well and melt down. So they thought about ways to keep the reaction under control and, if need be, to stop it in a hurry. The problem, Halban recalled, "preoccupied us very much . . . how do you stop the whole thing from exploding. We were sure of being able to make the reaction . . . but we didn't yet know what to do about stopping it. We discussed it every day."[9]

Help came from Felix Adler, a young, self-effacing Swiss theoretical physicist who had come to Paris in 1938 as a guest researcher in Joliot's laboratory, and who also happened to be a relative of Halban's. He timidly offered Halban the suggestion that the reactor might be stopped by cooling it. Halban told him that the idea was all wrong. But as they talked it over, Halban came to realize that temperature could be used to stabilize the reaction automatically. When a reactor heated up, all the neutrons in it would move faster. Since only slow neutrons are likely to cause fission, the number of fissions would level off, and the chain reaction would reach a steady state. Francis Perrin had the same idea independently about the same time.[10]

Adler and Halban also thought of ways to manipulate the speed of the reaction. Halban's earlier work had familiarized him with cadmium, a silvery element that had the handy property of hungrily absorbing neutrons. One could insert a sheet of cadmium metal into the reactor periodically for control: when the reaction went too fast, the sheet could be pushed in, and when it went too slowly, the sheet could be withdrawn again. Adler and Halban thought this process would have to be repeated as often as several times a second, fearing that once a reactor was the slightest bit above critical size, it would flash to a high temperature within seconds. In hindsight they were being excessively cautious, but in principle their method would work.[11]

By the end of April the French team had a vague picture of a nuclear power station. There would be a massive block of uranium, somehow combined with a moderator, stabilized at a high temperature, and provided with cadmium sheets as controls. Water or gas would circulate to remove the useful heat, and the whole would be surrounded by a radiation shield to protect the workers. Perrin pointed out that the critical mass could be reduced by surrounding the reactor with a heavy material, like iron or lead, to reflect escaping neutrons back inside. The plant could produce not only heat and power but also, as a side benefit, radioactive elements for medical use. All this was a sketchy but essentially correct outline of nuclear reactors as they are known today.[12]

One day around the end of March when these ideas were germinating, Halban, Joliot, and Perrin were eating dinner together when a question came up. They had their hands on something that might be of enormous economic importance; should they patent it? Joliot automatically opposed the idea, for Pierre and Marie Curie and other scientists had made a principle of not deriving personal income from their discoveries in pure science. This altruism was often mentioned by scientists who sought support from private or state philanthropy or who tried to get industry to acknowledge remunerable "property rights" in scientific discoveries. Scientists as a body must be rewarded, they said, since it would be scandalous for a scientist to sell a discovery for personal gain. In 1938 Joliot had written, "obtaining patentable processes, while always possible and legitimate, can in no way be considered the main aim of our work." Kowarski too, as he recalled, felt that taking patents would offend scientists; moreover in this case it would be futile, for if the reactor worked, it would be so important that patent laws could not control it.[13]

However, more and more scientists were going to the patent office, if not to line their pockets then at least to make sure that science got a share of the proceeds from research. Even Pierre Curie had taken out patents on laboratory instruments and made arrangements to exploit them. The attempts during the 1920s to set up scientific "property rights" had made many scientists think about the issue, but when this movement failed, private patents were left as the only way to make industry pay for using a discovery.[14]

It was up to individuals to do as they chose about taking out patents. Few universities in any country had a formal policy on professors' patents. The budget cutbacks of the Depression and increased concern with the effect of applied science on society led in the thirties to renewed longing for a better system. The United States National Research Council was tempted in 1934 by the idea of setting up a central agency that would administer patents in the joint interest of the universities. "The results of scientific work," the Council noted, "would thereby be protected also from socially harmful exploitation." Yet this idea too failed to take hold.[15]

The only scheme that had worked was a single private initiative, well known to nuclear physicists. In 1912 Frederick G. Cottrell, an American inventor, had set up a company called The Research Corporation, to which he gave his patents in the hope that the profits would be used to encourage invention. By the mid-1930s The Research Corporation was one of the main underwriters of American cyclotrons and other particle accelerators. Ernest Lawrence, for example, was aided by the corporation and gave it his patent on the cyclotron.[16]

In early 1934 Szilard applied for a secret patent on his idea for a nuclear chain reaction, hoping thereby to give scientists some control over future developments. At first he thought to assign the patents to an established laboratory but then he was attracted to the model of The Research Corporation. For the next two years Szilard tried to create a similar corporation run by prominent physicists, such as Chadwick, Cockcroft, Joliot, and Fermi. He hoped it would take out patents and raise money to finance his attempts to find a mechanism for a nuclear chain reaction. The patents might also, he wrote to Cockcroft, "be used by scientists in a disinterested attempt to exercise some measure of influence over a socially dangerous development."[17] But none of the scientists Szilard approached took any interest in his scheme, and by 1937 he gave it up.

Meanwhile the idea of patenting was spreading among physicists. In 1934 Fermi and his collaborators patented their method for producing artificial radioactivity with slow neutrons, since this method could create radioactive substances to replace expensive radium in cancer therapy. The idea of patenting a discovery had at first seemed strange to them, and they had done it only at the urging of their patron Corbino, the senator familiar with industrial circles. In April 1939, when the French began to think of patents, Fermi's friends were still negotiating with industry to squeeze a little money from their own patent. In Berkeley that same month, Lawrence was thinking of taking out a patent on production of radioactive substances by cyclotron bombardment.[18]

Szilard had applied the previous month for a patent on an "Apparatus for nuclear transmutation" using fission chain reactions. He felt he would need such a patent to tempt an industrial firm to support his fission research, in the event that he could not persuade private individuals, foundations, or the United States government to underwrite it. Like most universities at that time, Columbia left its researchers free to take out patents, so Szilard could act on his own.[19] His secret application contained in rudimentary form several of the ideas needed for a reactor, including the critical size, the moderator, and the neutron reflector. His conception, prior to and independent of the French work, was much less like the nuclear reactor as it has since developed. Nevertheless he was often very close to the French scientists' way of thinking, not only in working out key parts of the reactor but in looking to patents as a way to control the development of nuclear energy.

The French were not aware of all that their colleagues were doing, but they realized that the scientists of their day were searching for ways to exercise some control over the results of their discoveries. When Joliot and Kowarski objected to the idea of patents, Halban and Perrin pressed the point. Halban was at home in the German tradition of the professor who mixes with industry, for while he was growing up, his father had worked as director of an industrial laboratory. And Francis Perrin recalled a patent on ultrasonic devices, invented for detecting submarines, taken out in 1918 by his father's old friend Langevin and a collaborator. The patent was potentially lucrative, even though its imperfections had drawn Langevin into frustrating lawsuits which were still underway. Halban and Francis Perrin argued that the French had to take out patents less for personal benefit than for protection. Others, whether powerful industrialists or foreigners, might take out a master patent or key secondary patents, and thus the control of nuclear energy could fall into unwelcome hands. This argument convinced Joliot, a firm socialist and equally firm patriot, who had never opposed patents as such so long as they did not get in the way of true scientific research. A method would be worked out to channel profits from the patents back into science. Kowarski still distrusted the idea, but he had to go along once Joliot was convinced.[20]

The group would not only patent their ideas, Joliot decided, but would try to develop a nuclear reactor themselves. As Halban recalled, from the standpoint of pure research this undertaking was less interesting than other projects, such as Joliot's favorite, the use of radioactive tracers in biology: "By his own inclinations and by the tradition of the Radium Institute, he had little interest in any work that could lead him towards industrial development." To develop a reactor, it was necessary to learn only certain technicalities about how nuclei behaved, not the fundamental reasons and modes of their behavior, which continued to interest most physicists. But the argument over patents drove home the larger issue: nuclear energy was too important to ignore.

"Once Joliot decided to work towards an industrial application," Halban wrote, "he did it enthusiastically and even gaily."[21] The group agreed to push fast and hard. They needed more money and a larger stock of uranium, and they needed to inform some arm of the government of what they were doing. Although scientists in Britain, Germany, the Soviet Union, and the United States at about this time sent messages on fission to various bodies in their governments, a formal, written warning was not necessary in the close-knit world of Parisian science. Joliot simply talked the matter over with Henri Laugier.

Laugier, a physiologist with a passion for modern art, had entered the political arena as friend and adviser of Blum's foreign minister. He agreed closely on science policy with Jean Perrin, who picked him to become the first director of the CNRS, making him the executive officer of French science. An impetuous man of action, Laugier was struggling to bring what he considered the "runaway individualism" of professors under CNRS control. According to Kowarski, "Laugier was very idealistic, very progressive, and looked with great pleasure on any sign of anything flourishing in French science; and of course Joliot was quite a hefty sign of something flourishing. So he looked at Joliot, beamed at Joliot, with considerable pleasure."[22] When Joliot approached him, Laugier agreed to make some rearrangements in the CNRS budget, eventually freeing at least 50,000 francs to aid the Collège de France team. This was over a thousand 1939 dollars, enough to hire one or two technicians for a year. By the standards of the day it was a substantial addition to the disposable funds in Joliot's laboratory budget, which was already a large one for a French research establishment.[23]

Laugier discussed patents with the team, chiefly with Joliot, although Halban also took a keen interest in working out an arrangement. According to Halban, they reached a verbal agreement that the scientists would eventually assign their patent rights to the CNRS and that, if the CNRS drew any benefits from the patents, "at least half of these benefits should go to the furtherance of research and this money should be distributed according to the decision of [a] committee in which Joliot, Perrin, Kowarski and myself should have seats, with the right to nominate successors in case of our deaths." Thus the physicists renounced any large share of personal profit from their discoveries but arranged matters so that if a great sum of money flowed in, they would have a powerful voice in the councils that spent it. It was never specified whether or not they would have an outright majority on the governing committee; such details were to be worked out later.[24]

With some help from a patent agent, Francis Perrin hammered out the wording of the patent applications, and they were filed at the beginning of May 1939. There were three applications, one for the basic nuclear reactor, a second for the means of controlling it, and a third for a nuclear bomb. The applications were kept secret. This did not contradict the team's earlier decision to publish their discoveries, for the patents concerned industrial devices rather than purely scientific questions.

It is not known whether at this point Laugier or anyone else discussed the fission matter with other branches of the government or with higher authorities, but probably they did not. In France such matters were likely to be taken in hand by officials on Laugier's level rather than by a minister, since the minister and even the entire government might be replaced on short notice through a shift of parliamentary cliques. In any case, after the Munich debacle the government, disintegrating into a tangle of factions, had little attention to spare on such matters. Meanwhile government administration of science, still organized according to Jean Perrin's design, functioned as efficiently as ever. Laugier was the one responsible for scientific research, the committees stood behind him, and he had full authority to give his physicists all the help they asked for.

Money was not the only thing the team needed. In other countries scientists were trying to persuade their governments to locate and corner stocks of uranium. Here also the French could act more swiftly and directly than scientists elsewhere, because of an elementary fact of nuclear physics. Since uranium atoms gradually transmute into radium atoms, uranium ore always contains within it a minute amount of radium. In consequence, there had long been an association between the Radium Institute and the owners of uranium mines.

Almost as soon as she had discovered traces of radium in mine tailings, Marie Curie had remarked that her research, while purely scientific in motive, was likely to profit the mine owners.[25] This prediction soon proved true as the demand from doctors drove the price of radium to a great height, and the Curies established connections with French industrialists interested in marketing the element. Foreign uranium mines supplied the ore, but Paris remained a refining center, and the Radium Institute kept in close touch with the world market by certifying the strength of radioactive preparations and by studying chemical methods to produce them.

In the early 1920s a new force thrust into the trade, the Union Minière du Haut-Katanga. This powerful Belgian firm made its billions from the Congo's immense copper mines, but it had also discovered a lode of uranium in Katanga. As a sideline, the Union Minière began to mine the uranium in order to extract the radium associated with it. The lode was incomparably rich and the native labor cheap, so the Belgians cut the price from $100,000 to $60,000 a gram and quickly won a monopoly over the profitable radium trade. It was Union Minière radium, combined with beryllium, that produced the neutrons used to open up nuclear physics in many laboratories in the twenties and early thirties. Then a Canadian company discovered a deposit of uranium ore in the arctic and in 1935 broke into the radium market by cutting prices. In the confused price war that followed, a gram of radium sometimes went for as little as $12,000, which was probably close to its true cost. In 1938 the rival companies signed a cartel, establishing a price floor of $20,000 and giving the Union Minière the lion's share of the market. In 1939 the firm was still the world's largest producer of radium and, incidentally, of uranium.[26]

The company, having recognized that the use of radium could be expanded only through scientific research, had always been generous in helping the process along. They had a particularly old and cordial relationship with the Curie family. They used processes and standards derived from the researches of Marie Curie and her collaborators, supplied large amounts of radium and other materials at nominal prices to the Radium Institute and Joliot's laboratories, and kept informed about their research. The Perrins had also been in touch with the company, which in early April 1939 loaned Perrin's laboratory 200 kilograms of uranium oxide.[27] But the Collège de France team felt that they would soon need it by the tonne.

On May 4, the day when the last patent application was filed, Joliot wrote the Union Minière, "I would like to be able to talk with you about some important experiments which might interest your Company and which might be tried with its collaboration." He made an appointment for May 8 and went to Brussels to meet Edgar Sengier, the company's president, and Gustave Lechien, director of the radium division. Covering a blackboard with equations, Joliot explained the possibilities of nuclear power production and probably also warned Sengier that a bomb might eventually be built.[28] Sengier was a talented businessman, abrupt but not unfriendly. He had first made his mark as an engineer but had quickly proved to be an excellent executive, and his business sense, combining boldness and shrewd judgment, had rocketed him to the top of the Union Minière, where he made the company flourish as never before. When Lechien advised him that Joliot's predictions made sense, Sengier understood that his company held high cards in a new game.[29]

Joliot arranged for Sengier and Lechien to come to Paris and discuss the matter further. Meanwhile he inquired into the state of the Union Minière's stockpile of uranium oxide, especially its largest single pile. Probably he was curious to see whether the company had any heaps approaching the size that Francis Perrin had estimated would be the critical mass. The company did have many tonnes lying about in heaps nearly the size of Perrin's overoptimistic estimate, for uranium found only occasional use as a yellow coloring agent for ceramic glazes and was chiefly a by-product of radium manufacture.[30]

Sengier and Lechien came to Paris on May 13 to meet the physicists and Laugier. Sengier recalled that Joliot, Francis Perrin, and Halban suggested that their work could eventually lead to an experimental bomb which might be tested in the Sahara. They drew up a draft agreement on the spot; the preface noted that the Union Minière had "a large stock of uranium . . . as well as technicians who can develop industrial installations, and the international organisation necessary . . . for patents or granting licenses," while the CNRS had the patents. The Union Minière promised to supply a total of five tonnes of uranium oxide for a preliminary reactor experiment. If this was judged successful, they would loan about fifty tonnes for a full-scale trial. The Union Minière also agreed to "put its Technical Division at the disposal" of the CNRS and to pay expenses up to one million francs. A syndicate would be formed to exploit the French patents, and the Union Minière would get about half the income (20 percent in France, 80 percent in Belgium and 50 percent elsewhere). The balance would go to the CNRS, presumably to be spent for pure research through the committee including Halban, Joliot, Kowarski, and Perrin. The CNRS would contribute the patents to the syndicate, and the scientists promised not to take their ideas elsewhere. This was still only a rough outline of an agreement, and it left the proposed syndicate obscure. It may not have matched Halban's hopes, for he wrote on one draft, in his characteristic bold scrawl, the comment, "Syndicat sans capital!"[31]

The matter could not be swiftly concluded, for such an unusual agreement between a private Belgian firm and a French state organization would have to jump a number of hurdles, legal and political. Since the team wanted uranium at once, Joliot, unofficially representing the CNRS, and Lechien initialed the contract as a "gentlemen's agreement." The four physicists assured the Union Minière that even if the CNRS could not fulfill its part of the agreement, they would consider themselves bound by it.[32]

In fact the convention was never formally signed. While various provisional drafts were passed back and forth through the summer, the CNRS could not agree on the precise form the arrangement should take. The question was set aside when war broke out, but the CNRS administrative council took it up again in April and May 1940 and considered various proposals, such as paying the physicists a substantial lump sum for their rights. These rights, even if they came to only a small percentage of the total, could be worth vast sums of money if the patents proved to be as important as everyone hoped. Reluctant to press blatantly for personal gain, the physicists continued to insist that a large part of the proceeds from their inventions be used, under their partial control, to advance science. No final agreement was reached before the German invasion of France interrupted the debate.[33]

The Union Minière honored the gentlemen's agreement and sent the French a hundred drums containing a total of about 5000 kilograms of uranium oxide (U_3O_8), an extremely dense, grimy powder resembling pulverized charcoal. It arrived at Joliot's laboratory around June 1, 1939. The firm also loaned Joliot a gram of radium for the experiments.[34] Sengier and Lechien kept in close contact and supplied further uranium and radium on request.

The French scientists' arrangements with government and industry, although largely verbal or even tacit, nevertheless worked smoothly. These arrangements were made possible only by the hard work of the scientists and statesmen of the twenties and thirties who had set out deliberately to create close contacts among science, industry, and the state. Exploiting their contacts with enthusiasm, the Collège de France team had won something that no other group of scientists then possessed: tonnes of uranium and all the help needed to make of this uranium whatever they could.

This result was not only unique but remarkable. Personal contacts and political groundwork alone could not have brought the French team all the money and uranium they wanted, for their success also came from scientific results. Laboratory measurements and administrative arrangements followed one another closely, each depending on the last. On one side the facts of nature, such as the number of neutrons emitted per fission, pressed inexorably upon events. On the other side an equally strong pressure was exerted by human forces, including the ideology that the French scientists had unfolded during their long fight for funds and position. For a generation the liberal scientists had believed that science would produce measureless practical benefits, and would do so best under centralized control exercised by themselves; they had insisted on entering the thick of politics. So it is not surprising that Joliot and his team, unlike most scientists in other countries, determined to take over the development of the practical applications of fission themselves.

Everyone's intentions were explicit. In their agreements with the Union Minière and the CNRS, the Collège de France team said in almost so many words: give us money, supplies, personnel, and authority for our research; guarantee that a share of the profits from our discoveries will be used for pure science, partly under our control; and we will find out how you may build devices that will greatly profit the company and strengthen and protect the nation. This was the bargain that had been implicitly offered by the representatives of science in countless idealistic writings and speeches; now it was abruptly condensed into a contract. It remained to be seen whether the Collège de France team could fulfill these promises.

The Laboratory
of Nuclear Synthesis

Frédéric Joliot, Hans Halban, and Lew Kowarski with a geiger counter, 1939

7

Fission Research on the Eve of War

The weeks between late March and early May 1939 had been hectic. Experiments to determine how many neutrons were liberated in fission, discussions about how to build bombs and power plants, worries over secrecy, debates on patenting, and talks with Laugier and the Union Minière had all piled one on top of another. Now the pace slowed. "You cannot imagine the overwork in the last few months," Halban wrote Placzek on May 9. "It hasn't done me any harm, but it does make me happy that we can pause for breath now." In the next two months Adler and Perrin worked out improved calculations for the critical mass while Halban, Joliot, and Kowarski carried out some minor experiments.[1] But their chief job was to plan experiments with the uranium oxide loaned by the Union Minière, to purchase and set up the big tanks that these experiments would require, and to wait for the materials to arrive.

The physicists had little help from people outside the Collège de France. There were only two other laboratories in Paris fully equipped for nuclear research, and both were concentrating on other subjects. One was the Sorbonne laboratory of Pierre Auger. Auger—a Normalien, Jean Perrin's former student, Francis Perrin's brother-in-law, and a summer resident of L'Arcouest—was on good terms with Joliot and his circle. But he had recently discovered giant showers of cosmic rays, with which he was now naturally preoccupied. Moreover, later in 1939 he was called to the CNRS to create a documentation service to improve scientific communication, another organization that French scientists had wanted for many years.[2]

The other laboratory was the Radium Institute. In the last few years it had lost Marie Curie and Joliot, and the new director, André Debierne, although well able to inspire his students, was deeply involved in idiosyncratic and fruitless radiochemical research and could hardly take the place of his two excellent predecessors. Irène Curie also kept to the laboratory's specialty of radiochemistry, studying, like Hahn and Strassmann, Frisch and Meitner, and many others, the nature of the fragments coming from fission. She and Joliot discussed the problem of fission chain reactions with students at the Institute, who during the summer of 1939 used some uranium oxide that Joliot had loaned them to try to measure the absorption of neutrons by uranium. Their results were unreliable. Other studies were done on a small scale in other laboratories. Two workers in Jean Perrin's institute had made early attempts to measure neutron emission from uranium. And Constantin Chilowsky, a Russian inventor who shared Langevin's patent on ultrasonic devices and who knew the Perrins, took out his own patent on nuclear reactors, suggesting that liquid metals could be circulated in them to extract useful heat. None of this work did much to help the team at the Collège de France.[3]

Nevertheless in fission research France was relatively strong. By the end of 1939 somewhat over a hundred papers on fission had been published in western Europe and the United States, of which about 40 percent were written in the United States, 25 percent in France—about a third of these by Halban, Joliot, Kowarski, Perrin, and their collaborators—15 percent in Germany, 10 percent in Great Britain, and the remaining 10 percent elsewhere.[4] These proportions would change a bit if papers published in the Soviet Union and secret reports written in Germany were included, but they reflect accurately the predominance of the United States and the strong French interest. Fission had attracted everyone's attention. But within France and outside, most of the research did not point toward a working nuclear reactor. Hundreds of physicists and radiochemists studied fission, but nearly all of them treated it as an intriguing problem of itself or as an aid to understanding nuclei, not as something that could soon be used in industry or warfare.

A number of theoretical physicists, following Meitner and Frisch, studied fission in terms of a theory of the way nuclei behaved. Fission proved to be an interesting confirmation of some parts of the theory. A dozen papers were written on the subject in 1939, including work by Langevin's son-in-law, the theoretician Jacques Solomon. But the definitive paper was produced in the summer by Bohr and John Wheeler at Princeton. Buried in their paper was a most important deduction. The uranium coming out of a mine consists chiefly of one isotope, called uranium-238 because one atom of this isotope weighs about 238 times as much as one hydrogen atom. The natural uranium also includes a sprinkling (0.7 percent) of a slightly lighter isotope with three fewer neutrons, uranium-235. Bohr and Wheeler found theoretical reasons to believe that in a sample of uranium it was the rare uranium-235 which fissioned; the more stable uranium-238 normally absorbed neutrons without splitting. Through 1939 only a few groups of scientists paid much attention to this prediction. Even at Columbia, where Bohr and Wheeler often discussed fission work, their theory was received with skepticism. Fermi character-istically refused to pay attention to the idea until he had seen good ex-perimental proof. Another Columbia scientist, John Dunning, began to look for ways to make such a proof, but it was clearly going to be dif-ficult.[5]

Experimental physicists and radiochemists studied many other aspects of fission. Two sorts of work were particularly popular, since they could be done in any fairly well-equipped laboratory and seemed likely even-tually to give some insight into the structure of nuclei. They both in-volved study of the medium-weight fragments, like lanthanum and barium, that resulted from fission. On the one hand, a scientist could study these fragments to find how much energy was released when a nucleus split and how the energy was divided among the various pieces. On the other hand, the scientist could study the nature of the fragments themselves. There turned out to be dozens of different ways a uranium or thorium nucleus could divide, producing a bewildering variety of iso-topes, each of which was radioactive and went through a variety of fascinating transmutations. This was the sort of work pursued by Hahn and Strassmann, Frisch and Meitner, the Berkeley scientists, Irène Curie, and many more in France and elsewhere. This kind of research frustrated James Chadwick when he reviewed fission work at the request of the British government in late 1939. He complained of a lack of data: "Very few experiments have been made which throw any light on the actual mechanism of the fission processes; they have been mainly concerned with the radioactive products of the fission—interesting and important, but apt to degenerate into a kind of botany."[6]

In both Germany and the Soviet Union some scientists were excited about nuclear energy, but in 1939 they did little experimental work. In Germany the War Office and the Ministry of Education independently and slowly set up fission research programs and tried to procure uranium. German physicists made some good theoretical studies of chain reactions but got scarcely any experiments underway before the end of the year. Soviet scientists, who had been alerted to the possibilities of fission early in 1939 by a letter from Joliot, were impressed by the work of the Collège de France team and by Francis Perrin's optimistic figure for the critical mass. By November 1939 they had finished a few studies of fis-sion, had developed some of the basic theory of chain reactions, and were planning experiments on neutron production in uranium with a moderator. But like the German experiments, these had probably not been started by the end of the year.[7]

In Britain there were few scientists concerned with fission, but some did try chain reaction experiments. After G. P. Thomson wrote about the vast potential of fission to the government, the Air Ministry invited him to discuss the matter. Although Thomson felt diffident, as he recalled, on "putting forward such a wild suggestion suited only for a paperback," the government asked him to look into it and toward the end of May got him a ton of uranium oxide from the Union Minière. That summer P. B. Moon, a physicist working under Thomson's direction, set up spheres filled with uranium oxide with a neutron source at the center. This experiment closely resembled the earlier ones of the French and American groups, but the British conceived a somewhat different approach. They tried a number of spheres of different sizes, ranging up to large piles made with sacks of the oxide, and in each sphere they buried a radioactivity detector, at a fixed distance from the central source, to measure the neutron density. "It was hoped that the density would become greater and greater for the larger spheres until it became infinite for the sphere of the critical size at which the chain reaction would begin. No such increase was found, however." Thomson and Moon varied the experiment in a number of ways, for example by mixing in water as a moderator to see if this would improve the results. In all these trials they found no signs that a divergent chain reaction could be sustained.[8]

Meanwhile Mark Oliphant, another leading physicist in Britain, was looking into the chances of producing chain reactions with metallic uranium, a rare and expensive substance. In August he wrote: "We have been carrying out some further work on the fission of uranium and other elements, but the results of our experiments are such that I feel very great doubt now whether it can ever be made to produce a chain reaction, unless the isotopes of the element can be wholly or partially separated . . . I find it almost impossible to give my mind to things other than my official duties in connection with air defence." Oliphant and Thomson, like many leading physicists in Britain, in these last anxious months of peace were working on the desperately needed radar system and had little energy to spare for the remote possibilities of fission. When the war began, Oliphant's group dropped fission research altogether.[9]

Fission work had also slackened in the United States, although it was being pursued by more people there than in any other country. Most American scientists gave little thought to neutron emission and chain reactions, which were outside their purely scientific interests. The group under Tuve at the Carnegie Institution, for example, continued for a while to think about chain reactions, and they had a large water tank, sources of radioactivity, and other things needed for experiments, but they were willing to let Fermi's group take the lead. Roberts wondered about propagating chain reactions without a moderator, made some measurements, and found that this could not be done in natural uranium. He dropped the problem, for he had much to do helping the others build a new cyclotron.[10]

Elsewhere in the United States some organizations kept up work on a small scale just to stay in touch with the possibilities of nuclear power. These included the Naval Research Laboratory, the General Electric Research Laboratory, and the Bell Telephone Laboratories.[11] These small and isolated groups each eventually contributed in one way or another to the development of nuclear energy, but in 1939 they were making little headway. Of all the scientists in America that summer, only the ones at Columbia did significant slow-neutron chain reaction experiments.

Szilard had managed to borrow some 200 kilograms of uranium oxide from Eldorado Gold Mines, Ltd., the Canadian company which was the Union Miniere's only serious competitor. At the end of May he, Fermi, and Anderson used the oxide to carry out an experiment quite similar to the one that the French team had done in April, studying the balance of neutron births and deaths in a uranium-water mixture; like the French, they were eager to learn whether a divergent chain reaction could "go." As Anderson later remarked, the Columbia and College de France teams were independently moving along the same path, except that the French were always "about a week ahead of us."[12] This experiment, like the similar one in Paris, turned out to be ambiguous. Fermi's group could tell that the sort of system they had would not sustain a chain reaction, but they could not tell whether a better system could be built.

In the Columbia group's experiments the uranium oxide was diluted with much less water than in the French experiments; they had one uranium atom for each 17 hydrogen atoms, while the French had one uranium for 65 hydrogens. At first Fermi's group assumed the chances of getting a chain reaction would be better at still higher concentrations of uranium. But Placzek, on one of his frequent visits, pointed out a difficulty. Placzek had a lively critical spirit—so critical that it may have held back his own scientific work—but beneath his sarcasm there was warmth and an unusual willingness to share his brains, to help friends who needed a problem analyzed. He had already played the role of helpful skeptic in fission work with Frisch, Bohr, and others. Now he reminded the Columbia workers that uranium greedily absorbs neutrons in a range of moderately fast speeds: resonance capture. Thus if there were a lot of uranium in the mixture, most of the neutrons would be captured while they were slowing down, before they could cause fission. On the other hand, if there were only a little uranium, most of the slow neutrons would be absorbed by the hydrogen in the water, since hydrogen had a small but not negligible ability to absorb neutrons. At an intermediate ratio of uranium and hydrogen it was likely that the two types of absorption together would use up neutrons faster than fission could produce them. The group at Columbia was left with the lame conclusion that "even at the optimum concentration of hydrogen it is at present quite uncertain whether neutron production will exceed the total neutron absorption."[13]

There was a small but significant difference between the similar experimental setups in New York and Paris, a difference that helped Fermi see a possible way around the problem. Rather than mixing their uranium oxide directly into water, his group had distributed it about their tank in some fifty sealed cans. While calculating the production of neutrons in the tank, Fermi noticed that this heterogeneous arrangement might be better able to sustain a chain reaction than would a simple homogeneous mixture of powder and water. He discussed this possibility with Szilard, who became enthusiastic. Although a homogeneous mixture had a questionable future, separating the uranium from the moderator might allow them to make a critical system after all.

Another way around their problem would be to find for a moderator some substance which was less liable than hydrogen to absorb neutrons. Carbon seemed the most likely candidate, and Fermi and Szilard considered measuring its propensity for neutron absorption to see whether this was low enough. They did not publish these vague ideas, so the French did not learn that the Americans were thinking along these lines, just as the Americans did not learn of the ideas in the French patents, and both were ignorant of the incipient British work. Caution was beginning to interfere with communication between the scientists in different countries, although still incompletely. Until this point the teams in New York and Paris had advanced along parallel lines, but now they began to diverge.

In early July Szilard wrote Fermi, "I have reached the conclusion that it would be the wisest policy to start a large scale experiment with carbon right away without waiting for the outcome of the absorption experiment . . . If we waited for the absorption experiments we would lose three months . . . perhaps 50 tons of carbon and 5 tons of uranium should be used as a start." But Fermi was still pessimistic about nuclear energy, and he was no longer interested in doing experiments alongside the unconventional and vociferous Szilard. For the rest of the year Fermi occupied himself with calculations in the theory of cosmic rays. Soon after getting Szilard's letter, he wrote Pegram: "I agree with him that the loss of time for a semi large scale experiment would presumably be considerable. Nonetheless I would feel much better at ease if it were possible to try a large scale experiment after having convinced ourselves that the chances of success are greater than we can estimate now."[14]

Szilard was in no position to proceed without Fermi, for although he had several experiments in mind, he had no funds to do them, nor even a laboratory, his permission to work as a guest in the Columbia laboratory for three months having now run out. Moreover, he needed tons of raw materials just to find out whether his ideas could possibly work. Szilard fought to get a few tons of graphite and uranium oxide, approaching the Union Minière, Eldorado Gold Mines, Union Carbide and Carbon (a large graphite producer), the U.S. Navy, the U.S. Army, and private individuals. But he could neither borrow the raw materials nor raise two or three thousand dollars to buy them; he was far from winning anything like the support that the French enjoyed.[15] So American experimental work on nuclear chain reactions languished and was not resumed until the following April. Leaving aside the ineffectual study under Thomson in London, through the entire second half of 1939 the French had the field of chain reaction experiments all to themselves.

The scientists in Paris learned of the problems of uranium-water mixtures at about the same time as the Americans, warned by the same man. Placzek had met Halban in Copenhagen and again on a stay in the Collège de France laboratory, and the two had become good friends and correspondents. In May Halban wrote Placzek, saying he was sorry it was so hard for a foreigner to get a job in France, for life was difficult in a country where theoreticians like Placzek were in short supply. It was beginning to seem, Halban added, that a chain reaction could be started, and it all depended on how the resonance absorption varied with the concentration of uranium in the water. Placzek obligingly sent back several pages of theoretical calculations on the optimum concentration of substances, showing the conflict between uranium's resonance capture of neutrons and hydrogen's absorption of neutrons.[16] Now the French scientists began to discuss the difficulty. How could they slow down the neutrons while keeping them from being absorbed by either the moderator or the uranium?

Kowarski recalled that a solution came to him while he was sitting in his bath, "which was definitely conducive to the idea of neutrons spreading around in a liquid medium." The trick was to keep the neutrons away from uranium while they were slowing down, for only neutrons with a particular range of speeds, neither fast nor very slow, were threatened by resonance capture. Kowarski pictured lumps of pure moderator buried within a mass of uranium oxide. Fast neutrons from the uranium would enter a lump, then diffuse about in the moderator and slow down past the dangerous speed zone without risk of capture, and finally reenter the uranium as slow neutrons and cause new fissions. This was the same idea that Fermi came to a little later after contemplating his experimental setup. Kowarski, however, was not yet completely at home with the data of nuclear physics and overestimated the chances that hydrogen would capture neutrons. He figured that most of the neutrons would be absorbed in the moderator before they could reenter the uranium. It seemed that there would be no advantage to separating the uranium from the moderator, so Kowarski let the matter drop after mentioning it casually to Halban. In August the subject came up again, probably because Francis Perrin and possibly Adler had independently mentioned to Halban the idea of such an arrangement. In the course of discussions the team realized it might be an improvement. But by this time they were already beginning experiments with the usual homogeneous mixture, and they stuck to it. A homogeneous system would be simpler to analyze, and once it worked, or nearly worked, they could try to tune it up by separating the uranium and moderator.[17]

These experiments would be on a grand scale, building toward the full-scale trial using five tonnes of uranium oxide that the French had contracted for with the Union Minière. Fortunately there was ample space in Joliot's Laboratory of Nuclear Synthesis at Ivry, with its huge hangar, overhead cranes, and other industrial facilities. The team moved its operations there in June. At Ivry they could draw upon a crew of technical assistants to help with their tedious measurements and to handle their hundreds of kilograms of wet uranium oxide, denser than solid granite. At Arcueil, a nearby suburb, they had the use of another facility. It was an annex of the Radium Institute, built in 1923 after persevering efforts by Marie Curie, where radioactive minerals could be treated in lots of several hundred kilograms at one time. In developing these facilities at Arcueil and Ivry, Marie Curie and Joliot had specifically intended to work toward large-scale practical applications of radioactivity; now, quicker than they had imagined, the chance had come.

The team planned to approach the five-tonne trial by a series of steps, beginning with a copper sphere only 30 centimeters in diameter filled with uranium oxide and water, then progressing to larger spheres. Halban was so confident a chain reaction could be started that he did not want to begin with too large a mass for fear it might explode as soon as it was put together.[18] The team planned to use a high proportion of uranium oxide to water. Each sphere would be filled with wet oxide and then sunk in a larger water-filled tank three meters in diameter. This tank would reflect neutrons back into the sphere to produce more fissions; more important, it would let the team measure the distribution of neutron densities not only inside the sphere but also in the water outside, so they could get a complete density-distribution curve, tail and all.

The team kept their work going by taking their vacations in turn, staying in close touch by letter. In the second half of July they began measurements on the 30-centimeter sphere, and while the results were inconclusive, the team remained optimistic about the chances of making a divergent chain reaction with some combination of uranium and water.[19] These experiments were no simple task. Trying to figure out what was going on in their systems, the scientists carried out the tedious density-distribution measurements in a variety of combinations: with the sphere empty, filled with water, with cadmium to suppress slow neutrons, and so forth. The resulting curves were difficult to interpret, for the production of neutrons in the sphere made the situation very different from classic density-distribution experiments; sometimes they got a "camel" curve with two humps. The members of the team, particularly Halban, made various attempts to understand theoretically what was happening in their experiments. The curves should give enough data to tell them whether or not uranium in water could ever sustain a chain reaction; the theory they needed to interpret the curves should be simple; at times Halban seemed almost to have the answer. But he never quite got a clear-cut result, for in fact the theory he needed, although simple, was subtle. It eluded him and his colleagues.

Just getting the curves was a full-time task with the unreliable geiger counters of 1939. Many physicists avoided such counters altogether, for although they could be powerful tools, they often mulishly refused to function. It was Kowarski's particular task to nurse them along. Their sensitivity tended to wander, and so whenever a dysprosium detector was put in front of a counter for a radioactivity measurement, the counter had to be immediately standardized with the aid of a constant radiation source. As another check the team decided to measure each of the numerous points on the curves twice, using different geiger counters. When everything was working well, it took a day or more to chart a single curve.

Besides this lengthy routine the team had a variety of experiments to do on the side. For example, Joliot poured water into a box of the uranium oxide powder and some time later asked Dodé to take a core sample and see whether the water was uniformly distributed. If the powder in the water settled to the bottom, the experimental system would not be symmetrical and the team's calculations would be faulty.[20]

In August it was Halban's turn to stay in Paris. He and his wife were expecting their first child that month, and meanwhile he had high hopes for another and more momentous type of birth. Although he could not be sure whether a chain reaction was possible, he was optimistic enough to advise Adler to stay in Paris another year "in case the uranium business comes off," since success would guarantee financial support. Halban wrote Placzek that the prospects of fission were even worth the risks of staying to be caught in the next war—"you'll laugh over so much enthusiasm—but perhaps I'm entirely right."[21]

Even though he was by no means a senior French physicist, Halban had the use of all the facilities of the Laboratory of Nuclear Synthesis, including the aid of two or three technicians. He began experiments with a sphere 50 centimeters in diameter. Joliot, taking his annual vacation at

L'Arcouest, fishing, sailing, and preparing his notes for the fall semester lectures, wrote that he was "impatiently awaiting the first results with the big sphere." Halban filled the sphere with 300 kilograms of damp uranium oxide powder, submerged it in the water tank, and began measurements. He wrote his father, "My neutronist's heart beats faster when I think that this semi-industrial arrangement is the consequence of the exceedingly humble researches I undertook" with others a few years earlier at the Radium Institute.[22]

The area under the density-distribution curve, representing the number of neutrons in the system, increased as Halban increased the amount of water he mixed into the uranium oxide. But he had to stop when he had poured in an amount equivalent to three hydrogen atoms per uranium atom, making a dense black mud. "Alas," Halban wrote his companions, "it is impossible to add more water, because the uranium doesn't want to drink any more (like a camel!). In any case, we're sure that the water is not perfectly distributed any more." With the maximum amount of water that the uranium oxide would soak up, Halban estimated that the number of neutrons produced in the uranium was roughly equal to the number of neutrons escaping through the surface of the sphere. Chain reactions were definitely underway in the system; fission neutrons were going on to cause further fissions; but too many neutrons were lost, and each chain died out after a few steps. Halban began to worry about impurities, realizing for the first time that the team might need very pure materials lest foreign matter absorb neutrons. At the moment he could do nothing about this problem, but he could go ahead to a still larger sphere. "It's all quite encouraging," he wrote. "One has every hope for the big chain."[23]

By about August 20 the team was ready to go to a sphere 90 centimeters across. Following a design that Francis Perrin laid out for Halban, this would be built by stacking boxes of uranium oxide into a crude sphere with the aid of scaffolding. Meanwhile the results of the experiment with the 50-centimeter sphere were written up for publication, this paper being the only one authored by Perrin as well as Halban, Joliot, and Kowarski. They noted that under their best conditions, "the production of thermal neutrons in the system is *more than doubled* by the introduction of uranium . . . we conclude that *secondary, tertiary, etc. fissions have taken place, and we are in the presence of a convergent chain reaction.*"[24]

Stated this way, as the first proof of a nuclear chain reaction, the experiment seemed exciting and attracted popular attention. But its scientific worth was ambiguous. A convergent reaction, in which the chains petered out, was a long way from a divergent one, in which neutrons would multiply without limit. When the publication reached America that winter, Szilard asked Fermi what he thought of it, and Fermi replied, "Not much." On the French side Halban, when he got a copy of the similar experiment that the Columbia group had finished in June, wrote Joliot, "I think it's a great exaggeration to publish such a modest result."[25] In truth little could be learned from either experiment until the theory of the chain reaction had been developed further. Once this was done, both experiments would prove to be valuable.

The Collège de France team was hampered by the need to create a new theory, by the need to do experiments on a grand scale, and perhaps above all by the need to hurry. For one thing, they were eager to finish the preliminary experiments demanded by their agreement with the Union Minière and press on to a full-scale trial. More important, throughout the year they had been awake to the terrifying political situation in Europe, pressed on them every day by newspapers and personal contacts with refugees from fascist countries. An American scientist who visited Joliot's laboratory that summer wrote home that "one cannot talk long with anybody in Paris . . . without coming upon a mood of deep pessimism. After two or three such conversations I have felt so dispirited as to meditate a prompt return home, and yet the fear seems to be not so much of some immediate crisis as of a future of some months or years ending in eventual war." Kowarski recalled that the pressure of events affected the team's work: "Obviously the Axis powers were on the march . . . We didn't know when we would have to interrupt our work and in what conditions. Therefore, we began to look for simple experiments giving simple results which would allow future action to be undertaken . . . We were more in a hurry than our colleagues, chiefly Fermi and Szilard . . . who therefore worked in a different style. Their experiments were more deliberate, conducted with greater precautions and also took more time. We had the advantage on them of speed in reaching experimental evidence." In order to proceed with their increasingly expensive plans, the French team had to produce the sort of results that would attract ample support. As a result, Kowarski recalled, "We began to worry more about our relations with the government than our relations with our fellow scientists."[26]

French military research had until recently been the job of the Office des Recherches Scientifiques et des Inventions, an organization descending from the Directorate of Inventions that Painlevé had set up during the First World War. Although leading scientists were associated with the Directorate during the war, in the 1920s most of these ties withered away and military research, turned over entirely to the Ministry of War, stagnated. In mid-1938 the Popular Front government suppressed the Office and turned its functions over to the new applied-research organization, the CNRSA. The armed forces, nursing deep suspicions of the Popular

Front, were reserved or even defiant toward the new agency. When the CNRSA chief appeared at the military research laboratories to take command, the former director refused to vacate his offices, saying he was responsible only to the Ministry of War and vehemently criticizing the CNRSA; it took several weeks of negotiations to displace him. The military promptly created their own Institut National des Recherches Scientifiques Appliquées à la Guerre.[27]

All these institutes, including the CNRSA and the CNRS, were supposedly held together by interlocking directorates and by a coordinating committee, the Haut-Comité de Coordination des Recherches Scientifiques, which included Irène Curie, Jean Perrin, and Laugier. But cooperation was limited, and the government made frequent and confusing changes in the superstructure without ever finding a workable arrangement. The most important reorganization, pushed through by Laugier and completed in late 1939, was the creation of a Centre National de la Recherche Scientifique (CNRS), including both the old CNRS and the CNRSA. Laugier hoped eventually to intermingle these two organizations, which still led separate lives, but his progress was slow. Joliot sat on most of the important committees in the various bodies, serving, for example, as vice-president of the Haut-Comité de Coordination.[28]

The CNRSA was supposedly in charge of scientific mobilization in case of war, but the new organization had scarcely begun to lay plans before the Munich crisis of 1938. After initial confusion the scientists decided that in the event of war they would turn their laboratories into semimilitary establishments under the CNRSA, with their personnel remaining on the job on Special Assignment from the military. Joliot, nominally a captain in the artillery reserves, was promised a Special Assignment in case of war, as were important laboratory staff like Savel and Moureu.[29]

For Halban and Kowarski the situation was more difficult, for neither was a French citizen, although both had applied for naturalization. Mayer warned the professors of the Collège de France in September 1938 that their "foreign workers would be notified *later on* that the laboratories will be closed to them from the moment mobilization is announced until the authorities make a decision concerning their fate." This threat was a response to a general French antagonism toward foreign labor, a feeling intensified in the Depression and embodied in stringent security laws in early 1938. Special efforts were made on Halban's behalf by Joliot, Laugier, and Halban's wealthy banker acquaintances, who secured his naturalization in January 1939. Joliot, Laugier, and Jean Perrin helped Kowarski to become naturalized in August. "I salute you as a compatriot," Halban wrote Kowarski. "I'm so pleased to see you're doing your six weeks of military service next March, I'll be ready to do it myself just to see you debating with the sergeant whether it's you or all the rest of the company that's in the wrong."[30] Halban was denied this pleasure by the outbreak of war.

On August 23 French foreign policy collapsed with the signing of the Nazi-Soviet Pact, which left Hitler a free hand in the West. Every group in France was consternated. Joliot, Curie, Langevin, Jean Perrin, and other intellectuals signed a manifesto expressing their "stupefaction" at Stalin's about-face.[31] On September 1 Germany invaded Poland, leaving France and Britain at long last no choice but war.

In the French laboratories, as in the country at large, there was none of the disruption of the Munich crisis, only a dull resignation to the unwelcome task. Halban, Kowarski, Joliot, and others went into the army while remaining in their laboratories on Special Assignment. The CNRSA requisitioned Joliot's establishment, but except for extra paperwork the administration continued much as before. Military uniforms appeared and guards were posted, while the workers set up a first aid station, distributed fire-fighting materials, and held gas mask drills. The windows of Joliot's laboratory were painted over because people were working into the night, and the staff were warned that "everything they learn in the performance of their duties constitutes a secret."[32]

Much of the prewar activity in the laboratory simply continued. Joliot's team felt that their work on the chain reaction was important for the war. Like most Frenchmen, they supposed that the conflict would be a long-term affair of blockade and trench battles like the First World War, so that a new source of industrial energy could eventually be important. Other research under Joliot, such as study of alloys for the Air Ministry and work on the production of gasoline from vegetable matter, also continued. But the CNRSA rarely asked Joliot's advice on what should be studied. Instead it sent down a shower of little problems, mostly for the laboratory's chemical section, facing Joliot with a pile of paperwork on such matters as explosives and poison gas defense.[33]

For all their efforts to create an efficient and centralized system, the French made poor use of their scientists during the war. The scientists themselves were unsure how to divide themselves between their traditional pursuits and war work. It was less that they were unwilling to leave pure science than that they lacked proper direction. According to a later report, while "trying to maintain their laboratories' activities, with tasks often far removed from the War or the National Economy," the scientists also "did their best to answer the questions which arrived in masses, piecemeal and disparate, from the administration of the CNRSA."[34]

Joliot remarked that despite attempts to create a unified administration for science, nearly impenetrable barriers separated military from civilian researchers and kept the university scientists' knowledge from being applied to the war. Mutual distrust poisoned relations between conservative generals or admirals and left-wing scientists. Langevin, for example, had been shunned by the military since his support for the Black Sea mutineers in 1920, and it was not until the spring of 1940 that naval authorities asked him to resume his work on detecting submarines. Perhaps the most serious problem was the military's concern for secrecy. Afraid to let secrets slip into civilian hands, the officers decided by themselves what their needs were and then, as Laugier complained, tried to "extract the implied problems of pure science and give them over, in their crystalline, unworldly purity, to civilian scientists to be studied as such, and without any consideration of their application to military technique." When civilian scientists offered to help the Navy study the grave problem of magnetic mines, Admiral François Darlan told them that it was none of their business.[35]

The same sort of difficulties held back wartime science in other countries. The British were the first to see through the problem, and even before the war they had worked out a solution. As Cockcroft wrote Joliot in February 1940:

There seems to be a very fundamental difference between our organization and yours, namely, that in England the scientists are in general working in groups inside Service Research Establishments. We have found that this has been of the greatest importance since only by that means can we be brought fully into touch with the whole of the problem. In this way there is a complete interchange of ideas between the Service representatives and the scientists.

In addition to our groups working in Research Establishments we have of course a number of groups working in universities on special problems, rather as you are doing in your laboratory. Arrangements are, however, made so that the leaders are able to visit the Service Establishments and discuss fully the problems they are working on.[36]

No such close contacts with the military developed in France. While most senior physicists in Britain labored on radar and the like, the French universities were rigorously excluded from such work. But this exclusion was no disadvantage for the Collège de France team in their pursuit of nuclear energy, for it left them free to follow a long-range project with no obvious, immediate military use.

The war affected the team in other ways. There was much inconvenience and red tape. Throughout Paris subway lines were closed down to conserve energy and windows were taped over for protection against explosions, while in the streets many people carried gas masks slung over their hips. Fearing German bombers, the scientists kept their families outside Paris; Joliot and Curie sent their children to L'Arcouest, while Halban and Kowarski, who kept their children closer, could still spend one or two nights a week with them. Kowarski, lacking the money to maintain two apartments, began to sleep at the Laboratory of Nuclear Synthesis.

Not everyone got a Special Assignment to stay in his own laboratory. For example, Dodé, who had helped the team in a number of ways, was sent to Toulouse to study the manufacture of explosives; and the Arcueil annex, where the team had planned to dry out and otherwise process its tonnes of uranium oxide, was closed down and its personnel sent elsewhere.[37] Many foreign workers left, among them Adler, who returned to Switzerland.

The team also lost Francis Perrin. He had not requested a Special Assignment, Perrin recalled, for in the crisis he did not wish to be sheltered in a laboratory but preferred to have a well-defined task of immediate value in the national defense. He was sent out of Paris as a simple lieutenant in command of a section of antiaircraft searchlights. In early 1940 the Army ordered him back to Paris to join a team doing research on radar, and he worked on antenna design. He met Joliot occasionally but did not keep in close touch with the work at the Collège de France. From September on, Halban, Joliot, and Kowarski worked without scientific collaborators.[38]

8

Conditions for a Working Reactor

About the time the war began the team at the Collège de France reached a new stage of their work. From the moment fission had been discovered, they had stepped from one experiment to the next without needing any deep theoretical analysis. The sequence was clear to them: verification of fission, discovery of neutron production, measurement of neutron production, and a series of trials with larger amounts of uranium. But they were coming to realize that this direct path petered out. Now they would have to develop some theoretical tools to guide them back to a profitable route. Neither Joliot, Halban, nor Kowarski was a theoretician, and since the onset of war they had been cut off from their collaborators, so they would have to try to work out some elementary but essential chain reaction theory by themselves. It was not the sort of theoretical problem that only the brilliant or expert can comprehend; rather, the team's difficulties would result from the problem's simplicity.

Throughout the summer of 1939 the team had been increasingly forced to notice theoretical problems, and with the help of Adler, Perrin, and Placzek they had come up with an approach that seemed straightforward, but had hidden difficulties. When they published their August experiment with a sphere of uranium oxide, they also tried to interpret it. As usual, they had measured the area under the density-distribution curve as an indication of the number of slow neutrons in the system. It seemed reasonable to divide this measurement into two parts: the number of neutrons internal to the sphere (which they labeled Q_{int}), and the number of neutrons which had diffused out and were external to the sphere (Q_{ext}). Each could be found simply by figuring the area under the corresponding part of the density-distribution curve, the part measured within the sphere for Q_{int}, the part in the water tank for Q_{ext}. The team now cast about for a way to combine these quantities to produce a single number which could show how close their system was to a critical mass. This number would be a manifest measure of what they had done and what they could hope to do.

Their choice for such a quantity was the number of neutrons produced in the tank as the result of a single fission provoked by a single primary neutron coming from the central source. At the very least the scientists expected that each primary fission would give $v \approx 3.5$ neutrons; but if there were a chain reaction, these would go on to provoke further fissions. As the system approached critical mass, the number of neutrons that could be traced back to a single primary fission would approach infinity.

Careful thought and simple algebra were all they needed to write down an equation that would give the number of neutrons per primary fission. It was not hard to see the sort of numbers required to build up this equation. There would be factors describing the distribution of neutrons in the tank, and Q_{ext} and Q_{int} gave all the information needed to assemble these factors. There would be the birth and death rates of neutrons arising in fission and undergoing capture. And there would be the rate of "emigration" from the neutron population, as some neutrons slipped away through the walls of the tank. After writing down an equation and plugging in the numbers, the team calculated that at least eight neutrons were ultimately produced from each primary fission. Since they knew that the average single fission produced less than half that many neutrons, they concluded that they had produced short chains of two or three fissions at least.[1] While they had not yet achieved a divergent chain reaction, they seemed to be getting close.

But the team had made an error of logic, similar to the one they had made in April which had led to their overestimate of v. Their equations did not take into account that some of the neutrons which escaped from the sphere might diffuse back into it, and that some of the neutrons produced in fission, as well as the original neutrons, might diffuse out. They simply ignored such possibilities as a minor effect, but in a chain reaction, where a slight gain or loss of neutrons can be multiplied manyfold, these factors made a big difference. Their system was worse than they thought.

Around September 21 Kowarski saw the error and redid the calculation correctly in an original and elegant fashion.[2] This work was the first clear sign that he was blossoming into an independent nuclear physicist, able to produce new theoretical ideas in a coherent mathematical form. In particular he made use of one of the basic equations of chain reactions, which the team had already produced in rudimentary form but whose value they had not yet appreciated.

One approach to theoretical physics is to play with symbols until a cluster of letters, originally written down as part of a larger equation, takes on an independent life and, after a series of steps, turns out to be meaningful in unexpected ways. As far back as May, Halban and Adler had recognized that the number of fissions per primary neutron might somehow be a key to determining whether or not a chain reaction was possible, but they had not seen how to develop the idea.[3] And the team's equation had been for the number of *neutrons* per primary *fission*. Instead of working with this equation, the team began to work more and more with the number of *neutrons* per primary *neutron*, which was simply the number of fissions multiplied by their familiar v, that is by the number of neutrons emitted in a given fission. The slight change stood for an important, if scarcely conscious, shift toward a better way of looking at chain reactions.

Although the team had given up the traditional picture of a beam of particles shooting through a screen, they had still tended to imagine a one-time process. It was as if they had been shooting a single fast neutron into their three-dimensional pinball machine, then watching as it caused a fission which spat out other neutrons which provoked further fissions, and so on. Now they had closed the circle, starting and ending with neutrons and making nothing special of the neutron that started it all. They saw their system entirely as an ongoing population with a cycle of births and deaths. The quantity needed to describe it was simply the number of neutrons produced per neutron, the "reproduction factor," called "k." If one typical neutron, bouncing about in the system, does not give rise on the average to one new neutron in the next generation, that is, if it fails to reproduce itself, k is less than 1.0 and the chain reaction must die out. But if the first neutron gives rise to more than one neutron, that is, if k is greater than 1.0, then each generation of neutrons will be more numerous than the last: the system is above the critical mass. It turned out that k could be calculated in a surprisingly convenient way, as a simple combination of a few numbers which the team could measure or estimate in a rough way.[4]

They wrote all this down in October, but they would not publish anything now that war had begun. Instead, the team arranged to deposit their paper explaining the theory of chain reactions in secret at the Paris Academy of Sciences, which accepts sealed notes (*plis cachetés*). The envelope was not reopened for many years.[5]

Some time later, they saw that an even more useful equation would be one that was not limited to a system of a particular size, like their spheres, but that described an infinitely large system. In such a system there would be no problem of neutrons diffusing in and out through the boundary, and the equation would become even simpler. The reproduction factor would now be k ∞, "k-infinity," describing the potential for a chain reaction of a boundless mixture of uranium and moderator. The advantage was that k ∞ could be figured out in exactly the same way as k, by making a few measurements and estimates for any small system and plugging them into a short equation. If $k\infty$ for the particular system were greater than 1.0, then the only thing that would keep the system from being critical was the escape of neutrons through the boundary, and a critical system could be built simply by piling on enough additional uranium and moderator. On the other hand, if $k\infty$ were less than 1.0 for some combination of uranium and moderator, there could be no hope of reaching a critical mass with that sort of system.

While the French were developing a theory, so were their rivals in other countries. The report of the team's August experiment was one of the last papers on chain reactions published in any country, for beginning in the summer of 1939, fewer and fewer nuclear physicists were willing to share their results with foreigners. Even on such a simple matter as this straightforward equation for critical conditions, the various scientists began to express themselves in divergent ways. Nobody surpassed the elegance and simplicity of the French approach until Fermi two years later.[6]

Although other physicists used other and often clumsier methods than the French, they were working with the same physical facts and could reach the same conclusions. Many wanted to know whether a working nuclear reactor could be assembled out of uranium oxide and water. Szilard, on receiving in early 1940 a copy of the last French publication, carefully studied and reworked the data. He concluded that it was quite uncertain whether a homogeneous mixture of uranium in water could sustain a chain reaction. But he was encouraged to find that the system was at least close, and he felt that by making some changes one might get over the hump.[7] In Britain Thomson no doubt saw the French publication, but he also drew his conclusions from the similar experiments done in his own laboratories by Moon. By early 1940 he had realized that a homogeneous mixture of uranium oxide and water could never be critical and had "pretty well come to the conclusion that as a conceivable war project the thing is not worth pursuing." Werner Heisenberg, who had been put in charge of German theoretical work, reacted much as Szilard did. When he saw the French results around the beginning of 1940, he noted that they reinforced his own calculations, which indicated that no simple mixture of uranium and water could be critical. But he felt there was hope of getting a chain reaction in other ways.[8]

The French experiment attracted so much interest because it was the only experiment, aside from Thomson's unpublished ones, which used a very high ratio of uranium to water. Since the French did not have to wait several months to receive the publication of their own results, they realized several months before anyone else that uranium and water would not do. Around September 21, when Kowarski corrected the error in the team's logic, he also showed in a simple way that the condition for a divergent reaction—that $k\infty$ be greater than 1.0— is mathematically equivalent to the condition that Q_{ext} be greater than 1.0. In other words, unless the area under the density-distribution curve measured in the tank outside any small sphere of wet uranium oxide were above a certain value, it would never be possible to get a critical mass no matter how much bigger a sphere was used.[9] The August experiment on the 50-centimeter sphere had given a value for Q_{ext} far below 1.0. If Kowarski's deceptively simple logic was correct, then the team had to give up their hopes of a uranium-water reactor.

The team were not entirely convinced. They went ahead with the experiment, planned with Perrin before the war, on a 90-centimeter construction of damp uranium oxide. On October 7 Kowarski predicted the results, and the neutron measurements, carried out over the following week, gave a value for Q_{ext} that agreed fairly well with his prediction and which was far below 1.0. This was still not quite the deathblow for the wet uranium-oxide reactor. The team had made measurements only for certain proportions of hydrogen to uranium; they had measured systems with up to three hydrogen atoms per uranium atom, and beyond this there were no measurements until the ratio of seventeen to one, tested in June by the New York team. The French used their equation for k, along with some new measurements of resonance capture, to study what might happen with intermediate concentrations. They estimated correctly that the optimum conditions for a reactor would be reached somewhere between two and ten hydrogen atoms per uranium. At this optimum there would still be too many neutrons lost to uranium and hydrogen absorption. "It is almost certainly impossible," they concluded, "to get a divergent chain in a homogeneous medium containing hydrogen and uranium in their natural isotopic composition."[10] It was going to be harder than they had expected to get a nuclear reactor going.

The team was determined to find a way around the problem. One possiblility, if Bohr and Wheeler were right in figuring that the uranium-238 that makes up the bulk of natural uranium does not normally fission, would be to concentrate the fissionable isotope uranium-235. The system of natural uranium and water was fairly close to critical already, and the French estimated correctly that only a modest enrichment of this isotope, from 0.7 to about 0.85 percent of all the uranium, would be enough to make a working reactor.

Separation of minute quantities of isotopes had been a popular research problem among physical scientists in the late 1930s; a review of the field listed some forty papers published in 1939-1940 on one separation method alone. But development of such methods had not reached a point where enrichment of tonnes of a heavy metal seemed at all feasible. The French team had the idea clearly in mind by October 1939, and Joliot later considered bringing a French specialist on isotope separation to Paris. He eventually ordered some apparatus to try doing large-scale separation by thermal diffusion. The team felt, however, that long before anyone could achieve a significant amount of isotope separation, they would get a reactor working by other means.[11]

They had several ideas to try. Their thinking was still partly influenced by the error in their analysis of the April experiment, for although they had caught the comparable error in their August work, they had not yet gone back to recalculate the earlier data.[12] Thus the team still believed that some three or four neutrons were emitted in an average fission. Moreover, their measurements of resonance capture were imprecise, and their crude estimate of the cross-section for the capture of slow neutrons by uranium was only half the true value. Since their system nevertheless did not contain a great many neutrons overall, they had to assume that they suffered from a heavy loss of neutrons through resonance capture. The team's error was not great, given the severe difficulties of making all these measurements, but it did suggest that a reactor could be built if only they could beat resonance capture. And it seemed that this could be done by separating the uranium from the moderator.

By the end of 1939 the idea of such a heterogeneous reactor had occurred independently to physicists not only in France and the United States but also in Germany, and possibly in Britain and the Soviet Union. The French pictured a reactor made of lumps of moderator embedded in a mass of uranium oxide, which was quite different from the picture held in Germany and America. German physicists imagined an arrangement of alternating layers of uranium and moderator, an idea they stuck to for several years. Fermi and Szilard also considered layers at one point but quickly realized that a better reactor could be made by using blocks of uranium embedded in the moderator in a three-dimensional lattice.

The decision whether to embed a lattice of uranium lumps in the moderator or, as the French saw it, a lattice of moderator lumps in the uranium simply depends on the relative proportions of uranium and moderator. But the two pictures can suggest different theoretical approaches to the problem. Halban, Joliot, and Kowarski were experimental rather than theoretical physicists, and with Perrin gone and the war in progress they had little chance of finding anyone to do the complex mathematics needed for a full solution. They never derived the optimum ratio of uranium to moderator for a heterogeneous system, which was a difficult task. They felt it would suffice to estimate that their lumps of moderator should be large enough so that the neutrons would not pass right through them before slowing down, and yet small enough so that the neutrons, once slowed down, would be likely to return to the uranium before being absorbed by the moderator. A few centimeters seemed about right. The

Germans and Americans, in contrast, were beginning theoretical and experimental investigations of the details of the distribution of neutrons within both the uranium and the moderator. German theorists stuck to the simple case of layers, but Szilard was already doing full three-dimensional calculations by the start of 1940, and over the next few years Fermi, Wigner, and many others worked out an elaborate theory.[13]

The French did realize that they needed detailed calculations of the behavior of a system of neutrons. When they could spare the time from their experiments, they attacked the problem vigorously. Halban in particular labored at theory, perhaps aided by letters from the ever-helpful Placzek. The difficulty was that each particular neutron had a unique history: perhaps it slowed down gradually through many small collisions, perhaps it lost most of its energy in a single collision, or perhaps at some random moment a nucleus captured it. Through the winter Halban and to a lesser extent his colleagues tried again and again to calculate the energy distribution and capture probabilities for the entire population of neutrons. The problem was deceptively easy to state, but the solution required unusual theoretical insight and ingenuity. Halban must have felt near the solution at times, but in the end he produced only page after page of half-finished calculations. The others did even less.[14]

Sometime in 1940, probably in the spring, Joliot posed the problem of neutron energy distributions to the doyen of French theoretical physics, Langevin. Langevin helped the team with some advice, but he could do little, for he had other heavy wartime duties. When he finally did come up with a solution, wrought with his characteristic thoroughness, it was two years later—too late to help the team.[15] All their theoretical work had done little to get them nearer to a nuclear reactor.

The team had more success with experiments. They decided to compare homogeneous and heterogeneous distributions, using the techniques that were now so familiar to them. Early in November 1939 they filled their 50-centimeter copper sphere with over 250 kilograms of uranium oxide and "larded" it with almost a thousand three-centimeter cubes of paraffin in a regular lattice. Paraffin, rich in hydrogen, would moderate neutrons in the same way water would. They then measured a rough density-distribution curve. Its area was not significantly different from the area of the curve they had measured with wet uranium oxide, and therefore they were as far as ever from a chain reaction. They continued these experiments through the middle of December, trying different proportions of hydrogen to uranium with no success.[16]

Then in January 1940 they broke away from their density-distribution experiments, which had become almost habitual, and tried a clever new approach. They could measure the effects of resonance capture directly by finding the radioactive isotope uranium-239 which was produced when a uranium-238 atom captured a neutron. If the heterogeneous system did decrease resonance capture, it would produce less uranium-239; the less of this isotope, the more efficient the system.

They set up their sphere in two configurations—a heterogeneous one, filled with dry oxide larded with paraffin cubes, and a homogeneous one, filled with wet oxide—and extracted samples of the irradiated uranium. Before the radioactive constituents could change, both samples were rapidly processed following a chemical procedure outlined by Joliot, and the radioactivity of each sample was studied. The experiment failed to give any quantitative result for the simple reason that there was no uranium-239 at all to be found in the sample from the heterogeneous system. But some uranium-239 was detected in the sample from the homogeneous system. Now the team had solid proof that by lumping the moderator heterogeneously, they had managed to decrease the resonance capture.[17]

This experiment increased the team's confidence that they could build a nuclear reactor. They had not found any spectacular advantage in their trial heterogeneous systems, but they were encouraged to continue working on homogeneous mixtures; they felt that once they had reached critical conditions with a homogeneous mixture, or come near, they could switch to a heterogeneous system for final improvements.[18]

They had also come to realize that hydrogen absorbs too many neutrons to work well as a moderator. Hydrogen was the standard element used to slow down neutrons in physics laboratories in the 1930s and was used in all the early experiments on chain reactions, but it was not likely to give the French their divergent chain. Could some other element do the job? A moderator must not absorb too many neutrons, and it must be a light element, for otherwise the neutrons will simply rebound off the moderator nuclei without losing speed. These requirements left only three possibilities: deuterium, beryllium, and carbon. Whichever was chosen would have to be unusually pure, for only a few parts per 100,000 of certain kinds of atoms, like cadmium, would absorb enough neutrons to damp out the chain reaction. In a sufficiently pure and nonabsorbing moderator a neutron could travel long distances, a meter or more; thus, to build a reactor that would keep most of the neutrons inside to sustain the chain reaction, the team would need large masses of moderator. But while pure hydrogen was cheap and easy to find in quantity, this was not true of deuterium, beryllium, or carbon. If the team wanted to keep up their swift progress, they would have to get these in bulk, and for this they would need more help from both government and industry.

9

Moderators and Ministers

Sometime in the early autumn of 1939 Joliot made an appointment to visit Raoul Dautry, the minister of armament, who was charged with organizing the supply of raw materials and finished goods in the French war economy. A graduate of the Ecole Polytechnique, Dautry had made his name building railroads during the First World War and revitalizing them after; to the end of his life the railroads remained his great love. He had a square, earnest face punctuated by short black eyebrows, expressing friendliness without intimacy and confidence without arrogance. Like most Polytechniciens, Dautry did not share the socialist views of Joliot's circle. In 1937, when the Popular Front nationalized the railroads, he resigned his position, partly because he judged imprudent the sudden imposition of the 40-hour week, partly because he felt his railroads would become overcentralized and bureaucratized. As a technocrat, he held himself apart from political controversy. But Dautry, who was acqainted with Langevin, Jean Perrin, and other scientists, did share with them and others across the political spectrum a belief deeper than party doctrine, a trust in the power of science and technology to transform human life. He too worried about France's industrial backwardness and felt that science was a key to overcoming it.

Energetic and scrupulously honest, Dautry was one of the few figures acceptable to all parties in France. In September 1939 he was chosen to head the Ministry of Armament, and during that dismal winter of war without battles, he was one of the most active leaders of the nation, never letting himself or others relax. "Do take the trouble to come see me," he wrote one manager. "I have the impression your factory is a Sleeping Beauty's palace." On weekends he would board a train and early next morning appear unannounced in his bowler hat at some factory, shaming or encouraging the personnel, untangling red tape so that raw materials and skilled personnel would find good use.[1]

Meanwhile he had also taken charge of scientific research, overcoming the minister of national education who tried to keep it under his own ministry. Dautry, believing what the scientists and their supporters had so often maintained, considered laboratories essential for armament. He was preoccupied, he later recalled, with the search for new weapons: "A long war is always won with a new weapon and I believed it would be a long war, since I believed it would be possible to create a spirit of resistance in the country." Now that Joliot needed increased support, Dautry was the obvious person to visit.[2]

The two had first met before the war when Dautry came by the Ivry laboratory for a tour of the giant accelerators, and they had both been members of the Haut-Comité de Coordination des Recherches Scientifiques.[3] But it was chiefly on the strength of his position as one of France's most famous scientists that Joliot received an audience with the minister. According to Halban's secondhand report of the meeting, Joliot explained the possibility that research on uranium could lead to a powerful new weapon, and also laid out the more immediate possibility that one could build a source of energy useful for propelling submarines and for fueling industry. France, which in peacetime imported nearly all of the oil and a third of the coal she needed, was now, Dautry knew, in serious danger unless new sources of energy could be found for her economy and her army. Indeed, one of Dautry's preoccupations in the two years between his retirement from the railroads and his elevation to minister of armament had been France's inadequacy in sources of electrical energy. He had also, according to Halban, read a bit of journalism reporting that the energy contained in a table would suffice to blow up the world. Struck by the possibilities that Joliot opened up to him, Dautry promised the Collège de France team all the support it needed.[4]

This was not the only contact between the team and the higher circles of government. The Commission (formerly Ministry) of the National Economy under Daniel Serruys also took an interest in uranium. The high commissioner's son, Jean Serruys, who worked for a rare-metals company that had supplied beryllium for Joliot's earlier work, was aware of the team's progress on chain reactions, as he recalled, from social contacts with Joliot at dinner parties and the like. When Daniel Serruys was named high commissioner shortly after the start of war, at his son's suggestion he set up a committee to oversee the supply of rare metals in general and uranium in particular. The committee, after some investigation, apparently agreed to leave the matter of fission in Joliot's and Dautry's hands. The most significant action which Jean Serruys recalls that the committee took was to give Joliot secret instructions, on at least two occasions around the start of 1940, to urge the Union Minière to keep its huge uranium stocks out of reach of the Germans.[5]

In Belgium, Sengier was getting similar warnings from the British government. In the autumn of 1939 he was also tentatively approached by

Szilard and later on heard semiofficially from an arm of the United States government. But the formal French warnings carried to him by his associate Joliot may well have made the greatest impression on Sengier. In 1940 on his own initiative he dispatched many tonnes of ore to a warehouse in the United States; a like amount was left in Belgium to be seized by the Germans. Sengier later turned the warehoused ore over to the American government with the understanding that it would be used for military purposes, and it became a cornerstone of the American fission program.[6]

While Joliot's talks with Dautry, Serruys, and Sengier proceeded, the Collège de France team were more interested in a scientific question: how strongly would carbon absorb neutrons? The best existing measurements dated from Halban's visit to Copenhagen in 1937 when, with Frisch and Koch, he found an upper limit of roughly 1×10^{-26}, measured in units of square centimeters, for carbon's neutron absorption cross-section.[7] If the carbon cross-section were near this upper limit, it would be difficult or impossible to make a nuclear reactor with a carbon moderator, but if carbon were a good bit less absorbing, as it well might be, then a carbon-uranium nuclear reactor could be built.

To measure the cross-section of carbon accurately, the team needed a large amount of pure material. The first substance they tried was solid carbon dioxide—dry ice. At the end of November 1939 Halban directed the construction of a stack of dry ice nearly a meter square and 1.7 meters high. He put in a source of neutrons near the base and hastily measured the neutron density at various heights in the frozen pile. As usual, the idea was to find the total area under the density-distribution curve, in order to determine how many neutrons were lost in the carbon. But even at the top of the stack the neutron density was still high, and it was impossible to extrapolate the tail of the curve accurately to distances where the density would be low. Halban therefore could not find the total area under the curve; his pile of dry ice was simply too small. He was unable to prove that the cross-section of carbon was much smaller than the upper limit he had determined in Copenhagen.[8]

From his early training in chemistry Halban recalled that carbon was manufactured in a very pure and dense form as graphite electrodes. Joliot accordingly paid another visit to Dautry and asked him to locate a large stock of industrial graphite. This was done swiftly, and shortly before Christmas Halban drove down to a factory near Grenoble along with Pierre Delattre, the team's young laboratory technician, and other assistants. They built a pile of ten tonnes of the heavy, coal-black graphite, put in the usual neutron source and dysprosium detectors, and measured the neutron density up to a height of 1.6 meters. This time the neutron density fell off enough to permit a reasonable extrapolation of the curve's tail, and Halban found a new value for the absorption cross-section of carbon: 0.6×10^{-26}. Because the graphite was not free of impurities, he knew this value was again only an upper limit and not the true value.[9] The new number was encouragingly low, but still not quite low enough to prove that a carbon-moderated nuclear reactor could work.

Meanwhile Halban tried to work out a formula expressing the abilities of elements of different masses to serve as moderators. Kowarski, after Halban discussed this problem with him, produced an approximately correct expression, which was verified around the end of November 1939 in some simple experiments by Delattre. Kowarski's formula showed the team that as a moderator, a light element has a strong advantage over a slightly heavier element. Carbon, which has a mass about twelve times that of hydrogen, would be far less effective in slowing neutrons than deuterium, which is a hydrogen atom with a neutron stuck on. Kowarski had to insist on the implications of his formula for months before he could convince Halban and Joliot that the team should center their attention on deuterium.[10]

Deuterium behaves chemically like hydrogen and can take its place in substances. For example, replacing the hydrogen in water with deuterium makes "heavy water," which looks exactly like ordinary water but is a little denser. Halban was already familiar with this substance, for in Copenhagen he, Frisch, and Koch had measured its neutron absorption cross-section. The absorption was remarkably low. Unfortunately heavy water was far more difficult to obtain than graphite. Common water contains only about one deuterium atom for every 6000 hydrogen atoms, and it took a large amount of electrical energy and a long time to separate a drop of the heavy water. In July 1939 Szilard and Fermi had considered using this substance but had quickly abandoned it when Szilard found that he could buy no more than a few grams.[11] The French, with their assurance of government support, were less easily discouraged.

The Copenhagen experiments of 1937 had used sixty liters of water containing 10 percent heavy water, borrowed from the Norsk Hydro-Elektrisk Kvoelstofaktieselskab. Norsk Hydro was a Norwegian industrial firm which produced heavy water as a side line at an electrochemical plant in southern Norway and sold minute amounts to the world's laboratories at high prices. It was the world's sole producer of heavy water in quantities greater than a few drops.

At the end of November 1939, about the same time the team began studying carbon, Joliot got in touch with Daniel Serruys' Commission of the National Economy. "Among the products of foreign origin which can be used in applications of interest to the National Defense," he wrote, "permit me to call your attention to heavy water." He asked the Commission to find out how large a stock was held in Norway and, if possible, to get an option to buy or borrow it. Two months later Joliot made contact with an officer in the Direction des Poudres, the branch of Dautry's Armament Ministry that was concerned with the manufacture of explosives. "Heavy water," Joliot wrote, "is of fundamental interest for the realization of the important experiments which we have been pursuing for some months." It would therefore be useful, he remarked, to try to stop shipments of heavy water from Norway to Germany.[12] These contacts with the Commission of the National Economy and the Direction des Poudres were limited to an exchange of information; when the time came for action, the team went directly to Dautry.

Sometime around February 14, 1940, Joliot took a report to the Ministry of Armament, signed by all three members of the team. The five-page report, after reviewing what was known of chain reactions, concluded that "A uraniferous mass in which such a phenomenon took place would act as a violent explosive, as a furnace of very high temperature, or as a source of motive power, depending on whether the release of energy were more or less controlled." There were two ways to reach these goals: by increasing the proportion of the rare isotope uranium-235 in natural uranium, or by making an appropriate arrangement of natural uranium and, in place of the hydrogen used in their past experiments, deuterium:

Method (a) (enrichment of the rare isotope in natural uranium) has already been envisaged by American, English and probably German scientists. However, changing the natural isotopic composition of an element presents very great difficulties . . . the transformation of the present laboratory trials into a semi-industrial method, and its application in a case as unfavorable as uranium, would require an outlay in installations, personnel, and above all time which we consider prohibitive.

Method (b) (replacement of hydrogen by deuterium), on the other hand, we consider to be almost immediately applicable.[13]

For their experiments, the team continued, they would prefer to use metallic uranium, in which there would be no slowing down of neutrons caused by collisions with oxygen atoms such as would occur in uranium oxide.[14] The metal had never been produced in bulk, but they confidently asked that 400 kilograms, worth some $16,000, be bought from Metal Hydrides, an American firm which was the sole producer. Above all they would need deuterium, and they had learned to their surprise that Norsk Hydro was holding in reserve in Norway nearly 200 liters of very pure heavy water, worth some $120,000 if it were sold at current prices. Evidently the company had held it back to avoid saturating the market, which was limited to a handful of nuclear physics groups. Except for a few grams scattered around various laboratories, this was the total world stock of heavy water. "We would need all of this stock," the team declared.

Meanwhile the Germans had also taken an interest in heavy water, and acting through the German corporation I. G. Farbenindustrie, a part-owner of Norsk Hydro, they asked the Norwegians to make them thousands of liters. Norsk Hydro put the Germans off and finally in early February 1940 refused either to send them a substantial amount at once or to increase production. News of the transaction, when it reached Dautry, helped convince him that he must act decisively. He determined to send a secret mission to Norway to bring back the heavy water. Since the project was enlarging beyond the bounds of his own ministry into foreign affairs, he informed the premier, Edouard Daladier, about the entire matter. Daladier agreed to provide every necessary support.[15]

By good fortune the majority interests in Norsk Hydro were controlled by one of the great French banking houses, the Banque de Paris et des Pays Bas. Dautry called in a young acquaintance of his at the Direction des Poudres, Lieutenant Jacques Allier, who in peacetime had been an officer of the bank and had a close relationship with the Norwegian firm. Allier had already sent Dautry information about Norsk Hydro's heavy water. The dapper bank officer, scarcely imagining he was about to be launched on the adventure of his life, met with Dautry, Joliot, and Dautry's chief of staff Jean Bichelonne on February 20. Dautry explained, Allier recalled, that heavy water was necessary for experiments on "the violent liberation of atomic energy, with effects infinitely greater than those of the most powerful explosives." Joliot vigorously seconded Dautry's explanation and emphasized that the French were in a race with German scientists. Then Dautry asked the lieutenant to arrange a mission whose success, he said, could have incalculable consequences: to take a crew of secret service operatives to Norway and bring back all the heavy water they could lay hands on. Allier volunteered for the mission.[16]

Dautry was worried by the presence of Halban and Kowarski in the midst of these delicate and secret matters. Halban was of Austrian and German background, while Kowarski was from Russia, which now, at the height of the Soviet invasion of Finland, made him as suspect as any German. At one point in the meeting Dautry apparently asked Joliot whether his two collaborators could be eased out of the work in favor of native Frenchmen; Joliot refused to consider this step, saying that they were indispensable.[17] Finally Joliot and Dautry agreed that until the heavy water was safely in France, Halban and Kowarski should be sent on "vacations," under close surveillance, to isolated islands off the coast of France. The two scientists subsequently agreed, and on February 23 each went off to a different island.

Enrico Fermi
in the 1930s

Leo Szilard in the 1930s

Joliot, Raoul Dautry, and Jacques Allier in Dautry's office, 1940

The Montreal reactor team, c. 1943: Bertrand Goldschmidt,
Jules Guéron, Hans Halban, Pierre Auger

Halban immediately ran into a problem. The military commander on his island had not been told why Halban was sent to him. With his German accent and unexplained mission, Halban found himself under house arrest with a guard outside his door. When the guard kept him awake by snoring, Halban threatened the commander that he would tele-graph the minister of armament and "refuse further collaboration" unless he at least got a guard who did not snore. It took him a week to straighten things out. Kowarski meanwhile rode about his island on a bicycle and read *Gone with the Wind*.[18]

After several briefings by Joliot, Allier departed with a team on February 28. The next day the French intercepted and deciphered a German cable to agents in Norway, naming Allier as a suspicious person who should be waylaid. He made his way safely to Oslo under an as-sumed name and there met Axel Aubert, the general manager of Norsk Hydro. Some of the French secret service people, suspecting Aubert of pro-German sympathies, were afraid that his firm might prefer to deal with Germany. The matter was considered so delicate that Allier's in-structions authorized him to make a deal which would allow the neutral Norwegians to share the heavy water between Germany and France. He also carried one and a half million Norwegian kroner. But when Allier carefully sketched out Joliot's reactor research and the valuable part that heavy water might play in it, Aubert replied with trust and friendship. He not only agreed to turn over his entire stock of heavy water but of-fered to reserve all future production for France. Now Allier felt honor-bound to reveal the military implications of the transaction. Aubert, Allier recalled, replied that in that case he would not charge a penny for the heavy water. If France should lose the war, Aubert con-tinued, "I'll be shot for what I'm doing today; but it's an honor to run that risk." A gentlemen's agreement was soon concluded. Aubert volun-teered that he could soon increase production tenfold to about 1.3 tonnes per year. Allier promised in return that France would protect Norsk Hydro's interests if nuclear power transformed the world economy.[19]

Allier's next problem was to spirit away the stock of heavy water, 185.5 kilograms, without running afoul of the German agents on his trail. The heavy water was removed from the factory at midnight, poured into twenty-six cans, each the size of a bucket, and driven over icy roads to Oslo, arriving March 10. Allier's team got seats on two planes, one flying to Amsterdam and the other to Scotland. The planes were parked side by side on the runway so that Allier's crew could drive up between them, out of sight of prying eyes, to load half of the cans of heavy water onto the plane bound for Scotland. The other plane then took off, ostensibly with Allier, who had boarded it ostentatiously, then switched planes. The Amsterdam plane was forced by German fighters to land at Hamburg and was searched. Meanwhile Allier and his heavy water arrived in Edinburgh, where the next day the other members of the team with the rest of the heavy water joined him. By March 16 they had quietly made their way back to Paris. Joliot and Dautry were beside themselves with joy over Allier's success.[20]

On March 20 Daladier resigned as premier, exhausted less by the war on the frontiers, which so far had been almost bloodless, than by the party warfare in Paris. French politics was in a shambles. The one major group that had steadfastly opposed the Nazis, the Communist party, had ignominiously endorsed the Nazi-Soviet Pact the previous August, had then refused to support the war, thereby losing many of its liberal supporters, and had finally been declared treasonous and forced underground. There were scarcely any coherent parties left.

On April 3 Allier paid the new premier, Paul Reynaud, a call. He described what Dautry and Joliot were doing about nuclear energy and said that state support of the enterprise would be needed for some years. Reynaud promised to give it.[21]

Although the secret was never spread widely, the work at the Laboratory of Nuclear Synthesis was now known by a number of key personages in the government—Reynaud, Daladier, Dautry, Serruys, Laugier, and some of their subordinates—who gave Joliot's team whatever they asked for. The heavy water affair caught the imagination of these personages, and the prospects for chain reactions attracted them. Dr. Robert Wallis, who was attached to the Commission of the National Economy, later stated that Daniel Serruys was "understanding enough to ask me to negotiate with the French colonial office to rent in the Sahara Desert a piece of land 100 kilometers in diameter to make the first experiments with an eventual (not yet realized) atomic bomb."[22] The ministers were apparently more excited about the possibility of a weapon than about a nuclear reactor. Dautry recalled that the government had asked Joliot to direct his work "less toward the utilization of radioactive elements for the production of energy for the use of industry in peacetime (a domain where, nevertheless, extraordinary perspectives could already be seen)," than toward development of a nuclear bomb.[23] The scientists themselves had a somewhat different picture of what they were doing, although they seem to have made little effort to explain this difference to their official superiors. Their job, they felt, was to start up an experimental nuclear power plant. After that there would be time enough to think about whether there was any chance of making the bomb Dautry wanted.

10

From Research to War Project

The laboratory with its jumble of apparatus does not often resemble the engineer's office with its drafting board and shelves of manuals, but the one can transform into the other. The Collège de France team, while still behaving like scientists, had taken the first steps in this transformation. They would be thinking less and less about how universal nature works, more and more about how a particular artificial device could be made to function. This was one reason they had to meet outside experts, from patent lawyers to secret service agents. To build a working reactor, they would need more experimental and theoretical scientific work, but they would also need still more money, more personnel, and above all more supplies.

As winter gave way to the particularly fine Parisian spring of 1940, they relaxed their efforts in the laboratory, for at the moment their work depended on the procurement of materials. In wartime France, waiting half-unconsciously for battle, there was little to spare. The team did have their heavy water, but only enough for a preliminary experiment. They hoped that this would give evidence to justify going ahead to a full-scale reactor. Even if Norsk Hydro expanded its production as Aubert had suggested, and even if the product could be brought safely to France, it would still take several years to get the tonnes of heavy water needed for a single reactor. In late March Allier wrote Dautry that it was essential to set up a French heavy water production factory, and Dautry let him look into the matter. There was little chance, however, that French production could begin soon.[1]

The Collège de France team realized that they might have to use some other substance as a moderator. They were not completely convinced that hydrogen was useless, thinking it might work in a heterogeneous arrangement, so they ordered a ton and a half of refined paraffin. Beryllium was another candidate, if an unlikely one, and already in December Joliot had written Daniel Serruys to ask whether there was any chance of getting the metal in bulk. But the team realized that the best alternative to heavy water was still carbon. Around April 1940 they ordered 530 cubes of very pure graphite, each ten centimeters on a side. Even with some help from the Armament Ministry to expedite the order, the graphite was not expected to arrive until June. Finally, whatever moderator they used, they would need uranium. In March the Union Minière agreed to send the team another three tonnes of uranium oxide and also to loan them ten grams of radium, an enormous amount, for their experiments. "Pursuant to the indications you have given me on the new uses of uranium metal," Lechien added, "we have decided to get in touch with Metal Hydrides and have them make a tonne of metal."[2]

All this called for a lot of money. Sometime that spring Joliot wrote a memorandum, apparently to Dautry, which spelled out the team's needs. Besides the heavy water and uranium, he pointed out, they would need five tonnes of graphite cubes and a tonne of metallic beryllium, as well as unspecified chemical products, large containers, and a two-tonne truck. In all, an extra 500,000 francs would have to be spent on matériel. This sum was not given him all at once, for it was not yet needed. But Joliot's ordinary laboratory budget request at the Collège de France for 1940, much larger than his 1939 budget, was granted, and in March the CNRS added another 50,000 francs.[3] More would be provided as needed.

Joliot also requested a modest enlargement of the team at an annual cost of 170,000 francs. He wanted the Arcueil annex of the Radium Institute reopened with three workers to treat the large masses of uranium and other substances. Six workers and technicians would be employed at Ivry in addition to the three already there helping Halban, Joliot, and Kowarski. Some of these people were requested by name. For example, Joliot wanted Bertrand Goldschmidt, a young student of radiochemistry from the Radium Institute who was working at a military gas warfare laboratory in Poitiers. In early May Joliot told Goldschmidt that he would soon be ordered to Paris to work on the purification of uranium.[4]

Throughout this reorganization the team continued their scientific work. For example, in late March, shortly after Halban and Kowarski returned from their island "vacations," the team measured the average distance a neutron would diffuse through paraffin. Another piece of work, which they had begun back in January, was a search for the delayed emission of neutrons after fission, such as Roberts and his colleagues in the United States had reported a year earlier.[5] The French team's experiments were hasty and inconclusive, the more so as they were interrupted by the heavy water affair, but the experiments did open an important new line of thought.

The team noticed that some radioactivity was still found in a mass of uranium days after the central neutron source was removed. It occurred to them that there was a chance that the resonance capture of a neutron by uranium somehow led to the creation of a new element, which might itself be liable to fission. But recognizing that their experiments were far from proving the case, they did not pursue the matter. They did not notice that the new fissionable element might be used to make a bomb. Being far less preoccupied with bombs than with nuclear reactors, the team simply supposed that the vaguely conceived new element would improve their chances of keeping a reactor running after some of its uranium had been used up.[6] They remained convinced that their proposed nuclear reactor, except for its indirect value as a tool for understanding fission, was of no importance for work on bombs. In the spring the team wrote down their ideas on the production of a fissionable element in a reactor, delayed neutrons, isotope separation, and various other possible, though in some cases impracticable, ideas for further improving the chances of making a nuclear reactor. These scientific reports were deposited as sealed notes in the archives of the Academy of Sciences, and two secret patents were written on the ideas.[7]

Of all the team's work, their plans for experiments with heavy water seemed the most important. They wanted to try both heterogeneous and homogeneous experiments along the same lines as the ones they had done with ordinary hydrogen. For a heterogeneous experiment they proposed to put their heavy water in aluminum boxes buried in uranium oxide, as they had done earlier with paraffin cubes. Their struggles to get this experiment underway were typical of all their problems in conducting large-scale experiments in the midst of a war. They had to write letters to a variety of firms to locate the aluminum, then secure priorities from the authorities, and then find a shop to make the sheet metal into cubes. Not until late May, after various false starts, did Joliot find someone who could provide aluminum sheets, but then the firm said it could not make delivery until September. Joliot had to go back to the Armament Ministry to ask their help in finding another source, reminding them that his experiments were "of an urgent nature."[8]

Meanwhile the team's plans for experiments on homogeneous mixtures of heavy water and uranium oxide ran into a problem. The great advantage of deuterium was that it did not absorb neutrons. Without risk of losing neutrons, then, the team could use a high ratio of deuterium to uranium, thereby giving the neutrons ample opportunity to bounce around in the heavy water and slow down before striking a uranium atom. The team would therefore be dealing not with uranium oxide dampened with some water but with a suspension of uranium oxide in heavy water. Preliminary tests, however, showed that the uranium oxide powder would slowly settle to the bottom of the sphere.

Joliot noted a solution on the back of an envelope: "We have to make the ball turn." Drawing on his talents as a mechanical engineer, he designed a motorized arrangement to keep the sphere rotating so that the suspended powder would stay uniformly distributed. He also began a frustrating search for still more aluminum and for someone to make it into a sphere. Meanwhile, a wooden hut was thrown up outside the buildings at Ivry. The experiments were to be conducted there to avoid contaminating the laboratories with radioactivity.[9]

In France that spring, Joliot's group and some others under Dautry were almost alone in doing effective work. The Army vegetated in the concrete caves of the Maginot Line, its high command fractious and unrealistic. By the night of May 9, as German troops were taking up positions to invade France, the nation was effectively without either a civilian or a military leader. A few weeks earlier the British scientist Bernal, in France on an official mission about explosives, had dinner with Langevin, who was "in an extremely sad and depressed mood," Bernal recalled: "He felt, and we were beginning to feel very much from our reception by the French military, that there was something very bad about the state of France at that time . . . he felt that at that time the country had already been betrayed . . . It was, I think, the lowest point in his life."[10]

Meanwhile, in laboratories around the world fission research was reviving. In each country two separate paths were becoming apparent: reactor development and isotope separation. Physicists everywhere gave some thought to isotope separation after September 1939 when Bohr and Wheeler published their argument that the only component of natural uranium which was likely to fission was the rare isotope uranium-235. In Britain the idea had been mentioned even before the publication of Bohr and Wheeler's paper, and Oliphant and Chadwick were thinking about it in the summer of 1939. Working under Oliphant in Birmingham was Frisch, who had managed to get an invitation to Britain and stayed on after the war began. Frisch was interested in isotope separation, for he wanted to get enough uranium-235 for an experiment to check whether this isotope really was the fissionable part of uranium. As a recent immigrant, he was not allowed access to military secrets, so most of the physics laboratory with its radar work was out of bounds, but Oliphant let him do isotope separation studies, which seemed of little immediate importance. Even before the war British physicists had been saying, with sour humor, that most of their number were working on war problems, leaving pure physics to the refugees. Frisch set up some apparatus in a small lecture room. His progress was slow, partly because he got little support—certainly less support, he wrote Meitner, than he would have gotten had there not been a war on.[11]

In the United States the idea of isotope separation had also occurred to several scientists, and John Dunning, excited by the possibilities, took over this part of the work at Columbia. He made a bet with the skeptical Fermi and Pegram that uranium-235 would turn out to be the fissionable isotope. In early 1940 a team under Dunning proved that uranium-235 was indeed responsible, and in the spring they published this result.[12]

Fermi and his collaborators, who kept in touch with Dunning's group around a lunch table at the Columbia Faculty Club, saw no particular reason to change their plans. From this time forward there was a divergence among the Columbia University scientists. Fermi's group continued to work toward a nuclear reactor made from natural uranium, while Dunning's group calculated ways to separate uranium-235 from natural uranium on a large scale for a bomb. Both investigations, lacking manpower and money, crept forward very slowly.[13]

In Germany by December 1939 Heisenberg and others recognized the value of isotope separation for both reactors and bombs, and from about that time encouraged research into the problem. By spring serious pilot studies were underway on a somewhat larger scale than in other countries.[14] Scientists in the Soviet Union were also engaged in such work.

Reactor development was going ahead much like isotope separation: in each country a few people pursued the problem uncertainly but unremittingly. During the winter of 1939-1940 German scientists ground out a number of theoretical studies. They had almost no uranium to work with until January 1940, and for months afterward had only about 150 kilograms of uranium oxide. Serious experiments on nuclear reactors could not start until the summer. These experiments, which were often remarkably similar to those performed by the French several months earlier, gave similarly inconclusive evidence. The Germans nevertheless felt that a reactor was probably feasible and wavered between the possible moderators, coming more and more to prefer heavy water.[15]

The Soviet Union's fission program may have been as active as the German one. In the spring of 1940 an official committee under the Academy of Sciences began to organize fission research, to oversee an investigation of uranium deposits in the Soviet Union, and to coordinate studies on isotope separation and the production of heavy water. The Soviet scientists seem to have stayed in the race until the German invasion of 1941 forced them to take up more immediate tasks.[16]

Through the second half of 1939 Szilard and his friends were searching for a way to get a large experiment underway to test the feasibility of graphite as a moderator for a reactor. When Wigner prodded him to make contact with the United States government, Szilard approached Albert Einstein, with whom he had worked before they both fled Germany, and got him to sign a letter to President Franklin D. Roosevelt. In response, Roosevelt set up a small committee under Lyman

J. Briggs, a cautious bureaucrat-scientist, head of the Bureau of Standards. Szilard's group apparently had nothing more in mind than to seek official support for his attempts to raise money from individuals, corporations, and foundations. But when the question of funds came up at the Briggs committee's first meeting in October 1939, the Army representative agreed to allocate $6000 to buy raw materials for fission experiments, which were to be conducted by Fermi. This modest progress was well below the level of Szilard's plans, and he continually urged Fermi to press on headlong to a full-scale attempt to build a nuclear reactor.

Szilard's anxiety increased in early 1940 when he read the last open publication of the Collège de France team, describing the experiments with a wet sphere of uranium oxide done the previous summer. As Szilard recalled: "I was able to conclude from it what I was not able to conclude from our own [June 1939] experiment, namely, that the water-uranium system came very close to being chain-reacting, even though it does not quite reach this point. However, it seemed to come so close to being chain-reacting that if we improved the system somewhat by replacing water with graphite, in my opinion we should have gotten over the hump."[17] Szilard therefore made still another prolonged effort to get money from private pockets, such as the Union Minière's, and to stir up the United States government. He worked out a scheme for an organization to control the development of nuclear energy. "It was assumed," he recalled, "that the scientists would have adequate representation with this government-owned corporation, and I proposed that we all take out patents for our inventions and assign them to this government corporation without financial compensation." Like the CNRS-Union Minière arrangements of the Collège de France team, which it resembled remarkably, Szilard's scheme revealed deep-seated ideas about the proper relations among science, industry, and the state. But being an outsider in America, he could get little done beyond attracting attention to the fission problem.[18]

At this point Szilard and most American scientists saw nuclear energy chiefly as a new source of controlled power. Like the French, Szilard's appeals for support stressed the potential of the reactor for propelling Navy ships. He did not forget the possibility that "explosions of extraordinary intensity" could result from fission. But he, and the other scientists even more so, thought like the French that this was a remoter possibility.[19]

Meanwhile fission research at Columbia waited on the delivery of four tons of graphite bricks, ordered for Fermi by Szilard at a cost of $2000 to the Army, and these did not arrive until the spring of 1940. Fermi, less hopeful than Szilard that scientists could build a reactor and less fearful that they could build a nuclear bomb, had decided that his group, before trying to get huge quantities of materials, ought to confirm that a reactor was feasible by measuring the neutron absorption cross-section of carbon. He loathed being in error and would never promise results until positive that he could deliver. In the meanwhile Fermi continued his theoretical studies of cosmic rays.[20]

When enough graphite was at last on hand, Fermi and Anderson began an experiment which was similar to the one that Halban had done back in December, a measurement of the neutron distribution in a large stack of graphite. The work at Columbia, however, went ahead in a more leisurely and painstaking fashion and was analyzed more thoroughly. Because Fermi, who was the only scientist of his generation to be both a brilliant theoretician and a consummate experimental physicist, explicitly calculated the effects of the finite size of the pile, his experiment was more sensitive than Halban's and thus needed less graphite to yield reliable results. The Columbia group had another advantage: graphite with fewer impurities, a substance Szilard had spared no pains to find.[21]

In May the Columbia group got encouraging results, and they soon found that the neutron absorption cross-section of their graphite was no greater than 0.3×10^{-26}, or half the French team's upper limit. This experiment was crucial, for it showed that a graphite-moderated reactor was likely to work. From this time forward the Americans were committed to exploring the graphite reactor, although they still hesitated to launch a crash program.[22]

Up to this time nearly all American work had been published, but now doubts about this practice again came into the open. Szilard had never given up hope of creating an international pact to circulate fission information under cover. But the Briggs committee remained all but inactive, leaving everything up to the physicists. As late as April 27, 1940, when the committee held one of its rare meetings, the only suggestion to emerge was that the scientists working on uranium get together and impose upon themselves whatever censorship they felt necessary. The government itself would do nothing.[23]

Szilard had already taken the single step that was entirely within his power: he withheld from publication a paper of his own. This was a report on the calculations he had completed in February 1940, after receiving a copy of the last paper published by the French, in which he concluded that there was a strong possibility of making a nuclear reactor work, if not with water then with graphite. Had this report been published, it would have been a great stimulus to nuclear reactor work in various countries. But when Szilard sent it to the *Physical Review*, he requested that printing be delayed until further notice.[24]

Now Fermi and Anderson finished their crucial experiment demonstrating that the carbon absorption cross-section was very small and that therefore a graphite-moderated nuclear reactor could probably be built. Szilard approached Fermi with the suggestion that the value for the cross-section should not be published. "At this point," Szilard recalled, "Fermi really lost his temper; he really thought this was absurd." But while Fermi stuck by his priniciples, Pegram had second thoughts, and finally he asked Fermi to keep his work secret.[25]

This decision came late, but still in time. German scientists, drawing on experiments of their own carried out later in 1940, wrongly concluded that carbon had a substantial neutron absorption cross-section. From that point on they abandoned carbon as a moderator and attempted to use heavy water, which they never managed to get enough of. Soviet scientists too did not at first seriously consider carbon as a moderator. And the French scientists were committed to deuterium. Anderson and Fermi's work could have put all these groups on a different track.

Theirs was not the only hole in the dike that had to be plugged. In late May, Louis Turner, a physicist at Princeton, sent Szilard a copy of a paper showing theoretically that when uranium-238 absorbed neutrons, there should follow a series of steps that would end in a new element. This he predicted to be fissionable—it was the element later named plutonium. Although Turner did not realize it, he had written the prescription for the easiest route to building a nuclear bomb. Szilard wrote back at once to say that a paper of his own was being kept secret, which implied that an official move was underway to withhold papers. He persuaded Turner to write the *Physical Review* and delay publication.[26] It was well he did so, for Turner's paper could have been an essential clue for the Germans and others.

Before more progress was made, the June 15 issue of the *Physical Review* appeared, containing a letter from Edwin McMillan and Philip Abelson, young scientists at Lawrence's Radiation Laboratory in Berkeley. The tedious investigations into the chemistry of fission fragments had at last yielded an important result. When they had bombarded uranium with neutrons, the Berkeley scientists had observed the production of a new element, heavier than uranium. This was the first and most essential step of the process that Turner had predicted should lead to plutonium. But Abelson and McMillan failed to see the connection between their work on uranium bombardment and the chain reaction problem.[27]

A word about the site of this work. The Radiation Laboratory was a friendly rival and in some ways a model for Joliot's laboratories. Like other nuclear physics institutions, while pursuing pure science, it kept an eye on the practical applications of research. Regarding the large-scale use of nuclear energy, for example, Lawrence had written in the early thirties, "I have no opinion as to whether it can ever be done, but we are going to keep on trying to do it." Lawrence's cyclotrons, which Joliot had copied, were built like Joliot's with support from foundations, industry, and government, and Lawrence used them to irradiate cancer victims and to produce radioactive elements on a scale rivaling the Union Minière. Yet the ideal of disinterested research was strong among Lawrence's subordinates, and the cyclotrons were often used to advance nuclear physics. Thus Abelson and McMillan were responding not only to Lawrence's search for new radioactive substances of commercial interest but even more to a purely scientific impulse when they identified an element heavier than uranium.[28]

Their publication brought down a flurry of protests which helped to settle the secrecy issue. From as far as Britain, scientists interested in fission objected to the publication of such revealing information. But the most effective action came from Gregory Breit, a physicist at the University of Wisconsin, who had known Szilard and Wigner for years and had been awakened by them to the secrecy problem. Around the beginning of June Breit found a way to impose complete secrecy. In a meeting at the U.S. National Academy of Sciences, he spoke out for censorship, and a committee on publications was appointed to consider the problem. On his own initiative Breit immediately began writing letters to journal editors, proposing that papers relating to fission be submitted voluntarily to his committee before publication. Although some eyebrows were raised, Breit recalled, the editors and other leading scientists agreed to the plan. Within a few weeks Breit had imposed total censorship on American fission research.[29] Following Szilard's lead, he had gotten round the obstacles to self-censorship, using a show of secrecy and authority to impose secrecy and authority. His job was made easier by the international situation, which in mid-1940 was obviously much worse than it had been a year earlier before the start of the war in Europe. Communication between fission groups outside the United States had long since closed down, and the Americans knew little about whatever work was being done in other countries.

In Britain, research on chain reactions had all but died by 1940. Thomson and Moon were "tidying up" their discouraging experiments with uranium oxide and water, concluding that bombs could not be made with these materials and that reactors were too far off to be a reasonable war project. Chadwick, on reading the Collège de France team's last published paper, concluded that there might still be a chance for a hydrogen-moderated chain reaction. He requested some uranium oxide so that one of his laboratory workers could do experiments, but he did not give the work much attention, and it went slowly. Frisch too requested some uranium oxide, thinking to stack up alternate layers of the oxide and paraffin in cheap baking pans. Meanwhile he was single-handedly pursuing his fruitless experiments on isotope separation.[30]

None of these little projects could have led very far. Then a new line was discovered, a line that nobody anywhere had yet followed. It swiftly led the British, and then the rest of the world, into the age of nuclear weapons. The line can be traced back to Francis Perrin's first attempts to calculate the critical mass of uranium oxide.

Rudolf Peierls, an émigré scientist at the University of Birmingham in England, had read Perrin's paper when it first appeared. Peierls was a man who loved mathematical problems, provided that they had some vague connection with real things, and he was struck by the concept of criticality in the French paper. He worked through the problem himself in the spirit of a mathematical exercise. By June 1939 he had produced a more refined calculation of the critical mass, based on Perrin's concepts, and he published his equations without believing that they had any practical significance. He knew that natural uranium would have a critical mass of tonnes at least, far too large for a usable bomb.[31]

Peierls' friend and close associate in Birmingham was Frisch, who more than anyone else in Britain was interested in uranium-235 and chain reactions.[32] Sometime in late February or early March 1940 an interesting question occurred to Frisch: what was the critical mass of pure uranium-235? Since the prospects of separating substantial amounts of the isotope looked dim, virtually nobody had asked this simple question. It was only necessary to put the appropriate numbers into Peierls' formula, and this Frisch and Peierls proposed to do. There was one number they needed, however, which nobody had measured, the cross-section for medium-energy neutrons to produce fission in pure uranium-235. It seemed reasonable, in the light of Bohr and Wheeler's theory, to guess that this cross-section was about equal to the physical area of the nucleus—that is, to assume that whenever a neutron hit a uranium-235 nucleus, it provoked fission, which was in fact not far off the mark.[33] Once this guess was made, it was trivial to plug the numbers into Peierls' formula and find the critical mass. "And we were completely staggered to find this was very small," Peierls recalled. The first rough calculation gave a value of less than a kilogram of pure uranium-235. A chain reaction could be propagated directly by fast neutrons within this mass; therefore, utterly unlike a reactor with its voluminous moderator and slowed-down neutrons, the lump of uranium-235 could explode violently. Frisch and Peierls next calculated the force of the explosion; the result was appalling. They immediately wrote a memorandum on fission for the British government, which set the government, and eventually much else, in motion. Unlike many of the ideas on fission, which sprang up in various places independently, the Frisch-Peierls calculation of the likelihood of bombs was unique: every nuclear weapon ever made can be traced to it.[34]

On April 10, 1940, Thomson, Cockcroft, Moon, and Oliphant met in London to discuss the uranium problem and, in particular, the possibility, discussed in Frisch and Peierls' memorandum, of separating enough pure uranium-235 to make a nuclear bomb. This was the birth of the official "Maud Committee" which directed British fission research for the next year. From the start the committee was in contact with France. Up to this time the British and French governments had freely exchanged information of military value, about radar for example, but Joliot, despite British inquiries about his work, had not yet informed them fully about his research into the chain reaction. Perhaps he, like the British, had not considered the subject of immediate and pressing importance for the war. But after the heavy water was brought to France, Dautry decided it was time to inform the British, and for this purpose sent Allier to London.

The lieutenant happened to arrive in time to attend part of the first Maud Committee meeting. He warned them about the German approaches to Norsk Hydro, and afterward supplied them with a list, drawn up by the Collège de France team, of scientists in Germany who should be watched. Many of the people on the list were in fact already working on fission. Another subject that came up was heavy water procurement. On his arrival in Britain Allier had heard the shocking news of the German invasion of Norway, and he feared that the Germans intended to capture the heavy water plant. He was anxious to find other sources. Upon asking the British about their chances of getting the substance from Britain, Canada, or the United States, he learned that they had already looked into the problem but had turned up scarcely one kilogram.[35]

The British, as Allier reported on his return to France, were skeptical about the chances of making a nuclear reactor, with or without heavy water. A month later Sir Henry Tizard, the British government's foremost scientific adviser, noted that "certain physicists say that the controlled disintegration of uranium is a scientific possibility, and the French are excited about it—I think unnecessarily." Tizard stuck to the view that "uranium disintegration is not in the least likely to be of military importance in this war." Nevertheless he felt that the matter was serious enough to warrant impeding the Germans from seizing uranium. Other British scientists, impressed by the possibilities for bombs as spelled out in Frisch and Peierls' memorandum, and perhaps encouraged by Allier's enthusiasm for nuclear energy and his warnings of German interest, pressed their research more vigorously. Unlike scientists in all

other countries, the British were now scarcely interested in power production; they were concentrating on isotope separation, the only thing
that seemed liable to lead to a nuclear bomb.[36]

In May 1940, less than a year and a half after the discovery of fission
and less than a year after most scientists had stopped publishing their
results, work on fission had taken noticeably different paths in the various countries. There were striking similarities between countries, for the
basic physical facts and the most important ideas were available, or
would soon be available, to scientists everywhere. Nevertheless their
opinions on the various possibilities had diverged.[37]

All the major countries except France were sluggishly starting to work
on isotope separation, although no country had yet made much progress.
This was perhaps the biggest difference between France and the other
countries. Serious work on chain reactions in France was restricted to the
Laboratory of Nuclear Synthesis, whereas in other countries several laboratories, including physical chemistry groups with an interest in isotope
separation, were involved. For example, in the United States during 1940
contributions to the work on fission came from Columbia, Princeton,
Berkeley, the Carnegie Institution, and several other places. In spite of
the generous support enjoyed by Halban, Joliot, and Kowarski, they
were not able to draw on scientific resources at all comparable to those
available in the United States, the Soviet Union, and the British or German empires. Neither did they have the industrial base; an isotope separation plant was beyond the means of the beleaguered French nation of
1940.

The French were somewhat ahead of the others in investigating nuclear
reactors, and they were unique in their preparations for important experiments on mixtures of uranium oxide and heavy water and on a
heterogeneous arrangment of uranium oxide (and later uranium metal)
with graphite. They were still several years away from a working reactor. They needed more theoretical work and were definitely behind the
United States and Germany in analyzing the details of neutron distribution in a heterogeneous system, although the French did have the most
elegant approach to the basic ideas of the chain reaction.

Their key problem, as the project moved out of the strictly scientific
domain toward engineering, would be the procurement of materials.
Despite the progress that the team had made, they would have trouble
getting either pure graphite or heavy water by the tonne. Scientific manpower was also in short supply. The French scientists, however, had a
close rapport with the upper levels of government, perhaps by virtue of
the same centralization of French science which had helped to restrict
fission research to the Collège de France. They had a high potential for
immediate help from state and industry, and they were more eager than

most physicists to call upon this help. It is entirely possible that, as Blackett believed, "Had the war not intervened, the world's first self-sustaining chain reaction would have been achieved in France." The war, while accelerating American efforts, impeded Joliot's group. Even Szilard would later admit: "If his work had not been interrupted he might have beaten us to it."[38]

Paris, Cambridge,
Chicago, Montreal

Joliot at the end of the Occupation

11

In Occupied Paris

On May 16, 1940, Dautry was making one of his visits to a factory when he received fearful news. The commander-in-chief had telephoned during the night to warn that the German attacks, begun less than a week earlier, had shattered the French army on the Meuse; nothing stood between the enemy tanks and Paris. Consternation spread through the government. Called to a cabinet meeting, Dautry saw bundles of secret documents plunging past the window to be burned on the lawn below. He was one of the few who kept a cool head, delivering a precise account of the explosives available to destroy bridges and fortifications. After some confusion the other government leaders overcame their panic and determined for the moment to stay in Paris. But Dautry had already telephoned Joliot that morning and asked him to transfer his work to a safer place in southern France.[1]

As soon as possible Moureu, the faithful subdirector of the chemistry section of the Collège de France laboratory, drove south with the stock of heavy water. Halban soon joined him. Because the countryside was in a state of alarm, Halban attracted suspicion with his German accent and manner, but the two finally established themselves in a villa near Clermont-Ferrand and began to fit it up as a makeshift laboratory. In Paris, Joliot and Kowarski arranged to evacuate the rest of the equipment, and early in June Kowarski loaded geiger counters, amplifiers, lead bricks, and the like on army trucks and led a group down to Clermont-Ferrand. Irène Curie, watching France collapse while Britain, the United States, and other democracies offered little help, wrote an American friend: "I feel bitter and cannot help it . . . The events of the past year have made it clear that fascism and communism are international. The fascists in different countries help one another, and so do the communists. If the democracies do not develop a notion of International Solidarity, they will certainly be destroyed."[2]

The German tanks gave Paris a respite by driving west to the sea, then turned south again. On June 10 the French government fled the capital amid a flood of refugees, and two days later Joliot evacuated the last movable equipment from his laboratories and burned documents under a sky shrouded with black smoke from oil refineries afire west of Paris. Soon he and Irène Curie appeared, exhausted, at Clermont-Ferrand. The team prepared to resume research. But on June 16 Lieutenant Allier unexpectedly drove up. He reported that Dautry had telephoned with the news that the Germans could not be stopped anywhere; the French scientists should make ready to flee the country with the heavy water. The team must move on, this time to Bordeaux, the last capital of the Third Republic.[3]

Dautry, working with his chief of staff Bichelonne, was trying to evacuate not only the scientists and their heavy water but also Joliot's stock of uranium oxide and indeed everything in France of conceivable military value. Dautry was prepared to advise the government that the nation could restore itself within a few years by withdrawing beyond the seas. The Army, he declared, must be evacuated whole to the French colonies in North Africa, along with "every Frenchman between 16 and 30 years of age" and whatever industrial resources could be brought along. But he never had a chance to submit this bold plan. With the French Army disintegrating and the populace taking to the roads by the millions in blind flight, many in the government and military had no stomach for further resistance. Marshal Philippe Pétain, who had already been sixty years old when he became a hero of the First World War, was brought in to head the government, and he immediately demanded an armistice. According to Pétain and many others, the greatest threat to France was not fascist conquest but Communist revolution, and the route to salvation lay through an orderly transfer of power—to themselves. As the government collapsed, Dautry went into retirement.[4]

On the night of June 16 Joliot had long talks with Halban and Kowarski, together and separately. With extraordinary prescience he predicted the future of the war. There would be an armistice in France, he said, and the nation would be subjugated so long as Germany and the Soviet Union remained allies. But that alliance could not last, and meanwhile within France resistance to the Germans would rise under new leadership. Joliot probably never considered leaving his homeland, to which he was unshakably attached. He meant to keep his laboratory functioning as a research and training institution to help French science live through the Nazi occupation, even if that lasted for decades, and he meant to organize resistance. He asked Halban and Kowarski to carry on the team's fission research abroad.[5]

In the morning Halban and Kowarski loaded up cars with their wives and young daughters, laboratory papers, and canisters of heavy water, and made their way laboriously westward, cutting across dark rivers of refugees fleeing south. They reached Bordeaux and fought through confused crowds to the provisional offices of the Ministry of Armament. Here they found Bichelonne, who ordered them to board a British collier in the harbor, the *Broompark*.

They were welcomed aboard by a tattooed man stripped to the waist, looking like a pirate, shouting orders and cracking jokes in fluent French with a raw English accent: the twentieth earl of Suffolk. The formidable Lord Suffolk had come over to rescue machine tools and anything else of value he could lay hands on, including scientists. He had raced around Paris and Bordeaux, armed with a letter of introduction from Dautry and two pistols, collecting industrial diamonds and the like. Halban and Kowarski were among the first to board the ship Suffolk had commandeered, but during the next day about a hundred more people boarded along with many valuables. A raft was built on deck to which the heavy water and diamonds were lashed, so that they might be saved if the ship were sunk.[6]

On June 18 Joliot arrived in Bordeaux. On meeting Bichelonne, he learned that the remnants of the government were awaiting Hitler's reply to their plea for an armistice on almost any terms. Joliot also encountered Lord Suffolk, who pressed him to flee to Britain. Joliot now went in search of Halban and Kowarski for a last meeting. Had he located and boarded the *Broompark*, Lord Suffolk would probably have forced him to stay until they sailed, but because German planes had bombarded the harbor, the ship had shifted to another dock, and Joliot never found it.

People who knew Joliot have speculated about what might have happened had he been brought away from France in 1940. Even without him Halban and Kowarski were to strongly affect the development of nuclear energy. With the aid of Joliot's international fame and formidable political skill, the French scientists would certainly have had a still greater influence abroad.[7] But Joliot had committed himself to a different politics, casting his lot with a French resurgence.

An armistice was signed on June 22, and Joliot, his wife, and Moureu went to a small town near Bordeaux to wait for things to settle down. Curie was beset by a respiratory illness which had bothered her intermittently before, and which stayed through the war. In the middle of July Joliot heard that the administrator of the Collège de France, Edmond Faral, had gone back to his post in Paris. Faral had already made his views clear in late May, after the German breakthrough, when he told his superiors that "purely and simply to abandon these establishments [of science and learning] without leaving any personnel there would be to deliver them to the mercy of the enemy." He also felt that the persons left in charge should have the highest possible stature so that they could deal with the upper levels of the enemy command; otherwise the minor functionaries left behind "would be shoved aside, and it would be better to send them off immediately to cultivate French fields than to deliver them up to the enemy for I know not what tasks." Joliot also received a letter in mid-July from Charles Maurain, who had himself returned to Paris: "it would seem to be useful that the laboratories be occupied by members of the staff, for reasons of security, and so that work can be resumed . . . The German authorities have shown particular interest in M. Joliot's laboratory at the Collège de France, and in the Ivry laboratory . . . Numerous German officers, whom one may suppose to be persons familiar with your work, have asked about you in various places." Joliot promptly wrote Faral that he was awaiting instructions and hoped soon to be authorized to return to his work at the laboratory.[8] The permission was granted, and he went back to Paris, joined by Moureu, Saval, and other laboratory personnel, although many were missing.

German officers were doing their best to reconstruct the French work on nuclear energy by inspecting the physical traces and questioning personnel. Having captured documents and sealed Joliot's laboratories, they were able to reconstruct the outlines of the work if not all its details and results. When they met Joliot they interrogated him at length. They were particularly eager to learn where his uranium and heavy water had gone. He told them that the uranium had been evacuated, he did not know to where, and that the heavy water had left Bordeaux by sea, suggesting the name of a ship that he knew had been sunk.[9]

The Germans also took an interest in the nearly finished cyclotron in the subbasement of the Collège de France Laboratory. There were no cyclotrons in Germany, and although Walter Bothe was building one in his laboratory at Heidelberg, it was years from completion. Since cyclotrons were highly useful tools for many branches of science, including nuclear energy programs, possession of such an instrument would mean a great deal to German physicists. The German officers decided to seize Joliot's machine as a spoil of war and pack it off, along with whatever other laboratory apparatus seemed valuable, to Germany.

The interpreter for the Germans' first conversations with Joliot was a physicist sent down from Bothe's laboratory, Wolfgang Gentner. This was the young man who, as a student under Joliot a few years earlier, had checked out the geiger counter when Joliot first noticed artificial radioactivity. According to Gentner, the reunion was distressing to both men. He lingered behind when the other Germans left and arranged to meet Joliot in secret that evening. At one of the students' cafés that lined the Boulevard Saint-Michel, Gentner told Joliot that his cyclotron was in danger of being seized as booty. Joliot decided it would be better to ask the Germans to use it in place.[10]

With the aid of Gentner's private information Joliot set out to negotiate an agreement. This was not difficult, for the physicists in Heidelberg warned that the French cyclotron could not be moved and got working in time to affect the war, which they expected to end soon. Moreover, it was German policy at this time to treat French leaders with respect in the hope, not unfounded, that many would collaborate in integrating France into a Nazified Europe. In return for his cooperation Joliot secured a promise that he would remain the director of his laboratory, that he would be fully informed of the work done there, and that the only projects undertaken would be fundamental research and not war work. Before the end of 1940, four German scientists, headed by Gentner, began to put the cyclotron in operation.

There was also a chance that the Germans would seek cooperation in nuclear energy work from the administration set up under Marshal Pétain in central France at Vichy. At the time of the Armistice several men in the Vichy regime knew all of the implications of the French program, including Allier, Bichelonne, and possibly General Maxime Weygand, who had taken over Dautry's duties. Some Information was spread further in August 1940 when Allier gave a circumscribed report on the French fission work before a meeting of ministers convened at Vichy to discuss administration of the armistice, a group including Weygand, Admiral Darlan, and Pierre Laval, the turncoat politician who would head the Vichy regime under Pétain and the Germans through most of the war. Laval could not have learned much, however, for years later, after it was all over for the Germans and Vichy, he was surprised to hear about the full importance of nuclear energy when Bichelonne finally told him. Nevertheless in August 1940 Laval was sorry that the heavy water had been sent to Britain, Allier recalled, for "if there remained in France anything whatsoever that could be useful to Germany, he proposed that it be turned over to the Germans in order to speed the military victory. But he was alone in his opinion," and General Weygand vehemently rejected any suggestion that France betray her former ally by helping the Germans learn about nuclear fission. Allier was soon put in charge of a subcommittee responsible for dealing with German demands for war materials such as explosives. In December 1940, he recalled, a German general asked him to find out what had become of the heavy water— "a material he said was essential for the conduct of the war, and which had been seized in Norway by a French officer whose name was not mentioned." Concealing his reaction, Allier took the request to a meeting with Pétain and some of his ministers. On their instructions he reported back to the Germans that the heavy water had been embarked at Bordeaux and might subsequently have been sunk. This was the last of the German inquiries, and Allier lived through the Occupation unmolested, as did Dautry.[11] In the archives of the Academy of Sciences the sealed notes deposited by the Collège de France team were never touched.

Joliot was determined to keep French science alive through the Occupation. More than once he was approached in secret by Allied organizations who wished him to leave France, if only temporarily, to work with scientists abroad; he always refused, saying he wanted to ensure the survival of French nuclear physics and the education of the next generation of scientists. For the first two years he was concerned chiefly with recreating instruments lost in the evacuation of Paris and training new personnel. Beginning in mid-1942 a dozen researchers at the Collège de France and Ivry began turning out studies of biological and chemical processes using radioactive tracers, and Joliot himself took part in biomedical research on subjects such as the interaction of iodine with the blood, which was the sort of work he had been about to start before he heard of the discovery of nuclear fission. All this effort was typical of French physicists during the Occupation, who managed to accomplish some research and education, although only with great difficulties and by keeping as far as possible from the Germans.[12]

If the German scientists at the Collège de France realized how important cyclotron studies of uranium could be, they did not try hard to circumvent their gentlemen's agreement to avoid military research, as Joliot confirmed by secretly inspecting their workrooms at night. It took Gentner's team a long time just to make the cyclotron go, partly because they had to make changes in the design, and partly because they had a hard time getting electronic parts. Once the cyclotron was going, they used it to study problems of fundamental nuclear physics such as the behavior of radioactive gases. Gentner and his teammates were not eager to do war work in any case, for they were more friendly with their French hosts than with the Nazis. Joliot later recalled that the German scientists had warned their French acquaintances who were being investigated by the Gestapo.[13]

When Bothe came on visits, he was less obliging; at one point he insisted on studying uranium. But the cyclotron never functioned properly when he was running experiments. Gentner recalled that Bothe's measurements of uranium cross-sections were of little scientific value, and that when a sample of uranium was irradiated with a neutron beam from the cyclotron and sent back to Germany for study—a process which in the hands of the Berkeley physicists had pointed to the discovery of plutonium—Otto Hahn complained that he could get more irradiation with a few grams of radium. By one account, the difficulty was caused by Joliot's chief mechanic. Passing through one of the upper floors of the building, he would occasionally turn off the water that fed the cooling system of the cyclotron down in the cellar. Thereupon the machine would overheat and break down, a common event with the early cyclotrons and one for which Bothe, as the operator, would sheepishly take the blame. Such little failures, together with poor progress in laboratories in Germany, had a large significance. While the Germans suspected that plutonium existed and that it might possibly be used for a compact and overpowering bomb, they never got any idea of the element's properties nor even any proof of its existence. They failed to push this line of investigation, which was the quickest way to a bomb. Bothe's chief return for all his efforts was some practical experience in building and using cyclotrons. But he did not get his own machine in Heidelberg operating until 1944, and then he used it largely for basic biological research.[14]

In retrospect it might appear that Joliot came dangerously near to outright cooperation with the German fission program. But he would not have recognized any danger, for like the Germans, he remained convinced that a nuclear bomb could not be built during the war. He and Gentner, who had become close friends, went so far as to talk over whether a bomb could be made. They concluded that it could not.[15] Joliot still held this view when he was interrogated, shortly after the liberation of Paris in 1944, by an American intelligence team. According to their report, Joliot was not aware that German laboratories had done serious work on chain reactions, and he doubted that they could have solved the problems. The team reported that Joliot also held "the prejudiced opinion that no good scientist can work successfully on such a vital problem if exclusively destructive aims are considered. These weren't exactly his words, he definitely believes in doing war work on things like explosives or radar, but he appears to have some special idealistic view of the [uranium] problem. His basic interest in it is its use as a super source of energy, not as an explosive."[16]

At the time most of the people who knew Joliot could not tell how peaceful his research was nor how little progress was being made by the German researchers at the Collège de France. Some feared that Joliot was wholeheartedly collaborating with the enemy. It did not help that he agreed in early 1941 to an interview by the collaborationist newspaper Les Nouveaux Temps, even though he used the chance to call for the removal of all class privileges and barriers to entering scientific careers. France's downfall, he charged, had been caused particularly by the elite graduates of the Ecole Polytechnique, "the grand abettors of our industrial bankruptcy and of our deficiency in war materials."[17] This was not an unusual view, for liberals had been criticizing the Polytechnique for generations, and since the defeat most Frenchmen had been brooding over their country's faults.

Paris under the Occupation was quiet and strange. There were few young Frenchmen on the streets, for many had been detained and put to work as prisoners of war in Germany while others had slipped away to avoid labor drafts. Many scientists and other intellectuals had vanished into exile, or into the south of France to live under the Vichy regime. Another reason that Paris was quiet was the scarcity of fuel. Joliot had wisely left his car in the south and went about on a bicycle, although even bicycle tires and replacements were hard to get.[18] As the war continued, rationing grew stricter and people thinner, while black market profiteers lavishly provisioned the restaurants and night clubs favored by the conquerers and their French toadies. In the streets German uniforms were inescapable.

The leaders in Vichy were not always subservient to the Germans, having their own ideas about the future of France. Some followed the antique vision of Marshal Pétain of a patriarchal fatherland founded on family and religious values. But this notion was gradually pushed aside by other, more up-to-date plans. Bichelonne was an example of the modernizing Vichy leaders. After graduating from the Ecole Polytechnique with the highest grade average attained there in a century, working on railroad reorganization, and moving up during the war to become Dautry's chief of staff, he was still young and dynamic when, a few weeks after the armistice, he was named secretary-general of the Ministry of Industrial Production. A fortnight later he was arrested by the Gestapo, imprisoned, interrogated, and charged with signing the order that sent Halban, Kowarski, and the heavy water out of France. French leaders rushed to his defense, protesting that Bichelonne could hardly be blamed for having done his duty before the armistice, and after a few weeks he was released.[19] In 1942 Laval made him secretary of state for industrial production.

Like many Polytechniciens, including to some extent his patron Dautry, Bichelonne believed in a disciplined (dirigée) economy. To him this meant that the captains of industry, "the essential agents of production," should be organized into all-embracing trade associations, committees of capitalists which would be gently coordinated by a government hierarchy above while supposedly engaging in a harmonious dialogue with their workers below. His concept of industrial concentration and planning was not, Bichelonne insisted, very different from the goal the "pseudo-liberal" politicians of the Third Republic had been aiming at.[20] But in fact his vision was opposed to the vision of Jean Perrin's circle in the most important ways. Bichelonne hoped to see managers promote social stability and orderly industrial growth, with society arranged for managerial convenience; Perrin hoped to see pure thought join with popular aspirations to propel society headlong toward finer styles of life, with the managerial class merely making industrial arrangements in the wake.

When Bichelonne and his colleagues tried to bring to life their particular ideas on social organization, they found themselves tightly constrained by the demands of the nation's German overloads. The people in charge of French industry were obliged either to commit millions of Frenchmen to forced labor in Germany or to keep them at home in factories which aided the German war effort. Although they tried to find ways to shelter their countrymen, they never found a solution.

During this period Joliot established a corporation, one of whose functions was to provide work certificates for young French scientists so that they would not be hauled off to Germany. But he had higher hopes for this Société d'Etudes des Applications des Radio-éléments Artificiels (SEDARS). Along with the company's manager, Professor Léon Denivelle, he tried to raise capital for work toward a future nuclear industry. Although he did not get very far, and it is not known exactly what transactions took place, the effort brought him into close contact with various industrial leaders and financiers who had lost none of their influence during the war and would lose none afterward. This was another stage in Joliot's education in the relationships among science, industry, and the political processes that embraced both.[21]

He also encountered Bichelonne. The secretary of state for industrial production had not forgotten what he had learned at Dautry's side, and he too was interested in the promise of nuclear energy. In the summer of 1943, following a meeting with Joliot which was probably at his own request, Bichelonne wrote Joliot a letter:

1. The French Government invites you to reinstitute immediately your studies on atomic disintegration, particularly through research on the concentration of uranium 238 [sic] in natural uranium . . .

2. The French Government puts at your disposal all necessary credits . . .

3. All raw materials will be furnished you for the development of the Ivry laboratory.

4. All collaborators liable to service in *Travail Obligatoire* [forced labor] whom you desire will be attached to your laboratories.

It is unlikely that Bichelonne meant this proposal to be a contribution to the German war effort. Like many others, he must have felt that nuclear energy could not possibly be developed in time to have any influence on the war. Probably here, as in his work to reorganize industry, he hoped to prepare a way for the renewal of French strength in the postwar world, meanwhile sheltering young men from the labor drafts. The bait was tempting, including everything an ambitious scientist could have asked for. Joliot replied immediately:

You have asked consideration of the resumption of the experiments, interrupted since the Armistice, relative to the production of reactions in explosive chains of Uranium, in view of the construction of steam generating stations . . .

I have had the opportunity several times to state my position to representatives of the German services and they know that I will refuse any participation in scientific effort of the German war. Furthermore no proposition of work of this kind has been made me before.

I believe that it would be practically impossible and moreover scarcely opportune to resume our old experiments and I would not want it. It is in the field of pure research and in creating the greatest possible number of research workers that I believe I am most useful.[22]

In short, Joliot suspected that work on nuclear energy might somehow aid the German war effort, and he refused. The matter ended there.

In March 1944 Joliot received a letter proposing another meeting, but he apparently did not reply. On a Sunday not long after, as he was fishing in the Seine, an official black Citroën pulled up and summoned him over. Joliot soon found himself, still in his fishing outfit, seated opposite Bichelonne. The young technocrat was deeply troubled. According to a secondhand account of the meeting, Bichelonne told Joliot that he had earlier "played the card of German victory" because that victory had "brought order" and would maintain it, but now with the Americans on their way to winning the war he still hoped that order could be maintained. Perhaps he wondered whether a place could be found for himself in the next regime. Joliot, deeply distrustful, left without reassuring the minister, and Bichelonne died not long after under surgery in Germany.[23] He could hardly have known that the man he had dealt with, apparently no more than a scientist who cooperated with certain industrialists and Germans, had for several years been leading a double life.

The first stirrings of French resistance had come when the students and professors of Paris returned to the university in September 1940 and began spontaneous, isolated discussions with one another. Joliot attended one of these early meetings, where letters from two militants were read and the possibility of intellectual resistance was discussed. Even to consider the possibility was unusual at this time, when most of the French, depressed and introspective, observing that Britain alone was still fighting, believed that the Germans had already won the war.[24] The right and center relied on Pétain's promises of a return to traditional values, while the left reverted to pacifism. Even the outlawed Communist party, cramped by the alliance between Stalin and Hitler, saw nothing to gain in helping the British capitalists.

The university community began to awaken in October 1940 when the Vichy regime, acting well in advance of overt prodding by the Germans, decreed the dismissal of all Jews from influential positions, including teaching jobs. The professors could hardly overlook the parallel with events a few years earlier in Germany, where such dismissals had been followed by mass deportations and killings, as would in fact follow in

France. Another disturbing event was the arrest of Langevin on October 30. For three days nobody knew what had become of him, and then people learned that he was being held as a common prisoner because of his antifascist work before the war. Langevin had been scheduled to give his first lecture of the year at the Collège de France on November 8, and the Germans, anticipating a demonstration, locked the lecture hall. Joliot unlocked the room and addressed an assembly of students and professors, announcing that until his old patron was released, he would close his laboratories to all workers, French or German. Three days later, on Armistice Day, there was a demonstration at the Place de l'Etoile; it had been planned for some time by student groups, but Langevin's imprisonment added greatly to its force. The police brutally dispersed the demonstrators with gunfire, signaling the end of open opposition to the Occupation regime. Following talks between Gentner and the German authorities in Paris, Langevin was granted a partial reprieve and was sent away to the town of Troyes to live under surveillance.[25]

Around this time several underground publications began to appear. Among them was *L'Université Libre*, whose first issue appeared shortly after Langevin's arrest and was largely dedicated to him. It was put out by a group of three people who included Langevin's son-in-law, the theoretical physicist Jacques Solomon. At some point Joliot, who knew Solomon well, began to associate with the group as well as with other early Resistance circles.[26]

Many of the Resistance workers known to Joliot, including Solomon and other *Université Libre* workers, were ardent Communists. The party had the advantage, added to its unrivaled discipline and revolutionary expertise, of having already been underground when the Germans arrived. Under the initial shock of the Occupation the Communists were as confused and incapable of action as other French groups, and until mid-1941 they opposed the war altogether, damning as one the British, the Germans, and Vichy. In May 1941 the party issued a call for the formation of "National Front" committees to oppose the war, the Nazis, and the Vichy regime. The clandestine Communist newspaper *L'Humanité* declared, "Today, while the bombs of [British] airplanes once again menace our cities and villages because Pétain and Darlan once again drag us into the war, the duty which is thrust upon us is the constitution of a broad NATIONAL FRONT TO BATTLE FOR FRENCH INDEPENDENCE."[27] When soon afterward Germany invaded the Soviet Union, the French Communists could turn to full support of the war against Germany. The Nazis and Pétain responded by denouncing any opposition, any spontaneous or isolated act of resistance, as part of a Bolshevist threat. The unexpected result was that many antifascist French, believing this propaganda, began to gravitate toward the Communists.

Although there were Resistance groups of every political stripe, for the next year or so the Communists were well ahead of all others in organization and combativeness. The National Front committees, hoping to unite all anti-Nazi elements, accepted members of many different opinions, even priests, although the leaders were mostly party members. The Communist party itself recruited new members from within the Resistance, for many saw no other place to turn now that French democracy had, as it seemed, ignominiously failed.[28]

Nevertheless both the Communists and their National Front committees attracted only a handful of adherents in 1941, for the great majority of the French, scarcely able to hope that Germany could be beaten, had no heart for mortal risks. That June a University National Front committee was formed in Paris. Joliot was among the non-Communist members.

On June 29 the French police arrested Joliot. They brought him to the Prefecture of Police, a disturbingly massive stone building at the center of Paris. Turned over to the Germans, he was accused of being a Communist and interrogated. Happening to meet another arrested professor in a waiting room, Joliot told him that he was confident the German scientists working at the Collège de France would intervene to secure his release, for they believed, as he did, that the brotherhood of scientists transcended national conflicts. In fact Gentner had heard of the arrest almost immediately from Irène Curie, and a few hours later, as he recalled, he was telephoned at his hotel by the German SS staff who wanted confirmation of statements about Joliot's work. Gentner insisted that Joliot was very important for the German team's work, which had the highest priority; moreover Joliot could be detained only on the authority of Gentner's superiors in Berlin. Joliot was released almost immediately. Gentner later recalled that throughout this period he protected Joliot as much as possible, and at one point he told Joliot that he did not want to know anything about clandestine activities.[29]

The dangers of Joliot's double game became clearer as old friends like Cotton and Borel were imprisoned for various periods simply because they had been strong antifascists in the thirties. But the greatest shock came in December 1941 when Fernand Holweck was arrested. Holweck, a prominent physicist, was a former student of Langevin and scientific collaborator of Marie Curie, and had worked in a corner of the Curie Laboratory for many years. In the second half of 1941 he became a leader of one of the isolated Resistance teams that had begun to operate, helping downed British aviators to escape from France. A police agent masquerading as an Englishman penetrated the group, and Holweck along with seventy others was seized. At the end of December his family was summoned to receive his corpse. Medical examination confirmed what was only too obvious: Holweck had been savagely beaten and scalded. His torture and death, the first in Parisian scientific circles, stunned the professors.[30]

Another blow fell at the end of February 1942 when many leaders of the Communist-dominated *Université Libre* group were swept up by French police. Solomon was delivered to the Germans, tortured, and on May 23 shot without trial along with other intellectuals. His mother and his wife, Langevin's daughter Hélène, were deported to Auschwitz, where his mother died and Hélène survived only by exceptional luck. The party organization was equal to such shocks: *L'Université Libre* continued publication without interruption in issues of 4000 copies.[31]

Immediately after Solomon's death Joliot joined the Communist party. He told Pierre Villon, the Communist chief of the National Front, that if they were found out, he did not want to be treated any better than Villon himself. Later Joliot would say, "I became a Communist because I am a patriot." Like many others in France, after joining the Resistance he had begun to appreciate both the party's skill and the patriotic zeal which had dominated its policy since the days of the Popular Front. Further, he later remarked, in the intimacy of dangerous secret meetings it was the Communists who gave the most satisfactory answers to the troubling moral questions that continually had to be faced.[32]

During this period the National Front committees proliferated, and in the northern Occupied Zone they gathered under a single Directing Committee. The president of this committee was Joliot. He worked under the overall leadership of Villon, a quietly intelligent architect who, like Joliot, was just entering his forties. Joliot applied his political talents to both the National Front as a whole and the subsidiary university group at clandestine meetings held irregularly on an hour's notice, one day in a university building, another in a private home or shop. Through 1941 the National Front had been more an expression of determination than a concrete organization, but it gradually found supporters and took on structure. It became the largest, best-organized, and most belligerent of all the Resistance groups, although it never reached its goal of encompassing the entire Resistance. During 1942 it made contact with other growing secret organizations, some of them quite conservative politically, and a loose union was formed which papered over the enmities between the groups. To the end of the war the National Front held a commanding position among these organizations.[33]

As president of the Directing Committee, Joliot was in one sense working like a politician in normal times. The most important decisions were political ones involving such matters as negotiations with other Resistance groups, manipulation of public opinion, and relations with leaders abroad. The Directing Committee, meeting about once a month, received emissaries sent in secret from London, issued public pronouncements by way of its various clandestine newspapers, and approved issues for Villon to discuss with the other Resistance organizations. For example, the National Front leadership promoted a movement to unify the various anti-Vichy groups under the exile government of General Charles de Gaulle. According to Villon, "Joliot took an active part in all these meetings by competently presiding over them, but also and above all by giving his opinion on each problem under discussion. I saw him from time to time between two meetings, especially to prepare for the meetings themselves."[34]

In another sense nothing could have been farther from peacetime politics. Joliot helped coordinate a confusion of shadowy and continually shifting structures, paying particular attention to the secret recruitment and organization of intellectuals. He received and passed along information on police activity, such as anticipated raids, and information for the Allies, such as the effectiveness of bombardment. And when the National Front had to execute traitors, the final decision was said to have been Joliot's. He also took on even more direct work. A great opportunity arose when Moureu left his position as subdirector of Joliot's laboratory at the Collège de France to become head of the Laboratory in the Paris Prefecture of Police. The police called on Moureu to examine explosives seized from Resistance workers, and Moureu passed on to Joliot not only the information but sometimes the explosives themselves, which were hidden at the Collège de France until they could be redistributed and used. And in early 1944 Joliot personally organized Langevin's escape from Troyes to Switzerland. In later years Joliot never publicly discussed his feelings about this work, but they were probably like those summed up by Georges Bidault, another Resistance leader: "Our losses were heavy. We constantly had to reassemble groups that had been dispersed, look for new men to fill the ranks, new leaders to take the place of old ones. Our path was strewn with tragedies. The endless disappearances, griefs and fears set our nerves on edge, for the front was nowhere and the enemy everywhere."[35]

By the end of 1943 Gentner was no longer able to protect Joliot, for the Gestapo had received reports questioning Gentner's loyalty, and he was replaced and brought back to Germany. Nevertheless during the spring of 1944, in rooms at the Collège de France neighboring the ones where the German scientists were at work, Joliot's subordinate Savel managed to assemble several clandestine radio transmitters, and soon after the Allies landed in Normandy in June 1944, people in several Paris laboratories began to manufacture explosives, all probably coordinated by Joliot. The danger of discovery must then have become very great. Joliot smuggled his wife and children to safety in Switzerland, and then, three weeks after the Normandy landings, he disappeared.[36]

12

Exiles and Industrialists

Halban and Kowarski reached London on June 22, 1940. They were armed with an order signed by Bichelonne as Dautry's deputy, directing them to "carry on in England the researches undertaken at the Collège de France." British scientists were eager to interview them, and in the next few weeks the pair shuttled about visiting leading physicists in London, Liverpool, and Birmingham. They considered themselves morally, and probably legally, bound not to withhold any information whatsoever. Halban had taken up residence in a fashionable London hotel, where over the next week he and Kowarski wrote down for the British a full report on the research they had done in France.[1]

Halban held himself responsible for safeguarding France's position in nuclear energy, and he felt this would best be done by penetrating the highest circles of British industry and government. Like Joliot, Halban could be charming when he chose, and he soon struck up acquaintance with a number of leading men. Persuasively and with utter conviction he argued the long-range importance of nuclear power. He repeatedly stressed the value of the French patents. At his request Kowarski drew up applications for British patents duplicating the latest French ones.

Halban and Kowarski wanted to go on to North America, where their work would be safe from a second interruption by invasion and where it could feed on an entire continent's industrial and scientific resources. Committed to the war against Germany, however, they did not want to move to the neutral United States, so they asked the British to send them to Canada and give them a small laboratory there with a few assistants. If the government refused, they were prepared to look to industry for support. Halban was interested in reestablishing contact with Sengier or Lechien, whose whereabouts were unknown, hoping to get some radium and the 50 tonnes of uranium oxide promised in the gentlemen's agreement with the Union Minière; he was also on the watch for other possible allies in industry.

British scientists, preoccupied with more pressing war work and only mildly enthusiastic about the chances of separating enough uranium-235 to make a bomb, did not have much interest in reactors. Nevertheless they were willing to support some research which might lead to a valuable source of power and which would meanwhile give information about basic fission processes. The scientists further noted that, although a reactor "would not be explosive in the ordinary sense . . . the radiations produced would be such as to render uninhabitable a very large area."[2] The Maud Committee therefore decided to keep Halban and Kowarski in Britain and allowed the pair, with other foreign scientists such as Frisch and Peierls, to join a Technical Subcommittee. It was this subcommittee, not the strictly British Maud Policy Committee, that became the main forum for discussions on fission. The Maud Committee's first plan for giving the French scientists a laboratory was to install them at Liverpool, but the city was being bombed so frequently—the windows of the laboratory were blown in day after day—that Halban and Kowarski were finally offered space in the old Cavendish Laboratory building at Cambridge. Most of the normal occupants had left for radar work or been evacuated to other laboratories, for Cambridge was in the path of the expected German invasion.[3]

Halban and Kowarski drove up to Cambridge on July 14, which required care since all the roads signs had been removed to confuse the German soldiers when they came. That night the pair were invited to Trinity College, where they went through the dinner rituals, Halban recalled, "as if nothing was happening in the world, everybody realizing that it might well be one of the last before the invasion." The next day they moved into small rooms in the aging brick laboratory and set to work.

Intending to carry out the experiment planned with Joliot in the spring, they began to repeat the lengthy process of getting a little uranium oxide, ordering the construction of a large aluminum sphere that would rotate mechanically, and assembling electronic apparatus. Gradually they acquired a little group of helpers. The work of setting up the laboratory, although feverishly pursued, went slowly through the summer of the Battle of Britain. Their aluminum sphere was finished in August, then had to be modified. A ton of uranium oxide powder provided by the British government arrived in October, and in November the group did a number of experiments on the way in which the powder settled to the bottom of the slurry of uranium oxide and heavy water. Meanwhile Halban managed to get a message through to Joliot, which was their last contact for four years. Halban's father wrote Joliot from Switzerland on October 31, "My son has asked me to let you know that he has been very well received in his new surroundings and that he works under excellent conditions."[4]

Late in November the Cavendish group began to measure density-distribution curves. They had an aluminum sphere 60 centimeters in diameter spinning at a moderate 20 revolutions per minute, filled with the uranium oxide-heavy water mixture. In the center they put a gram of radium in a beryllium block as a neutron source, and the whole was immersed in a tank. On December 16 the group had their results. More neutrons came out of the rotating sphere when uranium was inside than when it was not. More precisely, Q_{ext} and, equivalently, $k\infty$ were greater than 1.0, which showed that the chain reaction was potentially divergent. To make a critical reactor, all they needed was a larger sphere and more materials.[5]

This was the first experiment anywhere to give such a result, anticipating other workers by over a year. The Germans did not produce a similar experiment, also with heavy water, until February 1942, and Fermi built a potentially divergent graphite pile a few months later.[6] The Cavendish team's hurried experiments were less thorough than the German and American studies, but they revealed the essential truth: given enough uranium and heavy water, a nuclear reactor could be built.

Meanwhile a few people were beginning to glimpse dimly the lethal potential of such a reactor. Up to the middle of 1940 everyone had thought of reactors as something of long-range potential, an unlikely weapon for the current war. Halban and Kowarski had a vague idea that new sorts of fissionable material might be created by the intense neutron radiation within a working reactor, but they gave little thought to this possibility, although they mentioned it in their first report to the British. The idea became more plausible when one of the last published papers on fission research reached Britain, the June 1940 paper of McMillan and Abelson reporting that neutrons from the Berkeley cyclotron could change uranium-238 into a new element, the first step on the way to fissionable plutonium.[7] A number of scientists independently guessed what this might mean. In the United States Fermi and Szilard, after seeing Turner's unpublished paper on plutonium, had realized already in May that the flux of neutrons within a nuclear reactor would turn uranium into plutonium by the kilogram. Unlike uranium-235, which is chemically identical to uranium-238 and therefore can be separated from it only with enormous effort, the plutonium could be extracted by chemical treatments. A few people in the United States now recognized that nuclear reactors could be used to manufacture material for nuclear bombs. But unlike the British, the Americans did not realize how few kilograms were needed for a critical mass, and their vague thoughts did not reach the higher levels of government. A physicist in Germany, Carl-Friedrich von Weizsäcker, had a similar idea in July 1940, but he too failed to stir up his government, and the Germans never seem to have realized how relatively easy it would be to use a reactor to turn out enough plutonium for a bomb.[8]

In Britain the idea was discussed from the time the paper of McMillan and Abelson arrived. Egon Bretscher and Norman Feather, young physicists working near Halban and Kowarski in Cambridge, developed the concept and presented it to the Maud Committee at the beginning of 1941, the same time as Halban and Kowarski reported their successful heavy water experiment. But Halban, Kowarski, and their colleagues were only beginning to suspect that their reactor could become an engine of war. Given a working reactor, they agreed that they might be able to produce a "super-explosive." But they calculated that to turn out a sufficient quantity of plutonium, they would have to build an enormous reactor, and Halban could not see how such a reactor could be kept from melting down under its own production of heat. It might be possible to build a machine that generated tens or even hundreds of thousands of watts of heat and still remained cool, but only a reactor producing tens of millions of watts could make enough plutonium for bombs.[9]

Another reason that Feather and Bretscher's report on plutonium was not pursued was the progress being made toward separating the isotope uranium-235, progress which overshadowed reactor work. The British could leave plutonium aside, for they were becoming convinced that isotope separation would swiftly supply them with a nuclear explosive. Long after, Chadwick said: "I remember the spring of 1941 to this day. I realized then that a nuclear bomb was not only possible—it was inevitable. Sooner or later these ideas could not be peculiar to us. Everybody would think about them before very long, and some country would put them into action . . . And there was nobody to talk to about it. I had many sleepless nights. But I did realize how very, very serious it could be, and I then had to start taking sleeping pills. It was the only remedy."[10]

Although the race for a nuclear bomb did not seem to depend upon reactors, Halban and Kowarski were determined to continue their work, sure that it would be a source of important scientific information and eventually of industrial power. By now they knew in a rough way how to build a reactor, given enough heavy water. But it would take large industrial resources to produce tons of heavy water, and the British hesitated to commit themselves when they were already hard-pressed to get uranium-235 separation underway.

Graphite was a familiar alternative. Already in October 1940 Halban had ordered a thousand cubes of it, each ten centimeters on a side, in lieu of the cubes left behind in France; in early 1941 he ordered a thousand more. Unfortunately the suppliers, as he noted, were "not able to guarantee complete absence of certain impurities which do not endanger the ordinary technical application . . . but which would make the product completely useless to us." In April 1941 he reported a disappointing carbon cross-section of 0.5 x 10^{-26}, whereas Fermi had obtained 0.3 x 10^{-26} with his graphite almost a year earlier, and Szilard was hard at work developing ways to produce still purer supplies. Halban and Kowarski felt unsure whether a carbon-moderated reactor could ever be made to work. To get the answer, the British government would have had to stake large industrial resources on an effort to turn out tens or hundreds of tons of extremely pure graphite, and in the midst of war they could not afford the gamble. They decided to leave pursuit of graphite reactors to the workers in the United States.[11]

Halban was prepared for these problems. From the beginning of the fission work in Paris he had consistently looked for industrial support, and through 1941 he pushed again in this direction. On January 8, 1941, right after the meeting of the Maud Subcommittee at which he and Kowarski explained their heavy water experiments, Halban went off with another subcommittee member, Roland Slade, who was research manager of Imperial Chemical Industries (ICI), the largest manufacturing firm in Great Britain. Slade had not always been confident that pure science was worth his firm's money, but for a few years during the 1930s ICI had supported a number of refugee scientists, including Szilard, with the result that Slade had been exposed to Szilard's nuclear chain reaction ideas as early as 1936. Now he listened with interest as Halban tossed out various ideas, such as that a reactor might be built with compressed hydrogen as a moderator. This suggestion clearly lay in ICI's line of work, so Slade called in Michael Perrin, an officer in the company's research division. Perrin (no relation to the French family) was an expert on high pressures who had made his name in 1935 by developing polyethylene, working unofficially after hours because the company had banned his hazardous experiments. Slade and Perrin agreed to study not only high-pressure problems but also the procurement of very pure carbon and other matters. Meanwhile ICI was involved in various other studies for the Maud Committee, for example on the production of uranium.[12]

Halban embarked on an intricate series of negotiations with ICI, looking to get the company's support for reactor research without betraying the rights of France and the old CNRS-Union Minière arrangements. With his usual optimism and energy he looked beyond the war, hoping to direct and exploit nuclear energy in peacetime industries. Patents might be the lever that would control developments. Halban continued to apply for patents on any new ideas for improving reactors that the team came up with, including Kowarski, Bretscher, and Feather as authors of patent applications.

In the spring of 1941 the Ministry of Aircraft Production, which was supporting Halban and Kowarski's group, began, like ICI, to negotiate with them for rights in their patents. They reached no agreement, particularly because Halban insisted that the government promise to keep him on as director of any research on nuclear reactors. Halban's talks with ICI were more successful. His enthusiasm for nuclear energy and his belief in the power of the French patents penetrated the ICI hierarchy all the way up to the chairman, Harry Duncan McGowan.

Lord McGowan was a stout, bald, cigar-smoking capitalist, a Scotsman whose "rather coarse heartiness," according to the historian of ICI, "served the purposes of a powerful mind and dictatorial personality." He had employed these in rising from poverty to the directorship of Nobel Explosives, in spurring this and other British explosives and chemicals companies to merge and form ICI in 1926, and then in personally taking control of ICI. McGowan was mainly concerned with the construction of ever-larger combines and the management of industrial power, but as a chief of the modern chemicals industry, he necessarily supported research and was familiar with the idea that scientific advances, expressed in practical matters through patents, could reshape industry.[13]

While Halban spread his ideas before McGowan and others at ICI, Kowarski remained as skeptical as ever about the strength of arrangements with industry based on patent rights; he was not involved in these negotiations. In Kowarski's eyes:

Halban's far-seeing plan was to establish himself in the eyes of the British industrialists as the coming man of this new force of nature . . . When Halban wanted to impress the government, having the purpose of getting access to big resources, he also at the same time had to impress the industrialists by appearing as one of those practical German wizards. ICI wistfully envied I. G. Farben in those days; see how reasonable the Germans are, what good cooperation there is between German scientists and German industry . . . For Halban the patent applications—not the patents themselves, but the applications—were important in order to situate him as a man they could understand . . . The patent application as a hallmark of practical-mindedness, of belonging to that particular brand of scientist, was more important to him at that moment than having the patents themselves.[14]

Halban, taking a bold gamble that nuclear reactors would work, was persuading others to join the game, including Slade and Michael Perrin. By midsummer 1941 they had sketched out an agreement, to which Kowarski acquiesced: the ICI staff would offer McGowan a plan to spend a quarter of a million pounds of ICI's own money over the next two years on a nuclear reactor research team under Halban, and the French would give ICI a share in the patents. McGowan agreed to take the chance.

Now ICI made a proposal to the government. Since reactors were not thought to have immediate military potential, ICI expected that the government would not wish to spend much money on them. Yet "the importance of the subject is so great," ICI believed, "that its successful development will make possible the reorientation of world industry." ICI reiterated to the Maud Committee that "this work should be started at once by British interests so that the British Empire will not be penalized after the war by the fact that other nations have solved or made great progress in the solution of the problem."[15] The firm therefore offered to take over Halban and Kowarski's entire team and all reactor research as a private venture.

This proposal, which dominated the view of the two French scientists, was only part of a larger scene which by now included hundreds of engineers, scientists, administrators, and industrialists on both sides of the Atlantic. Many other, sometimes more significant developments were taking place in both the British and the American nuclear energy programs. As the Maud Committee deliberated the ICI proposal to take over reactor work, they were also writing reports to the British government covering all the possibilities they could see for nuclear fission, and these reports gave Halban and Kowarski's work second place.

Detailed studies in various laboratories had satisfied the Committee that bombs could be made out of uranium-235, that they could be made within reasonable time and cost, and that their strength would be devastating. The first of the Committee's reports, dated June 12, 1941, recommended that work on such bombs be given highest priority. ICI meanwhile proposed to help build the necessary isotope separation plant, hoping to finish it before 1944. In a second report, dated July 15, the Committee discussed reactors. They were discouraged by the difficulties of producing heavy water and of building and controlling such an intensely hot and radioactive device. So despite the hazy prospect of plutonium, the Committee concluded that "the scheme requires a long term development and we do not consider that it is worth serious consideration from the point of view of the present war." But this conclusion did not mean stopping Halban and Kowarski altogether, for the Committee agreed that reactors could eventually be of great moment and might meanwhile produce scientific information that would help the work on bombs. The Committee made no recommendations about what role ICI should play in reactor work. Since, as the Committee put it, "Drs. Halban and Kowarski have done all that they can with the supplies [of heavy water] which they brought to this country," the Committee advised that "they should be allowed to continue their work in the U.S."[16]

Halban and Kowarski welcomed this recommendation. Knowing that they could not do much more in Britain, they were growing anxious to transfer their work to North America. ICI concurred, for Lord McGowan, who had risen by forming national and international cartels, believed strongly in international cooperation. The sharing of technical information, patents, and certain markets with the American du Pont corporation was a cornerstone of ICI policy, so it seemed likely that with du Pont support, Halban's team could be set up in the United States or Canada under ICI's authority. This sort of arrangement suited McGowan's plans for ever larger and more efficient combinations, which he felt would both aid the defense of the British Empire and lead to an expansion of industry and prosperity after the war's end. Even in 1942, during the desperate battles in Russia, North Africa, and the Pacific, he took a long view of ICI's prospects, stressing such postwar considerations as world markets and cooperation with du Pont in their exploitation.[17] The vision shared by many of Halban's scientific colleagues—a better world created through science, across national frontiers, applied by large-scale industry—matched McGowan's ideology with a precision that, given the scientists' long association with industry, was not entirely accidental.

The Maud Committee's reports, with ICI's proposals attached, were sent up to the Ministry of Aircraft Production for consideration in July 1941. The Ministry agreed with the Maud Committee that the reactor project was of long-term interest only and should be transferred to the United States or Canada, and believed that this should be done through ICI. Now ICI offered the Ministry an agreement: the corporation would reimburse the British government's expenses so far on the Cavendish team and would give the government free rights to exploit any patents for the armed forces, but otherwise would control all the research and patents themselves, taking on Halban and Kowarski as ICI employees. Thus a three-way arrangement among government, industry, and the scientists was taking shape. ICI prepared to hire staff for the project.

The Ministry meanwhile forwarded the problem for consideration to a higher level, the Cabinet's Scientific Advisory Committee, chaired by M.P.A. Hankey. Lord Hankey was already familiar with the plans through various contacts, including a luncheon meeting with McGowan. A panel of the Advisory Committee, after deliberating on fission, endorsed most of the Maud Committee and the Ministry of Aircraft Production's conclusions—but turned down ICI's proposal to take over reactor work. The potential of nuclear power was so great, the panel concluded, that "the matter should not be allowed to fall into the hands of private interests." The British government should keep control, hopefully in collaboration with the United States and Canada.[18]

By late September 1941, when the Advisory Committee's report was finished, Prime Minister Winston Churchill and the chiefs of staff had already agreed that Britain must try to develop nuclear bombs. Now the government organized a new department, code-named the Directorate of Tube Alloys, to handle all fission work. ICI's abilities were not forgotten, however, for the department would be headed by men borrowed from the company. The director was Wallace Akers, ICI's veteran research director, a man pleasant to deal with yet full of drive. He brought with him his assistant Michael Perrin. The Directorate of Tube Alloys, like the Maud Committee which it superseded, was overseen on the cabinet level by a senior minister, Sir John Anderson. A tall Scotsman with the "air of a slightly sad bloodhound," Anderson had been trained as a chemist before he went into administration and rose to become lord president. He had briefly been on ICI's board of directors in 1938 and would rejoin it after the war.[19]

A number of British scientists were upset by these rearrangements. It seemed that ICI, put out at one door, had come in by another. "I can see no reason whatever," Oliphant exclaimed, "why the people put in charge of this work should be commercial representatives completely ignorant of the essential nuclear physics upon which the whole thing is based." It was not that Akers, Perrin, or anyone else showed signs of putting ICI's interests ahead of the government's; it simply rankled that the academic physicists were displaced by engineer-administrators from a private corporation.[20]

As the British scientists argued about the reorganization of fission research, Halban and Kowarski found themselves in opposing camps. In November 1941 Oliphant wrote: "I have had a talk with Halban and Kovarski in Cambridge and find that Halban claims that the greater part of the reorganization is due to his own conversations with McGowan. Halban is adopting a very high and mighty attitude and feels convinced that the boiler [reactor] problem will now receive adequate consideration. Kovarski I find is in a very delicate position with regard to Halban, whom he feels has abandoned his attitude towards him of colleague and adopted a new one of paid assistant."[21] For some time Halban and Kowarski had been following separate paths. Halban was spending a good part of his time either in negotiations with leaders of science, industry, and the government or in administrative work. More and more of the scientific research proper at the Cavendish was falling into Kowarski's hands. For the first time Kowarski found himself an accepted and respected scientist, directing an important team and talking on equal terms with scientific leaders. He developed a deeper insight into the experiments and a closer rapport with the assistants than did the often absent Halban. But Halban assumed overall charge of the project and continued to direct the details. Kowarski's feelings for Halban, which had combined friendship with a respect that bordered on envy, were now more and more dominated by a well-founded belief that Halban was thrusting him into the background.

The two nevertheless held down their personal disagreements and continued to work together for some months, pursuing slow neutron research ancillary to reactor development. Their team was growing, until by mid-1942 it employed a dozen scientists and eight assistants. But the reorganization of British fission work separated them further, for Kowarski sharply disagreed with Halban's political moves. Ill at ease with industry, Kowarski had been suspicious of the attempted ICI takeover of reactor work, but it was the relations among Halban, the British, and the growing American fission program that brought his objections into the open.

The fall of France had shocked the United States and Britain into close cooperation, and they had soon begun to exchange scientific information of military interest. Ideas on reactors were one example. In October 1940 Cockcroft attended a meeting of the Briggs Uranium Committee and told them, among other things, about the work done by the French team at the Collège de France. A little later a copy of Halban and Kowarski's report to the British on the work done in Paris also reached the United States. When the Cambridge team succeeded with their December 1940 heavy water experiment, the news crossed the Atlantic almost at once. Cockcroft wrote the British liaison officer in North America that "Halban has obtained strong evidence that the D_2O_1U [heavy water-uranium] slow neutron reaction will go . . . He is now proceeding to try carbon. Can you now press Urey to develop his ideas for producing D_2 [deuterium] in quantity." There was a chance, Cockcroft continued, that a reactor might produce a new fissionable element. "It would seem therefore that the uranium investigation deserves more vigorous attack in N. America than it is now receiving."[22]

This was only one of many times the British prodded the uncertain and sluggish American program. Through much of 1940 and 1941 it was only in Britain that a number of leaders realized that they could build nuclear bombs in time to affect the war. The Maud Committee's report of mid-1941 recommending the development of bombs, promptly relayed to the United States, along with personal contacts helped persuade the Americans. "As far as I know," Fermi later recalled, "the emphasis on the development of high explosives and the first approximately correct estimates of the amounts of materials needed came from the British group." The pressure increased when Pegram and Harold Urey of Columbia visited Britain in October 1941 for full discussions. After meeting a number of engineers and scientists, including Halban and Kowarski, they went home more convinced than ever of the urgency of fission work. All these reports and visits influenced a series of decisions by which the Americans gradually gave this work higher and higher priority. Had the British not taken up fission in earnest in 1940 and 1941—as the French had urged them to do since the first meeting of the Maud Committee—and had they not then spurred the Americans to act, it is likely that no nuclear weapon would have been ready before the end of the war.[23]

Authorities in the United States welcomed the information from Britain, and in October 1941 President Roosevelt wrote Churchill to express the hope that there would be an extensive fission program, conducted in close coordination or perhaps even jointly by the two nations. Although this proposal fell in with prior British ideas about getting the work done in North America, Anderson, Churchill, and their associates now hesitated. They feared that the project might not be kept secret if they shared it, and they could not bring themselves to place the proposed British isotope separation plant and reactor team in American hands, risking the loss of Britain's independent stake in the future of nuclear energy. They allowed the tentative chance of fusing with the American program to die away.

The British nevertheless expected to keep close ties with the United States, so everyone was pleased when several leaders of the British fission program were invited to cross the Atlantic for extended discussions. Halban was among these leaders, and he went over in early 1942 as part of a mission headed by Akers. Having a weak heart, Halban could not fly at high altitudes and went by ship through the Battle of the Atlantic. He hoped soon to be in command of a large program to build heavy water reactors under British control in North America.

As usual, Halban got on well with some of the leaders he met. For example, Urey, leader of the American heavy water program, found him delightful. Urey and others liked the idea of bringing some of the Cavendish team over and putting Halban in charge of a heavy water reactor program in the United States; at the very least this move would bring the world's only significant stock of heavy water within reach. But Halban had less success with others. His ICI connections were not only useless but suspect in the United States, where the scientists and administrators who oversaw fission work meant to prevent any one company from becoming predominant in the field. The Americans were particularly wary of ICI's close attention to possible postwar industrial advantages.[24] Halban's Germanic background and manner, as well as the varied nationalities of his team, awakened other fears. He was no longer in an environment where he could prosper.

Further, having risen to a position in the upper circles of British policymaking, Halban was naturally unwilling to come to the United States as a simple research worker under American control, signing over all patent rights and personal autonomy; but these were the only terms suggested to him by Arthur H. Compton, head of American reactor work.[25] This was not the sort of cooperation the British had hoped for. They were unwilling to see their reactor team dissolve, and Halban himself flatly rejected suggestions that he go to work as Fermi's subordinate.

James B. Conant, a scientist in Washington, D.C., who through the summer of 1942 had overall responsibility for American fission work, including Compton's reactor team, noted that "the fundamental difficulty is that of having a group of English scientists working here without being essentially responsible to anyone in this country. In my mind, it would be very much better if they were to work in Canada where there would be no question of difficulties of 'extra territoriality.' " Moving to Canada had always been a possibility in Halban's mind. There his team could have access to all the resources of North America while remaining under the control of Britain and himself. By midsummer he was preparing to lead a group to Canada, expecting to start up a pilot heavy water reactor and only later, from a position of strength, to merge with Fermi's team.[26]

Much more than reactors were involved in the relations between Britain and the United States. The British mission had returned home convinced that their allies were rapidly outstripping them not only in reactors but also in isotope separation and that any advantage Britain might have won from its early start would soon be lost. A feeling grew that no matter what was done about reactors, the British isotope separation plant should be spliced into the American program. But when the British approached the United States with this proposal, it was their turn to be put off, for the Americans, with their program swiftly gathering speed, felt less and less need for British help.

Kowarski and others in Cambridge were distressed by these developments, particularly by the idea of a reactor team in Canada. They felt that France and Britain could maintain a place in the advance toward a reactor only if the Cavendish team immediately went off to join Fermi under American control. Thanks to their head start, they might have had a chance to join Fermi, had not Halban and the British leaders thrown the opportunity away. On top of this frustration Kowarski felt a more personal pain: "to work at last under Fermi was my most cherished dream," he recalled, "and from mid-1941 to mid-1942 I was helplessly watching this dream vanish." He strongly objected to the proposed move to Canada; as Halban had remarked in 1939, Kowarski always felt free to tell the other soldiers that everyone except him was starting on the wrong foot. The two teammates were beginning to part ways. While Kowarski, having little influence on the men in power, could only direct his hopes toward advancement as a scientist, science was merely where Halban's ambitions began.[27]

Halban, said a colleague, "had a single goal: to direct the team which would be the first to create a divergent chain reaction." It was not simply a matter of personal ambition, for recalling what Joliot had told him on their last night together in France, Halban considered himself charged with the care of France's interests in the revolutionary field of nuclear energy. Peierls, who knew Halban well during this period, recalled that Halban seemed ambitious not so much to advance himself as to get things done, to make things happen. Probably Halban, convinced of the promise of the reactor and regarding it as his own child, could not disentangle the interests of his project and his country from his chances for personal success. One day workers in the Cavendish started a game of making up famous last words for people they knew, and for Halban they hit on, "Get me God on the phone—and make it priority!"[28]

Halban's ideas had become quite clear once the new Directorate of Tube Alloys had taken over from ICI the task of negotiating French patent rights, that is, once Akers as government servant and his superiors had taken over from Akers as ICI employee and his superiors. By now Halban and Kowarski were coauthors of six patent applications in Britain as well as the original five in France. These applications, covering various ideas about reactor design, isotope concentration, and plutonium, were far from describing a working device; nevertheless Akers, Anderson, and their associates thought that such patents might be keys to legal control over the postwar nuclear energy industry, and they wanted the British government to have them. But the government was in a difficult spot, for through 1941 Halban and Kowarski had still not signed an agreement. It was true that they were obligated to Britain for their support, but the British were well aware of their own moral obligations to the French. If Halban and Kowarski had not come to Britain in June 1940, there almost certainly would have been no British reactor program at all. The negotiations were thorny, and at times it seemed Halban would simply refuse to assign any part of French rights to the British. But obviously the British would not give him a big reactor team until the matter was settled. Finally in September 1942 Halban and Kowarski signed an agreement with the government.

In this document they promised to work exclusively for the Department of Scientific and Industrial Research, which had overall responsibility for fission work, "wherever the Department may from time to time direct and will carry out as directed by the Department work within the field [of nuclear energy] and assign to the Department the sole and exclusive rights in any further inventions"—Halban for a period of five years and Kowarski for two. They were guaranteed good salaries, Halban's twice Kowarski's, for this period. The pivotal clauses divided up patent rights between France and Britain. The British would be given world rights outside France to the most important patents applied for in Paris, while the French would get rights in France to subsequent reactor patents. Halban, Joliot, Kowarski, and Francis Perrin would jointly share 12.6 percent of all profits from the patents. Presumably Halban and Kowarski expected to share their profits with the CNRS or a like

French agency, under their 1939 arrangements, to aid scientific research.

Everyone understood the weakness of the arrangement: Halban had no proof that he was authorized by Joliot or anybody else in France to make a deal. General Charles de Gaulle's Free French exile government was also kept out of the matter, for Halban and Kowarski felt that their orders from Dautry made them independent French representatives. In August 1940 they had informed de Gaulle's group that they were working under the British, but they had not placed themselves in any way under Free French authority and they were acting on their own. In the final patent agreement there were clauses stating that Halban "hereby affirms that he was instructed verbally by Joliot to deal in such a way as he might think best with all Joliot's rights" in the French inventions, that he "is convinced that he will be able to arrange eventually the proposed exchange of patent rights and that he would make every effort to do so." Kowarski was not mentioned in these clauses.[29]

Halban's long negotiations with ICI and the government, and the resulting contract, were practical matters full of legal details; but they were also an expression of the scientists' vision, an ideology congealed into a concrete bargain. Like Joliot in his discussions with the Union Minière and the CNRS, so Halban with ICI and the Department of Scientific and Industrial Research was saying: Give us the authority and the means for our work, give scientists a cut of the future profits, and we will put the corporation and the state in a better position to defend themselves and grow strong.

13

A Few Tons of Heavy Water

Halban's team, as they prepared to move to Canada, were bent on constructing a heavy water reactor there, which meant they had to get tons of heavy water. Since 1941 a division of ICI had been studying the problem and designing an enlarged, improved version of the Norsk Hydro facility. There was no reason to doubt it would work, but the projected cost, including the drain on Britain's supply of electrical power, seemed prohibitive.[1] Halban's success would therefore depend on two matters beyond his control: whether the Americans would manage to produce heavy water by the ton, and if so, whether they would hand it over to his team.

The Cavendish team was only one strand in a much larger British project, and the British project was more and more overshadowed in every area by American work. In this vast and complex history, the links with French work showed most clearly in the strange career of heavy water in the United States. As soon as news of Halban and Kowarski's Cavendish Laboratory experiment reached America, Urey, the discoverer of deuterium, took what he called a "fatherly interest" in heavy water. When a British liaison officer visited him at Columbia in January 1941 to explain the Cavendish team's experiment, Urey at once began to study ways to produce heavy water by the ton. The subject had been brought up before in the United States, but only casually, and heavy water had been all but forgotten until the news from Britain spurred Urey into brisk action. Briggs remained lukewarm. "There is one feature of the heavy water program that disturbs me," he wrote Urey. "If we keep jumping around we are not going to get anywhere. I think that the heavy water project should be looked into carefully, but I would be sorry to see it interfere in any way with what you now have on the books."[2] Urey was in charge of studies of uranium isotope separation, which in America, as in Britain, was looking more and more like the quickest route to a weapon. Heavy water could only get whatever attention was left over.

Urey nevertheless gave the problem as much time as he could, helped by Hugh Taylor, a Princeton physical chemist. They started various people thinking about heavy water production and soon came up with several competing methods, all of them feasible and all of them expensive. Urey talked the problem over at Columbia with Pegram and Fermi, who conceded that heavy water was bound to be useful, if only as a coolant for a graphite reactor. But the always careful Fermi would not admit that deuterium could be a better moderator than graphite until he had seen exhaustive proof.[3]

The Americans asked their British colleagues to review Halban and Kowarski's experiment, and Chadwick, one of the most meticulous of physicists, went over the Cavendish team's notes in detail and interviewed one of the assistants. He concluded in May 1941 that although there were a few ways the work might have been improved, "My final impression is that the experiments were carried out very carefully . . . In my opinion the results can be accepted with confidence and I think their claim that the fission chain is potentially divergent is fully justified."[4]

This never quite satisfied the Americans, for it was not enough that a system be potentially divergent. The experiment yielded $Q_{ext} = 1.06$ with an uncertainty of ± 0.02; the system was therefore potentially divergent, since this was certainly above 1.0—but just how far above? If the numerical result were exactly right, then one could build a reactor with roughly ten tons of heavy water. But if the true Q_{ext} were only slightly lower, if it were only marginally greater than 1.0, a reactor might require hundreds or thousands of tons of heavy water. The number was a little technicality, yet a difference of a few percent in its size might determine the success of great enterprises. Q_{ext} could not be measured much more accurately except in a far larger tank full of heavy water, and the British had to admit that without further study the Cavendish measurements could not be considered conclusive.[5]

That same May a committee met under Arthur Compton to review American fission work. They felt that Halban and Kowarski's experiment "was very encouraging, and has apparently convinced the British physicists that the chain reaction can be produced by this method. The experiment is not however clear cut, and needs repetition with perhaps a ton of heavy water before the conditions for the chain reaction itself can be definitely formulated." Nevertheless heavy water was potentially "the most promising method now on the horizon," although it might not be the first method to give results. The committee advised setting up a pilot facility to produce enough heavy water to make conclusive experiments.[6]

In July 1941, Briggs agreed to give the Standard Oil Company 50,000 dollars to set up a pilot facility in Louisiana to study a sophisticated "catalytic exchange" process, while another 10,000 dollars would be spent to continue Urey and Taylor's studies of the necessary catalysts.[7] There was no guarantee that a proper catalyst could be found or that the process would work when applied on a large scale, so the Americans also pursued studies of the "electrolytic" method proven in practice by Norsk Hydro.

Halban's visit to the United States in early 1942 reinforced American interest in heavy water. But in the midst of their negotiations with the British the Americans came to a decision. Fermi, Szilard, and their colleagues had continued to pursue the graphite reactor at Columbia and then at the University of Chicago, to which the team had moved in early 1942 as the core of an enlarged team led by Compton. Their work was beginning to look very promising. On April 1, 1942, Briggs, Compton, and Conant, the top scientists in the American reactor program, met to thrash out the relative merits of carbon and heavy water. As Conant recalled:

This was shortly after Halban had sat with the S-1 [Uranium] Committee . . . and had argued his point of view. Taking up the cudgels for heavy water, I then argued strongly that we should go slow on the Chicago development and concentrate our efforts on the alternate method which would involve heavy water with the idea of producing "94" [plutonium] by that route. Dr. Compton convinced me that under the schedule then set up for heavy water from Trail . . . he stood a chance of getting through by his method many months before one could hope for success along the heavy water route. Dr. Briggs and I, therefore, were converted to Dr. Compton's view.[8]

Conant was now sure that the Franco-British stock of heavy water was not essential, and this was a factor in the Americans' decision not to bring in Halban on his terms. They also decided to make no great effort to produce tons of heavy water, for with the limited resources allotted their fission program in early 1942 it would take the Americans several years to make enough heavy water for a reactor.[9]

While deciding not to stake everything on heavy water reactors, the American leaders were also reluctant to stake everything on graphite reactors or isotope separation, for both of these might well fail. They therefore kept a heavy water program going as a "second line of defense." Urey and Taylor continued their work and made encouraging progress, although Urey's main efforts continued to be directed toward uranium-235 separation.[10] Contracts were let in mid-1942 for a small facility at Trail, British Columbia, where there was an existing ammonia plant which was the closest North American counterpart of the Norsk Hydro plant. The heavy water production, however, was to be done with an ingenious new scheme which Urey and Taylor hoped would combine the advantages of the Norsk Hydro process and their own catalytic exchange process. The facility was not meant to serve as a prototype for full production, but it was expected to produce enough heavy water for the intermediate-scale trials which were needed before anyone could have full confidence in heavy water. Thus the entire heavy water program rested upon the Trail facility. And serious problems soon developed there.

The first difficulty was low priorities, which plagued the entire American fission program during the hectic year that followed the United States' entry into the war in December 1941. The Trail heavy water production facility needed some of the same chemical engineering talent that was dedicated to the crash synthetic rubber program; to avoid delaying the rubber program by a few weeks, the Trail facility was put off for several months. Urey doubted that a short pause in rubber production would lose the war but felt that a long wait for heavy water might do so, since the Germans might be ahead of the Americans. But he could not get a higher priority for Trail. Moreover, once the facility finally began to operate in mid-1943, it worked poorly, and the heavy water took months longer than expected to begin dribbling out.[11]

Decisions could not wait for the heavy water from Trail. By late 1942 the American fission program had at last been given ample funds and was rapidly expanding in several directions. If heavy water were to serve as a real backup for graphite, this was the time to get full-scale production underway. Deuterium had looked the more important since mid-1942 when some physicists had raised the possibility that it might be used to make a "super" weapon, the so-called hydrogen bomb, which would be even more destructive than a plutonium bomb. Therefore the leaders of the American project, rejecting the various sophisticated but unproved methods which had been proposed for producing heavy water, asked the du Pont Company to design facilities based on the fractional distillation of steam, a crude and expensive method but one that would use simple and common techniques and equipment.[12] In response, in December 1942 du Pont offered three alternatives, ranging from a low proposal of two and a half tons of heavy water per month produced by a facility costing some 15 million dollars and requiring a 22-million watt power supply, to a high proposal of over nine tons per month at three times the cost in money and four times the cost in power.

While these were large plans, they were only a small and almost neglected part of the mushrooming fission project. Du Pont was also planning to build an entire city at Hanford, Washington, centered upon three huge graphite reactors housed in windowless concrete buildings 120 feet high. Meanwhile another city was springing up at Oak Ridge, Tennessee, where starting in 1943 a uranium-235 separation plant was built four stories high and covering more than 40 acres, the largest single factory built anywhere till then. A construction project of this scale could only be carried through by a large and experienced organization, which was more than the scientists could offer. Accordingly command of the entire fission program was transferred in mid-1942 to the United States Army, which assigned General Leslie Groves as project commander. A career engineering officer, builder of the Pentagon, Groves was a skilled organizer who would let nothing stand in the way of success. When he took over in September, he reviewed the entire program, including the heavy water question. In agreement with the scientists, at the end of 1942 he took the lowest of du Pont's three options. Compton's Chicago group estimated that under this plan a large heavy water reactor could not be built and be producing plutonium until the autumn of 1944, or half a year later than the Hanford graphite reactors, according to their optimistic estimate for Hanford.[13]

The reorganization of the project disturbed many of the scientists in Chicago and some elsewhere. Already in June 1942 they had objected when Compton had merely raised the possibility of bringing in a large industrial firm; now, with du Pont taking more and more responsibility and with the overall command of fission work given over to an Army general who had little sympathy for scientists' feelings, some of the physicists rebelled. Like the British physicists who had objected to the formation of the Directorate of Tube Alloys, the Chicago rebels were convinced that their work could go forward efficiently only if they themselves were in charge, not government bureaucrats, Army officers, or commercial engineers who knew nothing of nuclear physics. "It must be made abundantly clear," Herbert Anderson declared in September, "that all the details are in the hands of the scientific group which bears the responsibility and which alone possesses the knowledge requisite for taking action. Interference . . . can only result in a delay in the progress." In November Wigner, whose habitual politeness overlay strong opinions, maintained, "Our process has been invented by, and is known to, only scientists, and every step to put others in charge is artificial and deleterious." At meetings of the laboratory's Technical Council Compton, Fermi, Szilard, Wigner, and others debated how the work could be better organized. Should there be a new corporation? If so, how should government and industry participate? Might the work go under the auspices of the National Academy of Sciences?[14]

The scientists argued passionately, for they believed that they were in a close race with Germany. They calculated that the Germans, if they had not wasted time as the Americans had, could put heavy water production plants in operation as early as 1943 and might be able to drop a nuclear bomb, or at least a cloud of deadly radioactive dust, on London by the end of that year. The Germans' first target might even be Chicago. Believing that their freedom and even their lives were at stake, the scientists, especially the refugees among them, fought angrily against anything that seemed likely to impede their progress.[15]

Their chief cause of frustration in dealing with the Army was the secrecy and compartmentalization that Groves attempted to impose. Questions from old friends and fellow-workers sometimes had to be turned aside with embarrassed silence, evasion, or bare-faced lies. The scientists felt that they could scarcely function where communications, so essential to normal scientific work, were strangled.[16] Still they respected Groves' bulldozer ability to get things done, and they were resolved to give him his bombs.

Relations with du Pont were equally cooperative in principle and equally upsetting in practice. As one refugee scientist pointed out, large-scale industry requires a rigid organization with well-defined duties, while the pure scientist is "an individualist, who chose his profession . . . because he values his academic freedom more than a high salary." The du Pont engineers felt that the scientists failed to understand how much work and organization were needed for a huge undertaking like the Hanford reactors; the scientists felt that they were being unfairly thrust down by people wholly ignorant of nuclear physics, risking the project and the war, with their laboratories turned into a mere subsidiary branch of the du Pont operation. Through late 1942 and early 1943 relations between the corporation and the dissident group in Chicago worsened until the engineer in charge of du Pont's work on the project complained that the company's men were being "characterized as 'stupid' and 'soft' . . . The du Pont Company has been suspected of seeking to obtain a monopoly of nuclear developments for postwar profit and of 'stalling' the Hanford project as a means to that end."[17]

The scientists who felt this way soon found that they were not in charge, even in Chicago, and that Chicago was no longer the center of the project; there was nothing they could do to impose their ideas of how their work should be controlled. They reacted in different ways. Wigner politely but firmly criticized Groves to his face and at one point in early 1943 offered to resign. Urey, who was still at Columbia but in close touch with the Chicago team, vehemently insisted that the isotope separation plants at Oak Ridge would fail and in late 1943 gave up command of that project. Fermi avoided all talk of politics and in August 1944 left Chicago for Los Alamos, New Mexico, where a bomb laboratory had been founded. And Szilard grew increasingly outspoken until General Groves could scarcely tolerate his presence in the project.

Szilard had another problem: his patents. The Army insisted that all rights to inventions in the chain reaction field should be sold to the United States government. For a number of reasons Szilard balked at this, so Groves ordered Compton to exclude Szilard from work until the matter was settled. For the entire year of 1943 Szilard was not employed, until finally toward the end of the year he signed over his patent rights for a nominal sum. The patents, which Szilard had once hoped would help scientists keep a measure of control over the results of their research, had brought only personal entanglements. Most other scientists who had thought to patent their ideas had similar experiences.[18]

The question of who should control fission work caused so much difficulty because of specific technical problems at issue, some of which involved heavy water. Szilard and other scientists pointed to du Pont's plans for crude distillation facilities as an example of the way the project was being held back. Urey's team at Columbia and Princeton had spent much effort investigating more sophisticated ways to produce heavy water. According to Szilard, Urey's team wanted to discuss their process with some other chemical firm but the Army refused to permit such discussions on grounds of secrecy; this raised further suspicions that du Pont was seeking a monopoly.[19]

Another thing that upset some scientists was General Groves' decision to give heavy water a low priority. Because of the scientists' objections, the plans had to be reviewed several times during the early months of 1943. Urey in particular spoke up for his heavy water frequently and vehemently.[20] He wanted to take another look at the chances of building a quick, simple reactor with a homogeneous slurry of uranium oxide in a few tons of heavy water—a scaled-up version of the rotating sphere experiment carried out by Halban and Kowarski in 1940 at the Cavendish Laboratory. Halban, the principal proponent for the slurry scheme, was invited to a meeting in early 1943 to discuss it, but because at that time Britain and the United States had fallen into a state of mutual suspicion over their fission programs, the British did not send Halban.

In his absence Fermi and Urey met in New York in March 1943. They concluded, as before, that the Cavendish experiment, while it seemed to prove that a slurry system would work, did not rule out the possibility that such a reactor might require many hundreds of tons of heavy water; much more experimental work was needed before they could commit themselves. This caution was typical of the perfectionism which was proving so successful for Fermi.[21] When a few months later he finally got enough heavy water, produced at the Trail facility, to make some measurements, he found that Halban and Kowarski's result was essentially correct: deuterium absorbs virtually no neutrons and a reactor does not need an impossibly large amount of heavy water. But the news came too late to affect the American program.

On December 2, 1942, Fermi's team in Chicago had achieved with a graphite-uranium pile the world's first divergent nuclear reaction. Their success confirmed General Groves and other American leaders in the decision to go with graphite. Some scientists, while generally agreeing with this decision, distrusted du Pont's plans for the Hanford reactors on technical grounds; the flaws that they found in these plans added to their objections to the way all decisions were made. Compton asked for authorization to build a pilot reactor moderated with heavy water, as a step to designing full-sized reactors for insurance in case the Hanford graphite reactors failed. When du Pont declared that it could not undertake any more large tasks, Groves appointed a committee of engineers and scientists in August 1943 to review the controversy, and the group supported the plans for Hanford. But the committee also attempted to soothe the Chicago physicists by letting them build a pilot heavy water reactor.[22]

The decision to give deuterium a lower priority than carbon was a choice made by the United States alone. The French and, following their lead, the British chose to pursue heavy water, if only for lack of pure enough graphite. Physicists in the Soviet Union, before the German invasion of 1941 dispersed them to more urgent work, were also concerned chiefly with heavy water, believing carbon's neutron capture cross-section was likely to be too high.[23] In early 1941 the Germans too abandoned graphite, partly because they had made erroneous measurements of the carbon cross-section. Thereafter their uncertain progress was fed by the thin and vulnerable trickle of heavy water from the captured Norsk Hydro facility. But it was not the choice of heavy water which led the other programs to failure. If the Germans had been convinced that a heavy water reactor could quickly be made to work, if they had seen clearly how this could lead to the plutonium bomb, and finally if they had persuaded their political leaders to take the necessary steps, there is a reasonable chance that they could have built heavy water reactors in time to use plutonium bombs in the war.[24]

Despite its low priority, the Chicago scientists managed to build a heavy water reactor, near Chicago at Argonne, Illinois. But it took much longer than expected to produce the tons of heavy water they needed, for the production program stumbled time after time. The small Trail facility was so slow in getting underway that by the end of 1943 only a quarter-ton in total had appeared, and the planned half-ton per month did not flow until the end of 1944, a year later than originally hoped. Du Pont's larger production facilities, each a series of tall distillation towers, attached to chemical works in three separate national arsenals in the South and Midwest, were also disappointing. They were finished on schedule at the end of 1943, but because they had been designed at great speed and without pilot studies, they were afflicted with serious mechanical defects, power failures, leaks, and fouling. Despite dogged and ingenious efforts, the du Pont engineers could only get their facilities operating at about half the planned output. Even so, du Pont was turning out a ton and a quarter of heavy water per month by the beginning of 1944, and there was soon enough for the Chicago team's small heterogeneous reactor.

The world's first heavy water reactor, Chicago Pile 3 (CP-3), reached critical mass on May 15, 1944, when 6.5 tons were poured in, only three-fifths of the amount that had been thought necessary. Heavy water proved to be an excellent moderator, but it came too late. By 1945, when the Americans had enough heavy water to contemplate using it to make plutonium, they already had the Hanford graphite reactors in full operation.[25]

The influence of heavy water research on the American nuclear bomb project faded into the might-have-been. Fermi's team might have been less successful with their graphite reactors, for graphite is never more than barely workable as a moderator for natural uranium. The Americans would then have been prepared to use heavy water instead. If Groves had chosen du Pont's most expensive (9 tons per month) option as late as December 1942, and if the additional plant had been no worse than the plants actually built, there would have been over 25 tons of heavy water in the United States by September 1944, the month in which the first full-scale graphite pile started up. That much heavy water would have sufficed for reactors which could have produced enough plutonium for a few bombs by August 1945. There are many ways to build a nuclear bomb, and heavy water reactors are not the worst.

The du Pont heavy water production facilities cost over 25 million dollars, yet they were scarcely visible within the American fission project, which cost 1.9 billion dollars and monopolized some of the nation's scarcest resources of engineering, scientific, and managerial talent, materials, labor, and electrical power. In their inefficiency and futility the heavy water facilities may serve as an example of the project's waste. American leaders, convinced that they were in a mortal race with Germany, had good reason to explore a number of lines at once, however expensive; yet a little more money spent at leisure during 1939-1941 on research and development might have saved enormous sums later on, and might even have shortened the war. The value of wide-ranging studies at an early point is worth remembering when new mammoth projects are touted as the solution to this or that national problem.

The Chicago scientists knew that they were not being used to the full. By 1944 other groups of scientists at Los Alamos and elsewhere were taking up the most difficult remaining work on making bombs. Those at Chicago, more than people in other parts of the fission project, had the time and the will to think about what they had done.

Before they ever heard of uranium fission, most of these scientists had shared an idealized picture of science as a high-minded search for truth, transcending national borders, which would automatically impel humanity toward a more just and abundant society. Some, such as Jean Perrin's circle and Szilard, had been concerned that this could happen only under the proper social conditions, but even they had thought of science as something above politics, something which simply required the right sort of society to release its benevolent action. When they began working on nuclear weapons, they felt no contradiction, for they saw nothing but evil for the world until the defeat of Nazism. Now, with Germany clearly about to be defeated but work on the bomb progressing faster than ever, they began to suspect that the conditions under which science would be beneficial for humanity were not so simple. They began to grope toward a more sophisticated politics.[26]

In June 1945 Szilard and others attempted to approach President Harry S. Truman to question whether the bomb should be used against Japan. When the Army expressed annoyance, Compton wrote them a memorandum explaining why his scientists were so presumptuous as to stray outside their technical duties:

The fact is that the scientists who were responsible for initiating and developing this project have felt that its control has been taken from them, that they are uninformed with regard to plans for its use and its development, and that they have had little assurance that serious consideration of its broader implications is being given by those in a position to guide national policy. The scientists will be held responsible, both by the public and by their own consciences, for having faced the world with the existence of the new powers. The fact that the control has been taken out of their hands makes it necessary for them to plead the need for careful consideration and wise action to someone with authority to act. There is no other way in which they can meet their responsibility to society.[27]

Years of abrasive daily contacts with industrial, military, and government people in connection with fission work had spurred the Chicago scientists to break down the traditional barriers separating pure research from social action. Many scientists elsewhere were also starting to burst these barriers, and none more thoroughly than the French.

14

Failures of Secrecy

In the fall of 1942 Halban was in Canada, hoping to create a program which would be controlled only by Britain and Canada while drawing on all the resources of the United States. He would not find it easy to get American heavy water, still less American knowledge. The technical problems of building a reactor, great though they were, would be small compared with the political problems.

First, the Canadian government would have to be won over. Sir John Anderson had opened discussions with the Canadians in August 1942, and they had been receptive. Halban's personality and enthusiasm helped. As one Canadian leader noted after a meeting, Halban "was quite optimistic about the power possibilities and it does not seem as inconceivable as we first thought."[1] Negotiations between the British and Canadian governments ended in an agreement that Canada would administer a substantial reactor team, built upon the Cavendish group, under a joint British-Canadian policy committee, with Halban in charge of the research; Canada would also give a share of the personnel and expenses. The project was well launched.

Halban, learning toward the end of 1942 that the team in Chicago had built the world's first critical reactor, must have been bitterly disappointed to be beaten by Fermi. But Halban knew that his proposed heavy water reactor might turn out to be more efficient, and in the long run more important, than Fermi's graphite pile. With his usual energy Halban set out to build up a team.

There was an immediate difficulty: Kowarski. Around the end of 1942 Halban and Akers offered Kowarski a position in Canada under conditions that would give him noticeably subordinate status and responsibilities. The bond between Kowarski and Halban, severely strained during the past year, now broke forever. Kowarski decided that rather than accept Halban's conditions he would stay in Cambridge. Although Kowarski, no less than Halban, could be brutally direct, the members of the Cavendish team had developed a close working relationship with him and had come to respect him. Unhappy with Halban's actions, they declared that they too would refuse to go to Canada. Senior British officials and scientists stepped into the dispute and finally persuaded most of the rebels that it was their duty to begin work in Canada at once, with reassurances that a suitable position for Kowarski would meanwhile be worked out. In the end he was left in Cambridge with a few helpers, and for the next year and a half pursued slow-neutron researches of minor importance. Halban, who felt that Kowarski had deliberately instigated the Cavendish rebellion and who had also grown tired of Kowarski's emphatic arguments over policy, was glad to be free of him.[2]

The reactor team moved to Montreal at the beginning of 1943. The core of workers who came over from Cambridge were not enough, however, and Halban looked around for senior scientists. Some valuable Canadians joined him, but not enough, nor were British physicists, deep in their own war work, available. Halban had to turn to people of other nationalities, and the most significant of these were French.

Many French scientists had taken refuge in North America. For example, Henri Laugier was there; he had been in London exchanging scientific information when France collapsed and had decided not to return home. Within a few weeks he came to the United States, where eventually he became executive vice-president of France Forever, a large American organization which supported General de Gaulle and his Free French.

Laugier was joined late in 1941 by Francis Perrin, who had managed to persuade the Vichy regime, then officially neutral like the United States, to let him leave for New York. There he took Urey's place teaching chemistry students at Columbia. He also became involved in exile affairs as a member of the executive committee of France Forever—another step in his political education. In 1943, when de Gaulle formed a Consultative Assembly for the Free French in Algiers, the executive committee of France Forever nominated a lawyer associated with the right to represent on the Assembly the French residing in the United States. Perrin, he later recalled, objected: "I said across the table, I don't think the French Resistance will accept him. Nevertheless the committee decided to send his name. And a telegram came back: We won't accept . . . Then they said, Well, you guessed . . . the spirit of the Resistance; you must go." So Francis Perrin left for Algiers, where he had an interview with de Gaulle and circulated among the Resistance leaders smuggled out of France. (Joliot, although among those delegated to Algiers, was unable to make his way there.)[3]

In the meantime Jean Perrin had followed his son to New York, where despite his advanced age he lectured, made radio broadcasts, and wrote articles on behalf of France Forever, as well as helping to organize the

French University in New York. He died in April 1942, having kept up this vigorous activity, so characteristic of his spirit, to the end.

The Perrins and dozens of other French scientists were aided by Louis Rapkine, a fiery young biochemist who before the war had worked on a CNRS salary at the Institute of Physicochemical Biology on the Rue Pierre-Curie. Happening to be in London on an official mission in 1940, he began to bring French scientists out of their fallen country, acting "without any authority other than his own character." He helped some forty scientists make their way to the United States and find positions there. In time the Free French confirmed Rapkine in his self-appointed position as the leader of French scientists in exile.[4]

Halban and Kowarski continued to have a somewhat distant relationship with the Free French. In December 1942 Kowarski accepted a passport from them, and in January 1943 he went on their register of French in the Cambridge area, thereby placing himself in a loose way under de Gaulle's authority while not directly enrolling in the Free French army. At the same time Halban approached the Free French delegation in New York and signed a specially adapted agreement.[5]

The United States government, however, mistrusted de Gaulle and held aloof from all scientists associated with the Free French. To their dismay the exiles in the United States thus found themselves restricted to teaching and to propagandizing for France Forever. Francis Perrin, for example, who was anxious to take up military work relating to nuclear physics, was forbidden entry to the laboratories. The only French scientists allowed to do military research were those who applied for United States citizenship, but the great majority expected to return to their homeland. In 1943 Rapkine struck an agreement with the British government, which had more confidence than the American government had in the French. Small groups of scientists under Free French authority moved back to Britain to study military devices and operations so that they could be of service to the growing Free French armed forces. Not all of them went, for by this time Halban had recruited some of the best for Canada.[6]

One of the first important physicists Halban found in the United States was a familiar member of the L'Arcouest circle, Francis Perrin's brother-in-law, Pierre Auger. In 1939 Auger had begun to work under the CNRS to improve communication of research results, and after France was occupied, he continued this work while resuming his research and teaching at the Ecole Normale. One day when Auger returned to his apartment, he learned from the concierge that the police had just left after trying to find him: evidently they had discovered that he was connected with the *Universite Libre* group. Like many others, he had a small suitcase ready, and he caught the next train into the unoccupied zone of France. With

Rapkine's aid he obtained United States visas for himself and his family and left by way of Spain. Once in the United States, he joined the Free French and found a position at the University of Chicago where, excluded from the military research he would have preferred, he returned to the study of cosmic rays. Halban visited Auger in November 1942 and asked him to come to Montreal as head of an experimental physics division. Auger, who was aware of the progress his colleagues were making toward nuclear energy and was even allowed to visit Fermi's first reactor soon after it began working in the next month, was glad to join Halban's team.[7]

Another French scientist in Chicago was Bertrand Goldschmidt. This was the competent young radiochemist whom Joliot had started to hire just before the German invasion. Goldschmidt's older brother had made his way into the intensely competitive Ecole Polytechnique, but Goldschmidt was content to enter, ten years after Joliot, the School of Industrial Physics and Chemistry. He was fascinated there, he recalled, by organic chemicals, "their colors, their odors, the brilliance and delicacy of their crystals."[8] But in 1933 when Langevin recommended him to Marie Curie as an assistant, Goldschmidt, like Joliot before him, turned from his planned career to the study of radioactivity. After Madame Curie's death Goldschmidt pursued his own researches as Debierne's assistant at the Radium Institute through the late thirties, familiar with the work of Irène Curie, Joliot, and their colleagues, but not joining it.

After the fall of France Goldschmidt lost his job because of the law dismissing people of Jewish background from university posts and spent all his time looking for a way to escape France. He finally got a visa in early 1941 and took a boat which was carrying hundreds of Jews and other "undesirables," jammed together like cattle, to Martinique; he went on in May to Puerto Rico. There he confided to a fellow-refugee, the anthropologist Claude Lévi-Strauss, that he believed a nuclear bomb was a possibility, that the major powers were engaged in a scientific race, and that the winner would be sure of victory in the war. Eager to take part in this struggle, Goldschmidt made his way to New York and applied for his first papers. He tried to get work under Fermi at Columbia, but after some hesitation the overseers of the project decided that there were too many foreigners on the team already and turned him down.[9]

Goldschmidt now enlisted in the Free French. When Halban came through on his visit to the United States in early 1942, Rapkine put him in touch with Goldschmidt, and after negotiations the Free French detached Goldschmidt to work on radiochemistry under the authority of the British Department of Scientific and Industrial Research. In June 1942 Halban got Goldschmidt an invitation to work with Compton's team in Chicago, where Fermi, Szilard, and others had now gone. It had not been possible for Goldschmidt to join them as a Frenchman, but since he was now under British auspices and since the Chicago team were willing to share their knowledge of plutonium with Britain, Goldschmidt was given the job of learning everything he could about the new element's chemistry.

The subject was peculiarly important. As soon as a nuclear reactor started up, the uranium in it would be immersed in a flood of neutrons, and many of these would be captured without causing fission. Uranium plus one neutron would yield a new element, neptunium, which within a few days would transmute to plutonium. By now the scientists knew that plutonium could sustain a chain reaction even more readily than uranium-235. But first the plutonium had to be extracted from the spent uranium by a chemical process. This would not be easy, for only a few grams of plutonium per day would be produced in tons of uranium; it would take a major scientific effort to develop a workable extraction process.

Goldschmidt, besides being one of the few trained radiochemists in the world, was forthright and friendly. Glenn Seaborg, head of the Chicago radiochemistry group, was attracted by his personality and talents. "He was only supposed to be there about a week," Seaborg recalled, "but we found him so congenial and he liked it so much that he worked side by side with our chemists for two months [in fact three]." Goldschmidt first set to work with Isadore Perlman studying the extraction and nature of various long-lived fission fragments. This was precisely the sort of traditional, purely scientific work that Chadwick in 1939 had called "a kind of botany," but now that nuclear energy was turning into a matter of practical interest, these same studies could be looked upon as a compilation of engineering data. The Chicago team could learn how to extract plutonium efficiently from the used uranium reactor fuel only if they knew exactly what else they would run across in the fuel. As of May 1942 there were 64 known products of uranium fission; by a year later 110 isotopes—nearly all those of any importance—had been mapped out, and engineering of the extraction plant could proceed. Goldschmidt and Perlman's work was central to this program.[10]

Next Goldschmidt went to work directly with Seaborg on the central problem of determining the chemistry of plutonium. The Americans had already learned something by studying microscopic traces of the element, produced by bombarding uranium at length with the cyclotrons at Berkeley and at Washington University in St. Louis—the same sort of work that Bothe had tried and failed to do with the Collège de France cyclotron. The properties of plutonium turned out to be different from what scientists might have expected, and Compton warned Goldschmidt that the insolubility of plutonium fluoride was one of the great scientific secrets of the war. Goldschmidt was with the Chicago radiochemists in August 1942 when they isolated the first minute but visible sample of a pure plutonium compound.

At the beginning of November 1942 Halban asked Goldschmidt to join him in Montreal. Full of secret knowledge and the excitement of the race for nuclear energy, Goldschmidt agreed and was made a group leader in the chemistry division. He later remarked that when he had decided to specialize in the recondite and all but useless study of radiochemistry, he had never dreamed of being so much in demand—it was like buying a cheap mine stock and learning it suddenly had great value.[11]

Working with Goldschmidt in Montreal was another French scientist, Jules Guéron. Guéron was a Jew, born in North Africa but imbued with French culture. He specialized in physical chemistry, first at the Sorbonne with a CNRS stipend and then on the faculty of the University of Strasbourg. A short, active, intense young man, he was one of the first scientists chosen from those under forty years of age to serve on CNRS committees dispensing grants. When France collapsed in June 1940, Guéron, hearing de Gaulle appeal over the radio, obeyed the call and made his way to London to join the Free French. In 1941 he was detached to Halban and Kowarski's team in Cambridge.

A problem came up when the Cavendish team moved to Canada: the Free French insisted that they, rather than the British government, would pay all Guéron's salary and expenses. The French representative stressed his interest in the "Tube Alloys" work, saying he was duty-bound to make sure that the postwar French government would be represented in the field and in particular that it would get technical knowledge. Anderson refused to make any formal agreement about postwar arrangements, although he did acknowledge orally that it would be natural for the French scientists' contributions to be taken into account. The Free French prevailed to the extent that Guéron stayed under their authority and the British let him go to Montreal.[12] He played an important part there in helping to direct the chemistry division, becoming a popular leader with an interest in everything.

Halban did not recruit every available French scientist. There was Laugier, for example, who had been unable to get war-related work in the United States and so was studying wounds and blood transfusion in Montreal. Halban did not seek out the man who had once been his administrative superior. Now, as throughout the war, Halban insisted that he and he alone had Joliot's verbal authorization to negotiate about the French patents; to reinforce this, according to Goldschmidt, Halban "slowly tried to eliminate everybody who had dealt with the patents," keeping Laugier, Kowarski, and Francis Perrin at arm's length. On an airplane trip in the fall of 1942 Goldschmidt chanced to meet Laugier, who passed on the information that there was an unfinished but usable wing available in the sprawling Montreal University buildings. On receiving this news, Halban quickly pressed the wing into service for the reactor group's laboratories, while still making no direct contact with Laugier.[13]

Halban also took on a number of scientists who were neither French, British, nor Canadian, such as the nuclear physicist Bruno Pontecorvo. One of the original members of the group that Fermi had founded in Rome, Pontecorvo had gone to study under Joliot at the Collège de France in 1936, then stayed on after Italy enacted anti-Semitic laws. Pontecorvo was the one who first encouraged Halban and Kowarski to work together. After the German invasion Pontecorvo fled by way of Spain and eventually found his way to Montreal. Another refugee was Friedrich Paneth, a venerable Viennese radiochemist whom Halban made overall director of the chemistry division in Montreal. And Placzek came up from the United States to lead the theoretical physics division. Placzek was probably Halban's closest friend in Montreal, and married Halban's first wife, whom Halban had recently divorced before marrying a still wealthier and more charming woman. The senior physicists and chemists were joined by many other scientists, engineers, and technicians, chiefly from Britain and Canada. By early 1943 there were about a hundred people under Halban, roughly half of them professionals.[14]

Early in January 1943, just as the Anglo-Canadian-French group were finally assembled and ready to start work, they were dealt a crippling blow. Relations between Britain and the United States concerning not only reactors but every other aspect of fission work had deteriorated to a nearly total rupture. Underlying all the problems in the complex relations was the simple fact of relative strength. Encouraged by the British to proceed on their own, the Americans with their far greater resources had forged ahead until by 1943 their program was ten times larger than the British one and would clearly be the first to produce fissionable materials in bulk. By the time that the British authorities had resigned themselves to being only junior partners in the alliance, the American authorities no longer felt the need to share information.[15]

And there were two good reasons to withhold information. In the first place, the Americans were reluctant to give their allies secrets that could prove important in postwar industrial competition. They were continually reminded of this problem by the persistent British concern over patent rights and by the presence in the talks of men like Akers, an ICI director only temporarily working under his government's orders. Conant said that controversy might never have arisen if the British negotiators had all been scientists rather than ICI men, and if the scientists had generally had a stronger voice in British policy. In the second place, many

American leaders were afraid that information they gave the British might slip away to Germany or the Soviet Union. Conant reminded Vannevar Bush, a one-time engineer who was now the chief administrator of American wartime science, that knowledge of fission work "is a military secret which is in a totally different class from anything the world has ever seen if the potentialities of this project are realized."[16]

In December 1942 Bush wrote President Roosevelt that "(a) there would be no unduly serious hindrance to the whole project if all further interchange between the United States and Britain in this matter were to cease, and (b) there would be no unfairness to the British in this procedure . . . We are now just reaching the point where the advances are military secrets of the first importance." To avoid completely alienating the British, Bush and Conant proposed to allow interchanges of information with the Canadian team, but only on fundamental scientific questions relating to chain reactions in heavy water, an area where Halban's group might produce useful information. Roosevelt approved this plan. No engineering details, even on heavy water production and reactors, would be given the British and Canadians, nor would Halban's team be allowed further information on plutonium chemistry. Halban's supply of heavy water and uranium would be held to the relatively small amounts necessary for purely scientific research; since the Americans were planning their own heavy water reactor, they saw no reason to furnish the Montreal team with materials to build another. General Groves ruled that neither graphite nor heavy water would be sent to Canada except on the express request of the du Pont Company, which might find the Canadian experiments useful for its own reactor engineering.[17]

When the bitter news reached Montreal on January 29, 1943, Halban's group held a desperate council of war. They agreed that even if London and Washington could not work out an agreement, the Montreal team could and must go ahead on their own. The scientists discussed how to get heavy water and whether factories could be built in Canada to produce pure graphite. Meanwhile Goldschmidt and Auger proposed to pay a private visit to their friends in Chicago to see if they could learn something.

They went down one day in February. Because the two Frenchmen retained their badges as consultants from their earlier contacts with the Chicago laboratory, they simply walked in. They found a warm welcome, for the Chicago scientists, themselves smarting under Groves' directives, were not as hostile as their superiors had been. Goldschmidt described the result:

We left that same evening for Montreal, Auger with the basic constants of the pile and myself with two tubes, one filled with a portion of the fission products

I had isolated, the other with a few drops of a solution containing four micrograms of plutonium derived from the two hundred micrograms I had helped to extract.

The line had been crossed, my scientific career had led me to commit a political act; already the atomic race had begun—between allies.[18]

Goldschmidt's little tubes gave the Montreal radiochemists enough material for a good start in studying the extraction of plutonium. Auger's experimental physicists meanwhile worked on analyzing the behavior of neutrons in materials, particularly in large piles of pure graphite. During 1943 Auger's group was perhaps the most important part of the Montreal organization, for they needed to know much more about basic neutron physics before they could design a reactor. Meanwhile an engineering group attacked a different but equally important sort of problem: how to keep a reactor from melting down.

One must run a reactor for a long time at a high power level, not only if it is used to generate energy but even if it is used strictly to make plutonium. To make a substantial amount of plutonium, there must be a great many neutrons flooding the reactor; this means there must be a great many fissions; and each fission releases a definite amount of energy. Thus there is a more or less fixed ratio between the quantity of plutonium a reactor makes and the amount of power it produces: about one gram of plutonium per day for each one million watts (one megawatt) of heat.

This simple physical fact has consequences which reach farther than anyone can yet see. Some of the implications were clear to the scientists in Chicago as early as 1942, before any reactor existed, as Halban reported during his first trip there: "One should not in the future forget that . . . any industrial plant which is making use of the slow neutron boiler [reactor] for the production of heat and energy does at the same time produce 94 [plutonium] as a by-product. It has been recognized that this fact alone necessitates a state control for nuclear energy production."[19] In 1943 such problems could be left to the distant future. The Montreal group did not see plutonium as an embarrassing but inescapable by-product of an energy-producing reactor; they saw energy as an embarrassing but inescapable by-product of a plutonium-producing reactor. Their main concern was the technical one of designing a cooling system to get rid of the reactor's heat.

Now they had to call on the engineers. Certainly the nuclear reactor was a child of pure science, and many of the calculations and experiments needed to design it were the same sort of work as had been done for years in academic nuclear physics research. But when it came to the cooling system, physicists like Halban and Auger could not get far with their science, for they simply could not derive, from basic physical theory or laboratory experiments, enough reliable knowledge about the vastly complex interactions of a hot reactor and its coolant. This sort of problem was the domain of the engineers, with their different tradition of knowledge. Where the scientist beats a way to a single abstract and permanent truth, the engineer must choose, by past experience or intuition, between a number of expedients all or none of which may work. The reactor, for example, might be cooled by blowing or bubbling air or some other gas through the heavy water, or by running ordinary water through it in pipes, or by circulating the heavy water itself, or perhaps by some more exotic method, such as the use of liquid bismuth metal as a coolant, an idea that Szilard was championing in Chicago. Each possibility had to be thought through in detail, and then somehow one method had to be chosen. Until the reactor was completed, nobody could be sure of the result. In this sense the Canadian reactor would be an experiment, one which would tell the Montreal team less about ideal nature than about the practicality of their own designs.

Cooling was not the only problem which required the special knowledge and approach of engineers. Scientists are satisfied with an apparatus if it can be coaxed to work just long enough to give the data they seek, but engineers need a device that will work the longest time with the least trouble; their overriding concern is to make the best use of resources.[20] For example, the team had to study whether materials like uranium could be kept from corroding under the intense heat and radioactivity within a reactor. They soon began to recognize that a slurry of uranium oxide in heavy water, the system which Halban had been banking on ever since the 1940 Cavendish Laboratory experiment, would be problematic. Attention therefore shifted to heterogeneous systems such as the Americans were building. The physicists and engineers, working together closely, sketched out various designs. A reactor, for example, might be a deep tank of heavy water, penetrated by uranium rods and cooled by ordinary water circulating through pipes.[21]

While the nuclear physicists came from the academic world, the engineers who worked on cooling and corrosion were more closely connected with industry. This affiliation was to be expected, since the engineer's goal—to make the best use of large-scale resources in technology—is close to the heart of modern industrial organizations. Therefore, as the nuclear reactor moved from an idea in basic physics toward a practical device, industrial personnel had to be added to the academics. This relationship was suggested as early as 1939 in the Gentlemen's Agreement between Joliot's team and the Union Minière, which included a clause obligating the company to give technical assistance; it was visible throughout the contacts between Halban and ICI, which led the corporation to study reactor engineering and large-scale heavy water production; and it became almost too clear when the du Pont engineers took over reactor work from the Chicago scientists. It was not so much the industries' money or power that was being called on as their irreplaceable technical expertise. In Canada, reactor construction was similarly contracted out in 1944 to Defense Industries, a corporation whose key personnel had been drawn from Canadian Industries, a joint subsidiary of du Pont and ICI. Guéron recalled that some of the Montreal scientists, like those in Cambridge and Chicago before, resented the growing role of engineers and would have preferred to run the job themselves. Underestimating the problems of building reactors, these scientists suspected that the engineers were brought in more for political than for practical reasons.[22]

Halban's tasks were growing more and more complex. He was in a new sort of job for a scientist, spending some of his time on scientific questions of neutron physics, some on engineering, and a great deal on trying to procure materials, adjust salaries, and fix up administrative arrangements. He wrote the head of the Department of Scientific and Industrial Research that the diversity of the Montreal group "is quite a new experience for many of us. I personally never worked in such a big unit as ours is now, and certainly never had the experience of so many different specialists exchanging their views."[23] Halban also had to move on the rough ground of international relations, getting into disputes over how much power the Canadian government should have over the project and chafing at the United States' embargo of supplies and information.

As time passed, all this struggle began to look like wasted effort, and all the reactor design work like a futile academic exercise. With no significant help coming from the United States, it became clear that the Montreal group could not soon build a reactor entirely on their own. As the scientists recognized their impotence, they became increasingly demoralized. They began, as Goldschmidt recalled, to be less anxious to know about the results of scientific experiments than about the results of negotiations between British and American leaders.[24]

It was also becoming clear that Halban was more suited for directing a normal scientific team than for handling so many novel and thorny administrative problems. His dictatorial approach made him enemies in Montreal. In Ottawa the Canadian authorities too came to dislike him, finding his independence and his insistent demands for the reactor project a burden. And with the United States refusing to cooperate, his scientific talent, enthusiasm, and energy were wasted, like a powerful engine that had nothing to exert itself against. He had managed to win the full support of governmental and industrial leaders in Britain, as Joliot had done earlier in France. But because Halban could not make the same connections in North America, his dreams withered.

In the summer of 1943 the United States and Britain at last began to work out their differences over uranium fission. Meeting in Quebec, Roosevelt and Churchill signed an agreement promising, among other things, cooperation in war work on fission. Information and supplies would be exchanged, although the precise extent of such exchanges was left to be worked out later. To reach this settlement, Churchill made two important concessions. First, he agreed that the American President could determine Britain's share in "any post-war advantages of an industrial or commercial character." Second, he promised not to communicate information about fission work to any third country without American consent. He apparently gave no thought to the difficulties this promise might someday raise with France.[25]

Months were lost while promises of cooperation passed around through various levels of administration and specific plans were negotiated. By early 1944 some American leaders were still hesitating to help the Montreal group, feeling that it was oriented toward postwar development and would not contribute to winning the war. Groves was secretly withholding tons of heavy water from the group. Morale in Montreal was abysmal. Meanwhile cooperation grew in other areas, such as isotope separation and studies of how to make a bomb explode, and a number of British scientists were put to work within American teams. Chadwick, who had come to the United States in late 1943 as head of the British nuclear scientists there, now turned to the problems hanging over the Montreal group. He began to work out with Groves an agreement under which the team in Canada would be allowed to build a reactor with the rapidly accumulating supplies of heavy water.

In April 1944 Halban, feeling that the Americans suspected his ICI connections and German background, and having little taste for the administrative tangles that enmeshed him, agreed to step down as chief of the Montreal group. He would remain as head of the physics division, since Auger was leaving that post to return to France in the wake of the advancing Allied armies as director of higher education. Halban's replacement was Cockcroft, a widely respected and thoroughly British physicist. "I want to thank you," Anderson wrote Halban, "for the way in which you have accepted and in fact urged the plan to send Cockcroft to Montreal. Many people in your position would have fought against this." Replacing Halban as head in Montreal broke the Americans' last reservations, and it began to appear that the British and Canadians would get a chance to build their reactor after all.[26]

In July 1944 General de Gaulle spent a few hours in Ottawa. Auger, Goldschmidt, and Guéron had decided that the leader of the Free French should be told that nuclear weapons were a near certainty. He must be prepared to protect the small uranium deposits in the French colony of Madagascar and to resume nuclear research in France once their country was reconquered. Further, as Guéron recalled, the scientists felt that their information might temper the general's habitual defiance of the United States: "It was very naive, but we were very young at the time." They persuaded the Canadian delegate of the Free French to arrange a secret interview with de Gaulle, although they did not tell the delegate or anyone else what the interview was about. "It was Guéron," Goldschmidt wrote, "the only one of us de Gaulle had previously met, who had the honor of giving our message in a little chamber hidden at the end of a corridor which the General, forewarned, visited for three precious minutes." De Gaulle said he understood the message perfectly.[27]

On August 19, 1944, as German troops streamed through Paris in headlong retreat from the Allied armies, French policemen under Gaullist leadership seized the stone fortress of the Prefecture of Police. It was a signal that the more conservative Resistance groups had yielded to the Communists' demands for an insurrection. Confused struggles broke out across the city. Within a few hours Joliot appeared in Moureu's laboratory in the Prefecture with two valises containing acid and potassium chlorate. With two helpers Joliot went down into the cellars of the Prefecture, poured the Vichy prefect's private stock of champagne on the floor, and refilled the bottles with acid and gasoline. The bottles were wrapped with paper soaked in the potassium chlorate, then rushed to the upper floors. When German tanks assaulted the Prefecture, such homemade explosives knocked out at least one of them. The attack failed. Joliot soon reappeared at the Collège de France, a German revolver on his hip, and over the next few days directed the manufacture of explosives at various sites.[28]

The University National Front was one of the many groups which joined the insurrection. On August 20 they seized the Ministry of Education and installed Henri Wallon as secretary-general of the Ministry. Wallon was a Marxist psychologist who had been Joliot's colleague both as a professor at the Collège de France and as a clandestine leader of the University National Front. While skirmishing continued throughout Paris, Wallon industriously signed decrees which removed collaborators from important positions in the Ministry and replaced them with members of the Resistance. Like other groups, the Communists were using the uprising as an opportunity to get into position for the coming political struggles. Joliot was named head of the CNRS. Wallon held his post for only two weeks and then, with the Allied armies in control of Paris, was replaced by de Gaulle's minister of education. But the new minister confirmed Joliot in the position to which Wallon had named him.[29] For the next year Joliot devoted much of his time to reconstructing the CNRS.

Meanwhile he began to learn where fission research had led. On a visit to London he went to the center for French scientists that Rapkine had established in de Gaulle's headquarters and enjoyed an emotional re-union with many old friends. A few days later Auger and Francis Perrin arrived.[30] With all these contacts, it was not long before Joliot began to get hints of the vast fission program secretly underway in Britain and North America.

General Groves, who kept close track of the French scientists by hidden microphones among other means, began to become disturbed. Of all the scientists with whom the commander of the Manhattan Project had to deal, Joliot was in the position of greatest independence. Groves had distrusted Joliot since the liberation of Paris, when American agents interviewed the scientist and confirmed that he had been host to members of the German uranium project. It made matters worse that Joliot was building up a good picture of the Manhattan Project. "We felt he was evasive about his contacts," Groves recalled, "and we placed little faith in his statements." The general must have become still more alarmed at the end of August when Joliot publicly revealed that he was a member of the Communist party. Ever since Groves had taken command of the American work on nuclear weapons, one of his principle fears had been that Communists would give his secrets away to the Soviet Union, a fear which later turned out to be fully justified, not for French scientists but for British ones.[31]

Under orders from Groves an American intelligence team plunged into the area of southern Germany that had been assigned for conquest to the French Army, for reports showed that German fission research had recently been concentrated there. A few hours ahead of the advancing French troops, the American team reached the German laboratories and removed or destroyed everything of value. The team also tracked down three freightcar loads of Belgian uranium compounds that had been sent into France one jump ahead of the Germans in 1940 and had passed the war unnoticed; in the confusion of the Liberation the Americans confiscated the uranium at gunpoint and shipped it off to the United States. Groves' experiences with Joliot had convinced him, he later wrote, "that nothing that might be of interest to the Russians should be allowed to fall into French hands."[32]

Groves' worst problem was closer to home. The French scientists in Canada, who knew a great deal about reactors, were thinking of returning to France. Auger had already gone to Paris, and in October 1944 Guéron visited France with the permission of authorities in London. At this point the Americans, alarmed, asked for a full statement of the obligations that the British owed the French. Anderson replied with an aide-mémoire which astonished the Americans. For the first time they heard of the 1942 patent agreement signed with Halban and Kowarski and of the fact that the British tended to feel that Halban and Kowarski's help obligated Britain to bring France in as a partner in reactor work.

The extent of the British commitments became clear when Halban, on a visit to London in November 1944, insisted that he be allowed to visit Joliot. He was anxious to bring Joliot into the political discussions and also was interested in arranging to go back to work in France within a year or two. The Americans and some British leaders strongly opposed Halban's plans to visit Paris. But he pressed Anderson, who began to worry that the British would lose Halban's services if they did not treat him courteously. Anderson persuaded the American ambassador, who did not know of Groves' vehement objections, to agree that Halban could visit Joliot and give him some limited information.

Halban immediately went to Paris and gave Joliot a thorough overview of the progress made by the British-American fission project. Among the facts he probably told Joliot were that large-scale uranium isotope separation was feasible, that nuclear reactors worked, and that these could turn out fissionable plutonium—all high state secrets. But the main reason for Halban's visit was to persuade Joliot to accept the 1942 patent agreements; and Joliot sharply repudiated them. He made it clear to Halban, and soon afterward directly to Anderson, that he would refuse to consider any patent arrangements except as part of a general Franco-British collaboration in nuclear energy. Halban's claim that he had represented Joliot and French interests throughout the war was exploded.[33]

Now the whole delicate structure of negotiations among the British, French, and Americans fell to pieces. Groves and others in the United States were furious when they learned what Halban had told Joliot, for they felt that Joliot might immediately pass the information on to the Soviets. Roosevelt refused to let the French into partnership on any aspect of the uranium project; Churchill was determined to abide by the 1943 Quebec agreement in which he had promised the American President, without giving any thought to prior obligations to the French, that Britain would not share information with any third power. Groves noted that if Joliot reacted unfavorably to these decisions, "My recommendation from the security standpoint would be that Kowarski, Guéron, Goldschmidt and Halban be placed in confinement in Canada and not be permitted to communicate with anyone." Joliot angered people further when he visited Anderson in early 1945 and warned that unless France were allowed to cooperate with the United States and Britain, Joliot would advise de Gaulle to think about making an arrangement with the Soviet Union; the Soviets had already been approached and had shown interest. The threat had to be taken seriously, for de Gaulle had regularly played the Russian card against American and British threats to his sovereignty. Churchill bluntly told Anderson that if there were any risk of a deal with the Soviets, Joliot should be seized and detained for a few months.[34]

Eventually tempers cooled and an arrangement was made to handle the French scientists in Canada and their dangerous knowledge. The British and Americans would keep Goldschmidt, Guéron, and Kowarski at work in North America at least until the end of 1945, while allowing Goldschmidt a carefully controlled visit to France. Halban, whom the Americans now mistrusted completely, was removed from the Montreal project. He was still under contract to the British by his 1942 agreement, so he could be ordered to stay in North America, cut off from his former colleagues, until the end of 1945. His meteoric career in nuclear energy was finished. When Goldschmidt visited Paris in April 1945, he told Joliot that all the French scientists—Auger, Guéron, Kowarski, Perrin, and Goldschmidt himself—would refuse in future to work with Joliot if Halban were asked to join them.[35]

This period may have been the nadir of Halban's life. Yet he was not quite the tragic hero crushed by his own ambitions. He was a tough man, able to learn from experience, and he still had friends. He took up some basic research on neutrons paid out of his own pocket and done in his own home, first in Montreal and then in New York City. It was a stressful period for him with his new wife expecting a baby and his older daughter to care for, and he wished to go to Britain. The British knew that for all their difficulties with Halban they owed him a great debt: the reactor team at Montreal, created largely through his personal efforts, was giving both Britain and Canada a secure foothold in nuclear energy. Akers made inquiries and found that Halban would be welcome in the Oxford physics laboratories of F.A. Lindemann (Lord Cherwell), Churchill's personal scientific adviser. In 1946 Halban accepted a professorship. He soon proved that under proper circumstances he was a fine research leader, for he built up a team which did notable work in pure nuclear physics. To France his most important legacy was the presence of French scientists on the Montreal team. They had learned a great deal about nuclear reactors, and they could not be kept out of France forever.

After Halban was replaced as team leader by Cockcroft, it was possible for Kowarski and the scientists who had languished with him in Cambridge to come to Canada. When Kowarski arrived in August 1944, Cockcroft told him that the team would begin building a small heavy water reactor at once. Kowarski would lead the project. "I remained slightly gasping," Kowarski recalled. "It was a considerable change from my previous position. Cockcroft said, somewhat alarmed, 'Well, don't you like the proposal?' I said, 'Well, I think I can do it.' "[36]

The United States had at last agreed to supply the uranium metal, heavy water, and information for an experimental Canadian reactor. This device, called the National Research Experimental (NRX), was designed to produce tens of megawatts of thermal power. Meanwhile the Montreal team wanted to build a small reactor operating at zero power, which could be built much faster since it required neither a cooling system nor heavy radiation shields, and which would serve as a pilot for the larger project. This first reactor was Kowarski's responsibility.

He paid a visit to the Argonne laboratory near Chicago where the American heavy water pile was already operating and learned a great deal from its builder, Walter Zinn, and other American scientists and engineers. He then returned to Canada and directed the design of the pilot reactor, which he dubbed Zero Energy Experimental Pile (ZEEP). Kowarski was now the head of a compact team working at a task on the borderline between a physics experiment and an engineering project. "Of course," he recalled, "All pretenses to my past as a neutron physicist were by then totally irrelevant because the amount of information on nuclear physics that began to flow from America was so enormous that whatever my past contributions and whatever I could contribute any more didn't count."[37] Like Halban and many other scientists in North America, Kowarski had left behind the smaller, more intimate research teams which had characterized physics in the thirties. His scientific knowledge, ingenuity, and instincts were still called upon when he made a decision, but he also had to judge resources and people and to manipulate them on a broad scale.

Besides designing a reactor, the Montreal scientists were laying the groundwork for a much larger program. A Graphite Group continued studies of the alternative moderator. A Future Systems Group roamed farther afield, considering a great many types of possible reactors, some of which eventually were built. Equally important was the work of Goldschmidt, Guéron, and their teams in the chemistry division. The Americans still refused to provide any information on ways to extract plutonium from used reactor fuel. Unless the Montreal group could work out their own processes independently, their reactors would be able to produce only power and not plutonium. The Americans did agree in mid-1944, however, to furnish some uranium rods that had been irradiated with neutrons, and with this material Goldschmidt began to attack the problem.

The young radiochemist, still surprised by his rapid ascent, was maturing into a research manager, although he also took a personal interest in the scientific details of the work, the manipulation of chemicals that always fascinated him. He could have blindly mimicked the Chicago chemists' methods for extracting plutonium. But unlike the Americans, Goldschmidt did not have to rush at once into production, so he took time to seek a better method. His problem resembled the problems familiar to radiochemists around the time fission was discovered: given a broth in which uranium, various fission fragments, and plutonium were all dissolved, he had to find an efficient way to get out one substance only, the plutonium. The team ordered over two hundred organic solvents from the United States and systematically studied their reactions with the various chemicals in the broth. By mid-1945 they found a solvent which absorbed plutonium while leaving behind all the rest, and they began to sketch designs for a chemical plant.[38]

Meanwhile their colleagues in the United States completed their own chemical plants and other installations. Scientists at Los Alamos were working furiously on the problem that in 1939 had seemed to make a great release of fission energy impossible, the problem of keeping an exploding mass together long enough for the chain reaction to run its course. Reactors had solved this problem by dispersing uranium through a mass of moderator, so that the fissions progressed continuously and calmly; now the Los Alamos workers went the opposite way, devising ingenious methods to force uranium or plutonium into a dense, symmetrical sphere the size of a grapefruit, a sphere which, if conditions were precisely right, would release its energy all at once. At the end of June 1945 Chadwick told Kowarski that there "unfortunately was no longer any hope that the bomb would not work."[39] The Hanford reactors were beginning to produce kilograms of plutonium every week, and the Oak Ridge uranium-235 separation plants would also suffice. In August a bomb made of uranium-235 from Oak Ridge was dropped on Hiroshima and a bomb made of plutonium from Hanford was dropped on Nagasaki.

The sound struck her like a blow. She crouched together against the masonry and looked up . . . There was nothing else in the world but a crimson-purple glare and sound, deafening, all-embracing, continuing sound . . . She had an impression of a great ball of crimson-purple fire like a maddened living thing . . .

[A soldier who witnessed the bombing wrote,] "I do not think any of us felt we belonged to a defeated army, nor had we any strong sense of the war as the dominating fact about us. Our mental setting had far more of the effect of a huge natural catastrophe. The atomic bomb had dwarfed the international issues to complete insignificance. When our minds wandered from the preoccupations of our immediate needs, we speculated about the possibility of stopping the use of these frightful explosives before the world was utterly destroyed. For it seemed quite plain that these bombs and the still greater powers of destruction of which they were the precursors might quite easily shatter every relationship and institution of mankind."

These passages from H. G. Wells' *The World Set Free* show that in the calm of 1913 the crisis of civilization had already been clear to an intelligent person who thought about science and history. "The power to inflict a blow," Wells had perceived, "the power to destroy, was continually increasing. There was no increase whatever in the ability to escape." The time must come when "a man could carry about in a handbag an amount of latent energy sufficient to wreck half a city."[40] It had taken just thirty-two years to transform Wells' fantasies into fact.

Years later Guéron recalled the deep depression, mixed with elation, that fell on the Canadian team when the bombs were used and the war ended. But they quickly returned to work, for they could envisage their own NRX reactor—a huge cube festooned with wires and instruments and serviced by a busy crew, pouring forth new isotopes and energy, much like the nuclear installations that the team at the Collège de France had envisaged in 1939. If Wells' nightmare could be realized, then so perhaps could his dream of humanity after the atomic wars. Commanding nuclear-powered cars and airplanes, humanity might spread to every region of the earth, regrouping according to mutual interests and shared tasks. "Our cities now are true social gatherings," wrote Wells in his history of the future. "They tower amidst eternal snows, they hide in remote islands and bask on broad lagoons."[41]

Something approaching this was already appearing at Chalk River, an isolated and lovely river shore in the midst of the Canadian pine forest, where industrial contractors were building an entirely new town to support the two Anglo-French-Canadian reactors. ZEEP, Kowarski's pilot reactor, was designed along the same lines as the larger NRX, with rods containing uranium suspended in a tank of heavy water, except that the smaller reactor did not need a cooling system. A month after the war ended, ZEEP went into operation.[42]

This was the first nuclear reactor to operate outside the United States. The second was the graphite-moderated F-1 reactor built by Kurchatov's team near Moscow, which reached critical mass on Christmas Day, 1946. Although that success was chiefly due to the efforts of Soviet scientists and engineers, it may well have been hastened by information which came from North America. Among other sources, the Soviets were told a great deal by Alan Nunn May, a British physicist who had held a leading position under Halban in Cambridge and then in the Montreal laboratory.[43] The third reactor outside the United States was the Canadian NRX, which started up in 1947; the fourth was Gleep, a graphite-moderated, low-energy experimental pile in Great Britain, descending from ZEEP not only in name but in design, except for its use of graphite as a moderator.

The next country to complete a reactor was France. This was no easy task for a nation deeply wounded by invasions, cut off by walls of secrecy from much of the knowledge of fission held abroad, and far poorer in resources than the United States, the Soviet Union, or the British Commonwealth. But the efforts of a generation of scientists had given France strong scientific organization and expertise. The French determined not only to have a reactor but to found a broad nuclear program. Soon France would become the first case of a nation that, without the resources of a superpower, sets to building a mighty arsenal of nuclear weapons. For the French scientists, the harshest struggles—technical, political, and moral—still lay ahead.

The
Fort de Châtillon

Joliot speaking in public, c. 1950

15

Organization for Nuclear Energy

Scientists returning to Paris after the Liberation found Joliot changed—thinner, middle-aged, the mark of the war upon him. He was more intensely active than ever, above all in political work. Guéron later recalled how Joliot, released from the repressions of a life underground, could scarcely stop talking. Deeply involved in Communist affairs, he "expatiated on the fraternity and the closeness of relations in the Party, like a man who at last had found a family."[1]

Paris too was changed, with wounds that would fester for years: a scandalous black market, a ruined currency, a political life more embittered than ever. The struggles between the Resistance and Vichy, and within the Resistance, were not over. When the public was not devoting itself to scraping up some potatoes on the black market, it was avidly reading about the arrest and trial of tens of thousands of collaborators. Conservatives accused the Communists of using the Liberation to elbow into positions of power and to shoot their opponents, while the Communists pointed to the bureaucrats and businessmen who had profited from the German occupation but still retained their comforts and influence.[2] France seemed an unlikely place for a reorganized and renascent science, yet this was what Joliot set himself to create.

Since August 1944 Joliot had been director of the CNRS. With the war still in progress, Germany far from beaten, and French science lying in pieces, Joliot set out not only to restore research but to transform it. Partly as a result of his work French science began, as one historian has noted, "its greatest expansion and reform of institutions since the Revolution and Napoleon."[3]

This advance in the direction opened up by Jean Perrin and his group in the 1930s had even been aided by the Vichy regime. The Vichy ministers were no friends of the leftist scientists and their creations. But instead of suppressing the CNRS, they reorganized it by decree in 1941, abolishing the directing councils and merging into one monolithic structure the applied and pure research sections, which despite Laugier's efforts had until then remained separate, each with its own director and council, linked only by a distant consultative committee. The regime probably intended only to subordinate science to government and industry, but when Joliot took over he found that the changes suited him.[4]

Joliot's plans were influenced by contacts between scientists and Communists. The two groups overlapped, for a number of scientists, including Langevin for example, had become party members before or during the war. Joliot also drew on personal impressions gathered in the 1930s and again in June 1945 on trips to the Soviet Union, where he found the scientists to be numerous, generously supported, and highly honored. He believed that science in France should be directed, as in the Soviet Union, by a prestigious and competent central body; this would ensure that important fields would not be neglected and that useless duplication of research would be eliminated. Whereas the prewar CNRS had usually supported whatever respectable scientists and scientific work came along, Joliot's new CNRS would make plans and assign priorities for research. In particular, he intended to pressure scientists to maintain a flow of ideas from pure research into industrial applications. In return he hoped to expand science by mobilizing greater resources and by recruiting from every social class. All this, declared Joliot, was necessary if France were to restore herself to her former position in the world. "I will speak plainly," he said; "if the country does not make the effort required to give science the place it deserves, and to give the servants of science the prestige they require to have an influence, she will sooner or later become a colony."[5]

The defeat, occupation, and reconquest of France had temporarily broken up the power of the bureaucratic and directing classes, leaving the nation in a dazed condition where prostration mingled with enormous hopes. For the first time since the Revolution young people in their forties, like Joliot, predominated, and no change seemed inconceivable. Joliot was well placed to take advantage of the disarray. To the value of his connections in the Resistance he added the force of personal prestige, for his fame as a Nobel Prize-winning scientist and a Curie were now redoubled by the wartime deeds that won him the Croix de Guerre. It was reported that when he gave a speech in a working-class suburb the crowd suddenly rose and sang the "Marseillaise." When the CNRS budget for 1945 came under discussion, Joliot went to visit de Gaulle's finance minister, with whom he was on close terms. At the minister's side sat his director of the budget, one of those parsimonious bureaucrats who normally ran the French government while politicians came and went. After a few sentences explaining the importance of the CNRS for France, Joliot asked for a 200 percent increase in its budget. "Allocation granted," replied the minister, while the bureaucrat sat in stupefaction.[6]

Joliot quickly filled key posts in the CNRS with scientists who had worked beside him in the Resistance and who shared his vision of a socialized research organization. It was hard to find French scientists in 1944, for many were still living in exile abroad, serving in the military, being held as prisoners by the Germans, or simply sitting in the provinces with almost no way of communicating with Paris. Those whom Joliot was able to find, however, functioned surprisingly well. Throughout 1945 they worked feverishly, reorganizing the CNRS without bothering to wait for government authorization, which did not come until November. The activity of these veterans of the underground was thus virtually illegal, but they were accustomed to such situations.[7]

Joliot personally singled out a few hundred leading scientists, particularly younger ones and those whose research was not in the distant past, and organized them into a set of interlocking committees. As in the prewar CNRS and CNRSA, whose committees the antidemocratic Vichy regime had dismantled, these new committees were the heart of the organization and included representatives of academic, industrial, and governmental research. Within its field each committee had broad authority to spend funds, recruit new workers, and make arrangements with industry. Between the fall of 1944 and the fall of 1945 the number of people supported by the CNRS nearly doubled, to over one thousand.[8]

Not all of the reforms took hold, for the French system still suffered from inertia. As the CNRS expanded, it developed its own hierarchy and bureaucracy, which brought the usual disadvantages; parallel efforts by Langevin and others to reform the educational system also had only partial success. Nevertheless, the visionary program of scientific organization forwarded in the thirties was now embodied in a permanent, valuable institution, and France began to rise back to the first rank of world science.[9]

Meanwhile, in 1944 and into 1945 the war continued. The CNRS organized groups to do military research, while other groups were sent into Germany behind the advancing Allied armies, bringing back as war booty 250 tonnes of scientific equipment and much technical information.[10] Some of the confiscated instruments were earmarked for nuclear fission research.

In 1944 Joliot had been astounded when his colleagues told him about the gigantic scale of the nuclear enterprise in North America—factories with a hundred thousand workers, reactors producing hundreds of megawatts, nuclear bombs nearing completion. It was plain that France could not mount a similar effort, particularly while the war dragged on, but Joliot grew anxious to start up some kind of fission work. According to information reaching General Groves, in November 1944 Joliot discussed nuclear energy with de Gaulle.[11] Joliot also talked the problem over with his wife and with Auger and Perrin. They soon got in touch with Dautry.

The former minister of armament, having refused to serve Vichy, had retired to live quietly in a remote village. In 1944, on the night of the Normandy invasion, he came to Paris at de Gaulle's request to join the Resistance. After Paris was liberated he became minister of reconstruction and urbanism in de Gaulle's provisional government. Around the beginning of 1945 he learned from Joliot about what had happened in nuclear energy since 1940, and during the next weeks he kept in touch with the scientist. In March 1945 Dautry sent de Gaulle two notes advising him that an organization for nuclear research should be set up under Joliot, while arrangements should be worked out for cooperation with the British. But nothing happened. In May, Auger and Joliot approached de Gaulle directly and told him that some sort of French organization was needed in the field of nuclear energy. Again there was no result.[12]

Although these discussions produced no immediate action, they put the advocates of nuclear energy in a position to act after nuclear bombs obliterated Hiroshima and Nagasaki. Now de Gaulle took up the matter in earnest, and in late September 1945 asked Joliot to get together with Dautry and draw up a plan for a nuclear energy organization. Joliot and his scientist friends quickly went through five successive drafts of a proposal, the last of which they submitted on October 12. A commission was named to draw up an ordinance. On the scientific side the negotiations involved Joliot, Irène Curie, Auger, and Francis Perrin, and on the administrative side Dautry, a government councilor (*Conseiller d'Etat*) named Jean Toutée, and Allier, who had been in on the discussions from the start because of his success in obtaining the heavy water in 1940. Working at a speed seldom seen in government circles, they drafted the ordinance and basic law for a Commissariat à l'Energie Atomique (CEA). De Gaulle promulgated the ordinance on October 18, 1945.[13]

The preamble to the ordinance said only that the measure was demanded by "pressing necessities of a national and international order." It would be easy to suppose that de Gaulle wanted the new organization simply because Hiroshima and Nagasaki had proved that the new weapon would dominate world politics. Certainly this thought hung over everyone's head, and in later years de Gaulle said that he had created the CEA above all so that France could have her own bombs. "He was very interested in nuclear weapons," Perrin recalled. "He did not insist that it should be done at once; he saw that it was in a rather distant future."[14] The future of nuclear weapons was far from clear during the first year or so after the war, if only because it seemed possible that they would be outlawed by international agreement. In any case, for the time being such weapons were beyond the means of crippled France.

Almost as important a motive as weaponry for French nuclear fission work was Joliot's original motive, the potential of a new source of industrial energy. France had long been aware that among the major powers she was one of the poorest in native sources of energy, with limited coal reserves and less oil. Coal mines had been vital in the First World War, and some had been worked within gunshot of the Front. Herriot repeated a common theme when he wrote in 1919 that the development of a nation could be measured by the quantity of coal it consumed, adding the warning that France's reserves might be exhausted in less than a century. In the twenties and thirties Socialist trade unions attacked the French private electricity industry for relying on imported coal and oil. During the Second World War energy stayed in the fore, and even those who failed to see the role of oilfields in the grand strategy of the war, or the increasing place energy budgeting took in the allocation of resources for military production, could not ignore the everyday effects of fuel shortages at home. Even before the war Joliot's laboratory had been studying the extraction of fuels from forest trash, and during the Occupation wood alcohol was pressed into use for automobiles, while most people, like Joliot, went about on bicycle or on foot. In the early postwar years Europe remained desperately short of fuel, and the first winter after the war was as bad as the war itself. French factories ran on half-schedule; homes went unheated for lack of coal; Paris had electricity for only a part of each day.[15] It seemed that power from uranium could not come soon enough.

The war had made the importance of science for national strength plainer than ever. If nuclear weapons had terminated the war, radar had done as much as anything to win it, while synthetic rubber, penicillin, and other scientific developments had strengthened the Allies more than a dozen army corps. Now that peace had come, these developments could be turned to beneficial uses, and the public could hope for even more striking advances. Joliot, who was speaking in public more and more frequently, did not discourage such hopes. In early 1945, for example, at a speech in Orléans he stressed not only the Communist policies of the day, such as cooperation among parties and nationalization of industry, but also "the importance of science and technology for the war effort and for the renaissance of the Nation." The audience must have been particularly impressed when the electrical power for the entire city failed and Joliot spoke for over an hour in darkness.[16] Few would have disputed his belief that science and technology had essential work to do.

Such feelings not only created the CEA but gave scientists a strong place in determining its direction. At the top of the CEA was an Atomic Energy Commission, the majority of whose members were required by law to be qualified by their work in nuclear energy; in fact four of the six original members were nuclear physicists of the L'Arcouest circle. Placed in command of all scientific and technical work was a high commissioner, Joliot.

However, some of the government people involved in creating the CEA were reluctant to give scientists total control of so great an enterprise, if only because they were afraid that physicists would not necessarily make good executives. Therefore administrative work was given over to a division that was separated from scientific work, which had its own parallel division. Allier recalled that Dautry was largely responsible for this bicephalous structure, maintaining during the preliminary meetings that it would secure an efficient collaboration between scientists and administrators.[17] Convinced of the importance of nuclear energy, Dautry now left the reconstruction ministry and took over direction of the administrative division under the title of administrator-general and delegate of the government. In this position Dautry, who had no sympathy for Communism, could also serve as a political counterweight. In principle he was on about the same level as Joliot, and it was not clear who should prevail if they ever opposed one another.

Dautry thought that Joliot's high position in the program was necessary, since strong scientific leadership was indispensable, but he hoped to keep his Communist colleague confined to technical work. However, Dautry had already lost the first skirmish. When the ordinance establishing the CEA had been drafted, he had tried to insert additional delegates of the government to counterbalance the scientific commissioners, and had also tried to get the duties of the administrator-general defined before the duties of the high commissioner, giving his position symbolic precedence. But these changes did not make their way into the final ordinance.[18]

The potential for conflict was not much ameliorated by the appointment of a secretary-general who was supposed to serve as intermediary between the scientific and administrative divisions. By mutual agreement this sensitive post went to Léon Denivelle, a former collaborator of Dautry at the Ministry of Armament and a friend of Joliot since their cooperation on the abortive SEDARS company during the Occupation. In the first years of the CEA there did not seem to be any serious difficulty with all these arrangements, but beginning around 1949 the problems raised by the CEA's bicephalous structure had to be thrashed out in bitter struggles.

The decree establishing the CEA in effect sealed a bargain whose terms were expressed in the structure of the organization. The state, represented by the administrative division, agreed to endow the scientists with organization, authority, and money; the scientists, in their own division, promised to build devices that would meet France's pressing needs. They would also use some of the funds to support basic research. These arrangements were similar to the scientists' 1939 bargains with the Union Minière and the CNRS and their 1942 bargains with ICI and the British Department of Scientific and Industrial Research. However, this time they were making arrangements without relying on any private industrial corporation. As a Communist, Joliot could hardly have done otherwise, and since nationalization of industry was a policy of the new French government, it was natural that Joliot should get authority to build up a new industrial enterprise entirely under state control. But in comparison with the earlier state-science-industry tripods, the CEA was a precarious structure with one leg missing.

The newborn CEA was the first primarily civilian nuclear agency in the world. Following the wishes of both Joliot and Dautry, the agency enjoyed a degree of autonomy unique among French government oranizations. It was modeled to some extent on the recently nationalized Renault company, whose ordinance had also been drafted by Toutée, but unlike such a state corporation, the CEA did not come under any ministry. Neither did it have to submit its accounts in advance, as other government agencies did, to the restrictive inspection and control of the Finance Ministry. The ruling body of the CEA, the Atomic Energy Commission, had statutory relations with other bodies and included a representative from the military research establishment, but it reported only to the prime minister. During the first years of the CEA France would have a bewildering succession of prime ministers, few of whom took an interest in nuclear fission. The Atomic Energy Commission was left on its own.[19]

The Commission in turn was dominated by Joliot, who as high commissioner was formally charged with the planning and execution of all the CEA's scientific and technical programs. Joliot's position, enhanced by his personal prestige, had been raised still higher by Hiroshima, which made the public feel that nuclear physicists had touched on superhuman powers. Joliot not only took command of the CEA but retained a strong influence over the CNRS, whose direction he turned over to the biologist Georges Teissier, a loyal friend, Resistance comrade, and fellow-Communist who shared Joliot's views on how science should be organized.

The members of the Atomic Energy Commission were designated on January 3, 1946. Besides Joliot and Dautry they were Auger, Curie, Perrin, and General Paul Dassault, a relatively liberal representative of military research and former member of the National Front. The first meetings of the Commission, Auger recalled, were marked by "team spirit and mutual confidence . . . [they] resembled meetings of the directing group of a club planning the ascent of Everest or a trip around the world in a sailboat. Proposals which were very audacious for that era could be made and discussed without having their wings clipped by protocol or the budget."[20] But except for Auger, whose stay in Montreal had been relatively brief, none of the leaders of the CEA had any substantial knowledge of recent nuclear energy work. Even Joliot knew little more about reactors than he had known in 1940. The Commission could hardly design a detailed program until Goldschmidt, Guéron, and Kowarski returned from Canada. Some months earlier Joliot and Auger had sent them word through Rapkine that they were wanted back in France, and all three were willing to come.

Their return posed a problem, for Groves remained unwilling to let any information on nuclear energy escape to a foreign country. The Soviet Union was the general's chief concern, and he feared that whatever the French knew might leak in that direction. "You've heard about the Communist Party in France?" he asked a reporter. "I read the Communist Party in France is disconnected with the Russians but I notice they go to Moscow all the time."[21] He was less afraid that the French would mount their own program. For some time he had been gathering information which suggested to him that few if any countries would be able to compete with the United States. For example, in a May 1945 telephone conversation with a du Pont engineer Groves asked how long it would take various countries to duplicate the Hanford reactors. It would take the Russians a very long time, the engineer replied, unless they threw all their resources into it. "Now how about our pals the British?" Groves asked. "They could build it in time," he was told, "but they will not hurry . . . it would take them a couple of years." "How about the Frogs?" "If they had the manpower," the engineer thought. "They are terrifically weak on organization as you can never get two Frenchmen to agree. Technically they have very fine minds, individually," but it would be "damn near eternity" before they got big reactors. Groves and many other American leaders shared these prejudices. Joliot recalled that when he visited the United States in 1946, the financier and statesman Bernard Baruch told him that it was "madness to try to do atomic energy in France. A pile—two piles—you'll never get them in the state your country is in."[22]

In any case the French scientists in Canada had only limited knowledge of Groves's secrets and knew almost nothing of the bombs themselves. And he could not forcibly detain them. So Groves reached an identical accommodation with each of them, an all but unspoken gentlemen's agreement. The French scientists would neither publish their knowledge nor let it out of restricted circles in France; they would use it gradually, one piece at a time, as they needed it for their work.[23] They would need it sooner than Groves imagined.

The British still felt that they owed the French a debt, but they could not afford to antagonize the United States. Sir John Anderson officially informed the departing French scientists by letter that they must keep their nuclear secrets strictly to themselves, and they were warned to take no documents. Before leaving Canada, Kowarski asked Cockcroft how far this last rule extended. What exactly was a document? Were notes written from memory included? The Englishman, Kowarski recalled, "drew out of his side pocket one of his black books which were famous all over England and Canada. He had a very small and meticulous writing, and these black books, which were roughly of the size of a paperback novel, would contain amounts of information about which people talked with admiring incredulity. 'This could not be considered as a written document.' I thanked him for this information, and I didn't ask him any more questions."[24] When Kowarski returned to Paris, he brought along, in addition to his unusually powerful memory, a little packet of notes written in his own tiny, precise hand. Goldschmidt and particularly Guéron also brought back written notes in addition to a great deal stored in their heads.

Joliot and Curie were off on a skiing vacation when the "Canadians," as they began to be called, arrived one by one around the end of January 1946. On his return Joliot embraced them in an emotional reunion. The "Canadians" were immediately hired. Kowarski, by virtue of his success in building a reactor and his earlier collaboration with Joliot, assumed the broadest responsibilities as director of scientific services, while Goldschmidt became head of extractive chemistry and Guéron became head of general chemistry. These impressive titles stood for intentions more than realities: the personnel of the CEA then numbered perhaps a dozen people. Years would pass before there was any formal organization chart or complete separation of responsibilities.[25]

While retaining firm control over the program, Joliot left the day-to-day direction in the hands of his colleagues, for he had many other concerns. At first he dreamed of getting back to research. "At present I have a heavy task at the Commissariat," he wrote G. P. Thomson in March 1946, "but nevertheless I have the joy of passing part of my time in my laboratory." He wrote similarly to others throughout the year. But there was much to do before he could seriously begin research. The cyclotron laboratory at the Collège de France had been left in full working order by the German guest scientists, but wayward bombs had struck the Laboratory of Nuclear Synthesis at Ivry in 1942 and again in 1944. For years everything had been exposed to the rain, so all the apparatus had to be taken apart and cleaned, while broken or deteriorated parts were replaced. Not only the apparatus but the staff had to be rebuilt, for the younger scientists, even if they had been able to do research at all during the war, had learned none of the new techniques developed abroad.[26]

As a professor at the Collège de France, Joliot was also obliged to give

an annual series of public lectures, and his course on nuclear chemistry
was the College's great attraction of the 1946 season. Hundreds of Pari-
sians, old and young, turned out each week to watch the chief of the
"sorcerers who wield atomic energy and domesticate neutrons," as a journalist put it, scribble equations across a huge blackboard. And this was
only the beginning of his responsibilities. A flood of mail descended upon
him. Every month brought requests for advice, materials, information,
recommendations, and jobs; requests to write book prefaces, give lectures, join committees, accept honors, and give interviews. Joliot yielded
to many of these pleas. He took the time to make inquiries, on behalf of
one of his old teachers at the School of Industrial Physics and Chemistry,
into the fate of a young Jewish deportee, and he intervened with a ministry on behalf of a fishing village near L'Arcouest which was not receiving its ration of butter and lard. He answered crank letters seeking
advice on strange theories, and pathetic letters seeking help treating
cancer. Nothing shows more clearly Joliot's renown, and his conscientious response, than his daily mail.[27]

But perhaps Joliot's first concern, which would come to possess him
more and more, was the part he might play in promoting world peace
and in doing away with the nuclear weapons he had helped create. Scientists of the previous generation had rarely stepped into the arena of
international politics. But Joliot saw that his prominence, and the weapons that science had created, gave him new possibilities and responsibilities for action. He not only served as scientific adviser to his government
in international negotiations over the control of nuclear weapons but
also went forth on his own to address the world public.

In July 1946 Joliot helped found the World Federation of Scientific
Workers (WFSW), which carried on the work begun in the 1930s by
leftist scientists like Bernal and Langevin. Joliot became the first president
of the WFSW, a post he held for over a decade. The WFSW was quickly
identified as a left-wing organization and could not unify scientists as
Joliot would have wished, but it did serve as a forum for work to prevent
nuclear warfare and promote more beneficial uses of science. Those who
helped create the WFSW, Joliot remarked, strongly expressed a feeling
that they could no longer remain silent: "Many scientists rightly think
that the abuses of science can be prevented: they do not wish to be the
accomplices of those whom a bad organization of society permits to
exploit the results of their labors for selfish and harmful ends. It is undeniable that a fit of conscience has seized the scientific world and that
the scientists' sense of social responsibility is growing and becoming more
defined every day."[28] It was indeed a movement that crossed national
and ideological frontiers. On the day the WFSW first met, American

scientists like Szilard, in newborn organizations like the Atomic Scientists of Chicago, were in Washington fighting passionately and successfully to remove the control of nuclear energy work from the United States Army and to put it into civilian hands.

Joliot's leadership of the WFSW, an organization that hoped to bridge East and West, was in accord not only with the general sentiment among scientists but also with the policies of the French government. De Gaulle had always balanced the Soviets against the Anglo-Americans to help maintain French independence. Although he was now gone, having resigned abruptly in Janurary 1946, his successors also worked to keep France from getting frozen within either of the two spheres of influence that were beginning to solidify. A "tripartite" government was formed by the three strongest parties—the Communists, the Socialists, and the Mouvement Républicain Populaire (MRP).

The Communist and Socialist parties were headed by men from the prewar era, Maurice Thorez and Léon Blum, whom Joliot knew personally and agreed with on many points. The MRP was a new Christian-Democratic party descended from Resistance groups. Like the Radicals in the Third Republic, the MRP in the early Fourth Republic held a commanding position in the center but was immobilized there, its leaders intelligent, well-meaning, and in the end ineffectual. The MRP's leader and France's prime minister for the second half of 1946, Georges Bidault, was a robust, dynamic liberal and sometime history professor who said he preferred making history to teaching it.[29] In 1943 he had been elected president of the National Resistance Council, the body which had loosely coordinated all French Resistance groups. He had met Joliot during the struggle, and the two had become friends.

Joliot and Bidault, both ardent nationalists, shared the hope that France could keep her independence between the rival blocs of East and West. For Bidault and his political companions this meant day-to-day efforts, on the one hand to block the Communists from gaining too great an influence in the coalition government, on the other hand to stand up against the United States on various foreign policy issues. Joliot's task was more circumscribed but looked farther into the future. In a world where nuclear energy was expected to become the backbone of industrial and military strength, he had to build an independent nuclear energy organization for France.

He was starting with very little. Some steps had been taken by the time the "Canadians" arrived in January. Provisional offices were in use, two extensive apartments on one floor of a building at 41 Avenue Foch. The building had a cold and classical symmetry; as if to symbolize the split personality of the CEA, the apartment on the left was held by Joliot and his scientists, the one on the right by Dautry and his administrators. Dautry provided a few typewriters from American war surplus, and secretaries began to gather, soon crowding into the kitchens and bathrooms as the CEA expanded.[30]

The CEA had also begun the difficult search for raw materials. To build a reactor—the obvious first step for any nuclear energy program— they would need tons of uranium. Today this can be bought on the world market, but in 1946 it was almost impossible to find. During the war the United States and Great Britain had set up a trust whose goal was a world uranium monopoly. The Canadian mines were already under their control, and in 1944 the trust struck an agreement with the Belgian government-in-exile and the Union Minière, gaining exclusive rights to the ore of the Congo.[31] None of the other known deposits were accessible to the French, aside from doubtful traces in Madagascar. In its early meetings the Atomic Energy Commission devoted much of their time to this problem. They decided to train prospectors and send them hunting for uranium throughout France and the French overseas territories.

Obviously this program would not produce results for years, assuming that there was any uranium to be found at all. But by good luck the French already had enough uranium to build their first reactors. The bulk of the seven tonnes of black uranium oxide which the Collège de France team had borrowed from the Union Minière in 1939 and 1940 had been evacuated to Morocco in June 1940 and concealed. It was brought back to France. Better still, a railway car containing nine tonnes of bright yellow sodium uranate was discovered on a siding in Le Havre. This was one of seven cars that the Union Minière had hastily sent into France in 1940 a few days ahead of the Germans; the cars had separated and gone astray in the confusion. One of them stayed in Le Havre throughout the occupation, ignored by the railwaymen, who thought it contained some sort of yellow dyestuff of slight value, and left unmolested by the French, Germans, and Americans.[32]

With a total of sixteen tonnes of uranium compounds the French could pick up the thread where they had lost it in 1940. They did the same with the other essential ingredient, a moderator. The scientists resumed relations with the Compagnie Industrielle Savoie-Acheson which had provided the graphite for Halban's experiments in early 1940. But the company's graphite was too impure to serve as a moderator, and to produce tonnes of graphite of the exceptional purity needed would take years of development. Meanwhile the French returned to heavy water.

In his March 1945 note to de Gaulle Dautry had advised that France buy the heavy water borrowed from Norway in 1940. In April 1945 Allier met the president of the provisional Norwegian government in London and concluded such an agreement in principle. After Hiroshima and Nagasaki, Allier pressed the negotiations, and in May 1946 the Norwegians contracted to deliver their total production of heavy water up to five tonnes, or about two years' output, which was enough for a small reactor. Later, in 1947 and 1950, Allier negotiated further agreements which brought the total deliveries up to eleven tonnes. In return France promised Norway cash, thousands of tonnes of specialized metal products for the construction of Norsk Hydro chemical plants, and probably a look at the designs for a French reactor once these were drawn up.[33] The French were fortunate that they could start with heavy water for their moderator. Because its tendency to absorb neutrons is exceptionally small, heavy water allows a reactor builder to use less uranium, materials of lower purity, and a less exact design than graphite would demand.

In February 1946 the four scientist members of the Atomic Energy Commission plus the three "Canadians" were formed into a Scientific Committee. This group now developed the CEA's program, meeting each week to thrash out the technical questions. At its first meeting around the end of February the Scientific Committee decided "to build a heavy water pile for the following reasons: a) it requires less [uranium] metal [than a graphite pile]; b) it leaves greater latitude for the purity; c) Kowarski has just built one in Canada whose construction took eleven months." The pile would go beyond ZEEP by producing some heat, not exceeding 300,000 watts.[34] It would thus allow the French to experiment with some sort of primitive cooling system and would produce a few grams of plutonium for the chemists' experiments. The construction of this first pile was assigned to Kowarski. Goldschmidt was meanwhile asked to purify the CEA's stock of uranium compounds. Once purified, these compounds would have to be converted to metallic uranium. Guéron was asked to oversee this conversion, as well as the production of heavy water and graphite.

At the same time the Scientific Committee looked beyond its first reactor to a larger effort. If enough uranium could be found, Joliot envisaged a second reactor, air-cooled and graphite-moderated, capable of producing about six megawatts. Beyond this he sketched a vision of numbers of giant reactors producing a thousand megawatts apiece, capable of supplying much of France's electrical power (the reactors were rated by their thermal power, about a quarter of which could be converted into electrical power).[35]

This program had implications which were never fully discussed. A reactor which generates so much power must also generate many neutrons, and these will inexorably convert the uranium to plutonium, yielding roughly one gram of plutonium per day for each megawatt of thermal power. Given the necessary chemical plant and allowing for periods when the reactor is not operating at rated capacity, something over half this amount can be extracted. Thus even the second, graphite reactor that Joliot proposed would already yield one or two kilograms of plutonium a year, and only five to ten kilograms are enough for a nuclear bomb. As Kowarski later noted: "There was a curious ambiguity in [Joliot's] statement about what would be the aim of this next reactor. It sounded something like Hanford. And Hanford, of course, was for making military plutonium. Now, Joliot didn't say at all that we would be making military plutonium there, but then he didn't say anything at all about what purpose this second reactor was supposed to serve . . . Whether plutonium was to be civilian or military, that was sort of tactfully not mentioned . . . Even within the Scientific Committee."[36]

16

The Commissariat à l'Energie Atomique

On March 19, 1946, Joliot presented his plan to the Atomic Energy Commission, which duly approved it. The newborn CEA would set out to build a series of reactors. Clearly the scientific and engineering difficulties would be enormous, but before the scientists could even begin to overcome them, they would have to face another set of difficulties—of recruitment, organization, and relations with external groups—which were not only technical but also political in nature. Like a vigorous young plant growing among boulders, the CEA would be shaped, and sometimes misshaped, by the obstacles it met.

The elementary problem of housing was the first of the cramping difficulties encountered by the CEA in its first years. The scientists planned eventually to build a fine new research center for their work. Until this could be finished, the CEA would install itself in a disused fort. At the request of the Atomic Energy Commission, General Dassault had found a compound on the outskirts of Paris, a fortress complex built in the aftermath of the Franco-Prussian War. A somber construction of stone blocks, mostly underground, the Fort de Châtillon seemed less than appealing to the scientists. It was in use as a warehouse for explosives and as a prison for persons convicted of collaboration with the Germans; Kowarski recalled that when the scientists inspected the premises in March 1946, he saw blood from prisoners executed a few hours earlier.[1] But it was the best place to be had. The fort was turned over to the CEA, and in July the first handful of employees moved into its dank barracks.

This first team at Châtillon could scarcely have had a more difficult task ahead if they had been asked to build a nuclear reactor in the middle of a desert. Unlike its rivals in the United States, the British Commonwealth, and the Soviet Union, the CEA did not enjoy the full backing of a united and victorious power, but received the sometimes grudging support of a nation rent apart and economically prostrate. The smallest scrap of iron had to be requisitioned, and requisitions were not always honored. Gasoline and many other commodities were still rationed.

There was a severe Europe-wide shortage of machine tools. And the technical information possessed by the CEA on fission and its applications would scarcely have filled one book. Many nations today, even underdeveloped ones, are in a stronger position to build up a nuclear energy program than was France in 1946.

The most pressing problem was recruitment. The leaders of the CEA naturally found their first collaborators within their own circle of acquaintances. For example, the biological and medical service was led by a professor from the Collège de France who was also a director of the Radium Institute, and the search for uranium was assigned to Jean Orcel and Louis Barrabé. Orcel was a mineralogist who in the 1930s had provided Joliot with minerals for artificial radioactivity experiments; Barrabé was a friend of Joliot who had worked under him on the reform of the CNRS.[2] Both men had Communist affiliations.

In fact Joliot recruited a number of Communists and fellow travelers. He would not have picked a Communist when a more competent person was available, but there was no lack of scientists with leftist views in Parisian academic circles, still less in the circles that Joliot knew best. His hiring policies had other features not common in France. He would not tolerate the presence of anyone who had wholeheartedly collaborated with the Vichy regime. And he did not have any special sympathy for graduates of the elite schools, even the Ecole Normale. When it was hinted that Normaliens might be given preference for jobs, Joliot instantly scotched the idea.[3]

When they needed to recruit engineers, the scientists entered unknown territory. Before the war most French scientists had only slight contact with industry, and even Joliot's experience had been limited largely to recruiting low-level technical assistants and purchasing raw materials. During the war, while scientists in Britain and America were thrown into major industrial efforts, nothing of the sort happened in France. Goldschmidt, Guéron, and Kowarski had learned something about large-scale cooperation with industry as it was practiced in North America, but they found it hard to transfer their experience to their own country.[4]

Whereas many of the scientific elite in France belonged to the leftist circles centered on the Ecole Normale, most of the leading engineers and industrialists belonged to an altogether different segment of society. The French often contrasted the exuberant Ecole Normale with the Ecole Polytechnique, a school run with machinelike discipline by military officers whose traditional function was to train engineers for the officer corps and the civil service. The Polytechnique also proved a rich source of talent for the upper echelons of French industry. It was typical that the Polytechnique alumni were represented on the CEA at this time by a general familiar with engineering, Dassault, and a technocrat familiar with the military, Dautry.

Polytechniciens, along with French industrial and military circles in general, were traditionally connected with parties of the right. The brutal struggles of the extreme right against the Popular Front and the Resistance had widened the ancient breach between the conservative and liberal divisions of the French elite. Joliot and his colleagues lacked the traditional respect for Polytechniciens, and he was disinclined to hire them. Nor was it likely that many of them would be willing to serve if asked. According to Kowarski:

When you entrust this place [the CEA] with sums of money completely unheard of in scientific work in France, and which have to be spent according to a completely new set of hierarchies and rules, you begin to feel that you get away from the grooves . . . we were something outside of the system . . . We noticed very soon that people who already possessed positions in the grooved system were not very interested in working with us. Also, we certainly made a few mistakes, such as a typical mistake of an upstart in a position of power to surround himself with people not selected according to the usual rules . . . When I was looking for the chief engineer, trying to recruit some very reputable French star of engineering science, I found out that I couldn't find anybody. Everybody would say, very gratefully, that they had lots of things to do.[5]

It was certainly true in 1946 that French industry faced tremendous tasks of reconstruction. Industry could offer much higher salaries than the government, as after the First World War, and as the franc inflated, the CEA had to push to keep its pay scales at an adequate level. The fledgling organization could not even guarantee job security, and some applicants wondered whether the posts offered them would still exist after a year or so.[6]

The scientists solved their problem by skipping the middle level of management and recruiting directly from the crowds of students just out of school, who otherwise would have had to spend most of their lives working their way up. The young people, offered a chance for education and rapid advancement in the glamorous new field of nuclear energy, responded with enthusiasm. The CEA had its pick of the best—"talented adventurers," Guéron called them. In early 1946 the CEA had around a dozen employees; by the end of the year it had 236; and by the end of 1950 it had 1610. Among the 200 top people employed at the end of 1950, only seven were Polytechniciens, none of whom were at the highest level except those in the administrative division, while there were twenty people from the much smaller School of Industrial Physics and Chemistry. The bulk of the cadres had no education beyond a bachelor's degree in science (*licence*) or a common engineering degree.[7]

The CEA had to do its own advanced training, for often no other school taught the skills it needed. The recruits were set to studying the basics of nuclear engineering. Throughout 1947, one of them recalled, "we went ahead together deciphering and studying the notes Kowarski brought back from Great Britain and Canada, which at that time were the basic text for what was known in France about the physics of reactors." To supplement this "textbook," photographic copies of foreign publications were collected. The students gathered in the evenings and projected them with an old magic lantern so that they could read and discuss the articles together.[8]

The CEA's lack of equipment was not caused by a lack of money, for the government was generous. As Kowarski noted in 1948, "Whether it is the scare thrown by the bomb, or some expectation of a new era in technology, those on whom we depend for our budget and allocations give us practically everything the country can afford." What France could afford, however, was restricted by her size, the swift inflation of her currency, and the wartime destruction of much of her industry and communications. The CEA's appropriations, starting in 1946 with a lump sum of 500 million francs and rising to about ten times this amount in 1950, were only about a hundredth of the appropriations of the United States Atomic Energy Commission.[9] But the limit on the CEA budget was set less by the government's ability to supply money than by the scientists' ability to spend it. Until they had recruited and trained personnel, until they understood reactors better, and until French industry recovered, they could scarcely launch a major construction project.

In 1946 even simple laboratory glassware was scarcely to be found. A group that Goldschmidt established in the Ivry laboratory to study uranium compounds was forced to use a bizarre collection of vessels, reminiscent of an alchemist's equipage, purchased from an apothecary. Machine tools were also a rarity, but after some difficulties the CEA furnished itself handsomely with booty from the French-occupied zone of Germany. The American war surplus camps were also raided. Much of this equipment was picked up by the picaresque Roland Echard, whom Joliot seems to have discovered working as a mechanic in the town where Irène Curie was convalescing in 1940. Echard had become Joliot's devoted follower and worked his way up within the Communist party. Now he traveled the roads of Europe using threats, his connection with Joliot, and the great prestige of nuclear energy to snatch supplies for the CEA.[10]

But when it came to sophisticated instruments, much of what the CEA needed was simply not to be had on the continent. Scarce foreign exchange could be used to buy some items from abroad, but the CEA had to build many tools in its own shops. France had respectable traditions in several fields that were important to the development of nuclear energy, including electronics, specialty chemicals, and the treatment of radio-active materials, but the war had left these industries far behind those of the United States and Britain. The CEA had to do the job by itself.

For example, the CEA had a pressing need for electronic apparatus, such as geiger counters. The core of an electronics staff was the first group to enter Châtillon. Kowarski had found an enthusiastic head for this group in Maurice Surdin, an electronic engineer of Israeli origin who had worked under Langevin in the Collège de France until the war and then, working in Britain on radar, had become familiar with the latest techniques. Like other group leaders, Surdin had difficulty hiring established engineers away from industry, so he took on young engineers and physicists and trained them himself. He rarely succeeded in attempts to buy apparatus assembled to order by private firms; after one such effort Joliot commented on "the difficulty of carrying on such negotia-tions, making contracts and deciding whether or not to take patents." In the end nearly all of the thousands of geiger counters and other electronic devices used by the CEA in its early years were designed either by Surdin's group or at the Collège de France and were manufactured at Châtillon. French private industry failed to fill the CEA's needs for per-sonnel or apparatus so often that the scientists had reason to treat in-dustry less as a partner than as a mere supplier of raw materials.[11]

All these problems and policies were discussed at length in frequent meetings of the Scientific Committee, the true ruling body of the CEA. In some ways this group was a descendant of the circles that had built up the CNRS decades earlier. The Committee even met at L'Arcouest each summer to review its programs. But the members were of several distinct types. At the center was Joliot, firmly supported by his wife and, a little later, by two men he added to the Committee—André Berthelot, who had been his assistant at the Collège de France, and Joliot's old friend and schoolmate Pierre Biquard. These two, like Irène Curie, strongly sym-pathized with the Communists, so Joliot's political views were well supported within the Committee itself. Auger and Perrin, in contrast, al-though friendly to the left, avoided party politics; old companions and fellow graduates of the Ecole Normale, they resembled one another in their neatly trimmed beards, refined manners, and detachment from the daily quarrels of the CEA.

All of these people had responsibilities outside the CEA. Joliot, for ex-ample, was not only a full-time professor at the Collège de France in charge of extensive laboratories but also served on many committees, while Irène Curie's first loyalties went to the Radium Institute, which she now directed. The three "Canadians"—Goldschmidt, Guéron, and Kowarski—were distinguished from the rest by their separation from outside academic positions: their careers were chiefly within the CEA.

They were the only members of the Committee immersed full-time in the technical problems, and Joliot gave them broad authority. To their surprise he did not seem eager to discuss all that they had learned, in science or in politics, during their years outside France. Goldschmidt got the impression that "he couldn't stand the idea that he hadn't participated in all that. It was a very strange feeling. We were bubbling with desire to tell him a lot of things . . . He didn't want to hear it. Slowly he got interested."[12]

Among the CEA's scientific managers Kowarski was in the strongest position by virtue of his noticeably higher official title, his old and close relationship with Joliot, and his personal force. He had left France as little more than a postdoctoral student and had returned a veteran of political and technological struggles on the grandest scale. He was still driving and restless, made no less so by the slow failure of his marriage, which over the next few years would be replaced by another and more successful one. His mordant wit, precise logic, and control of details remained, surprising in such a massive body, and he seemed more confident than before. Nevertheless, finding himself in an exposed position in the risky new CEA, he could never feel entirely secure.

The "Canadians" were set apart from Joliot's circle not only by their full-time commitment to CEA work and their experiences in the war but also by their lack of sympathy for the Communist party. Kowarski, while inclined toward the left, during the war had acquired an unconcealed admiration for British ways of life and thought; Goldschmidt had married an English Rothschild and was at home in financial circles; Guéron voted somewhat left of center but appreciated industry and scorned the Communist party. Even among the "Canadians" themselves there could be friction under the stress of the CEA's problems, as when Kowarski or Goldschmidt criticized Guéron's performance of his jobs, while he felt that they were choosing only the tasks that attracted them and leaving him to struggle with the rest.

With all these divergent personalities on the Committee, among them some like Irène Curie and Kowarski who not only called a spade a spade but felt free to analyze the spade's defects, the meetings tended to be long and occasionally marked by forceful discussion. But the talk was businesslike, and open quarrels were rare. Decisions could usually be made by consensus, although nobody doubted that the final word was Joliot's. The scientists had common aims and simply avoided such divisive issues as politics. They had enough work with their technical and administrative problems.

To outsiders the Scientific Committee seemed homogeneous and commanding. Their solidarity after they reached a decision and their unimpeachable expertise put the Committee's actions beyond dispute.

Dautry's administrative division had nothing to say about the technical questions which at this stage determined the CEA's program, and the full Atomic Energy Commission ratified the Scientific Committee's plans routinely.[13]

Dautry, who commanded respect in governmental and industrial circles, worked to secure a favorable environment for the fledgling organization; otherwise he was content to supervise the day-to-day administration of purchasing, personnel matters, and the like. This work was a challenge even for him, one of the most capable executives in France. The mushrooming CEA was not only a new organization but a new kind of organization, a legal singularity, rushing headlong into an unpredictable field. A young lawyer who was recruited in 1947 remarked that the administrator-general "had all the trouble in the world imposing a little order on the fantasy that reigned at the time. It was done, but with difficulty, the more so because he wanted to know everything, to be in touch with everything." Dautry's task was not helped by the lack of sympathy which sometimes appears between scientists and administrators, nor by his political distance from Joliot's circle. Nevertheless Dautry and Joliot, very different in personality but equally adept at diplomacy, avoided outright warfare between the scientific and administrative divisions. Dautry and the scientists regarded one another with respect and in some cases affection.[14]

The aging technocrat, more convinced than ever of the importance of nuclear energy, hoped to crown his career by giving France an efficient and successful agency. He tried paternally to make things easy for "his" scientists. Dautry's proudest statement was that "the scientific work and research have never been obstructed or retarded by lack of funds or delays in administrative procedures." In the beginning matters could be settled simply by stepping across from one apartment to the other at 41 Avenue Foch, and later Dautry took care that the administrative services evolved cautiously, conforming to the patterns laid down by the scientific and technical services.

Nevertheless his ultimate goal was a system which, like his railroads, would possess not only esprit de corps but discipline, not only technical virtuosity but faithful obedience to the orders of the government. He considered himself responsible for transmitting these orders. No doubt France had determined that nuclear science had "almost an absolute priority" and that the scientists must have all the means needed for their task. But the nation, Dautry declared, "certainly does not wish, as people sometimes propose, to be governed by scientists who are not specially trained for this task." He had always appreciated the managerial class. Sincerely concerned with problems of the masses, such as their need for better cities and public health, Dautry felt these needs could best be provided for by private enterprise, if need be rationalized under state guidance. In his vision of future society, workers and capitalists would each have a place, their conflicts reconciled through the mediation of experts—engineers obedient to their superiors and benevolent to their employees. In Dautry's view, science was only a tool to be used in the service of social goals set by others, for managerial and spiritual values, not scientific ones, must provide leadership. It did not suit him that scientists, for the time being, controlled the realm of nuclear energy.[15]

Even in foreign policy the scientists dominated thinking about nuclear fission. Joliot had high hopes that France could play a central role in Western Europe, and he undertook negotiations with Anderson, proposing close cooperation between France and Britain in nuclear energy work. In early 1946 he told the Atomic Energy Commission: "The smaller nations certainly would be well-advised to link up with us in this domain. My point of view is that we can help England release herself from the grip of the United States."[16]

This was not how the British saw matters. In November 1945 the leaders of Britain, Canada, and the United States had agreed to keep detailed information about industrial applications of nuclear energy secret, since "the military exploitation of atomic energy depends, in large part, upon the same methods and processes as would be required for industrial use." In July 1946 the United States Congress passed the MacMahon Bill, carrying secrecy a long step farther by cutting off even Britain and Canada from American nuclear information. The British fervently hoped that they would somehow be allowed to circumvent this ungrateful ban. Clearly, if Britain cooperated in any way with France, the United States would never consent to pass the British any information for fear that it might make its way over to France and ultimately to the Soviet Union. For Britain to cooperate with France would mean giving up on the United States, yet as a partner, the CEA—still minuscule when the French scientists went to Britain in May 1946 to negotiate with Anderson—could hardly take the place of America or even the hope of America. Nor could Joliot form a clear picture of the complex relationships that had grown during the war. The discussions with Britain over cooperation in fission work bogged down in the old problem of patents. The French maintained that their master patents of 1939-1940 would give them licensing rights and royalties for every nuclear reactor that would be built anywhere in the world, even though the United States, waving Szilard's patent and others, flatly rejected the priority of the schematic French patents. In any case possible gains from royalties were meaningless compared with the military implications of nuclear fission. After months of prickly negotiations the British turned down Joliot's proposals not only for patent arrangements but for cooperation in nuclear energy of any sort whatsoever. France was left on her own.[17]

All this was particularly frustrating to Joliot because he denied any interest in the military uses of nuclear energy. He wrote Cockcroft, "it is uniquely for beneficial applications that we are working. I still keep hoping that the scientists of various countries can soon combine their efforts along this direction." The hope of nuclear disarmament resulted not only from Joliot's deep personal distaste for war but from his nation's situation. In a world dominated by two or three superpowers with nuclear weapons, France and other smaller countries would be, as Goldschmidt later pointed out, not only discriminated against in peaceful industrial applications but in all ways reduced and impotent, "*déclassé*." If nuclear weapons could be done away with, France might still rate as a world power rather than as just another unimportant country. Beyond this the French scientists shared with many other physicists an urgent fear of nuclear war. For example, in May 1946 Goldschmidt suggested that the members of the Scientific Committee themselves translate, for the instruction of the French public, the book *One World or None*, a compilation of articles by Bohr, Compton, Einstein, Szilard, Urey, and others warning of the perils of nuclear armament.[18]

The scientists' feelings were echoed at the highest levels of government. On March 29, 1946, the Atomic Energy Commission proposed that a statement of France's peaceful intentions be made by the French delegate to the United Nations; the draft of the declaration was discussed and approved by Joliot and the others in the Scientific Committee meeting of April 5. In June the French delegate to the United Nations read the policy: "the goals which the French Government has assigned for the research of its scientists and technicians are purely peaceful. It is our hope that all the nations of the world may do the same as swiftly as possible." The French scientists continued from time to time to discuss mechanisms for nuclear disarmament in meetings of the Scientific Committee and to advise the government of their conclusions. One or another of the chief CEA scientists was always present and influential as an adviser to the French delegation at the United Nations meetings in New York, which were deeply involved at the time in nuclear disarmament negotiations. They all hoped that an international control system could be set up. This went on until 1948, when negotiations between the United States and the Soviet Union finally broke down.[19]

A poll taken in 1946 suggested that a majority of the French public felt their country should build nuclear bombs.[20] But these feelings had no immediate effect, for the CEA was restricted to doing what France could afford. "What results can we hope to achieve with . . . one percent of the American effort?" asked Kowarski in 1948. "It is obvious, to begin with, that it would be folly to divert any fraction of these limited resources towards military applications; in fact, no such intention has ever been harbored." It would have been premature even to discuss the matter when France did not yet own so much as one small reactor or uranium mine. Since in 1946 and 1947 it seemed that international controls must come sooner or later, the scientists often assumed that when France finally got a nuclear establishment, its activities would be strictly peaceful.[21]

But nuclear war, which some thought imminent, could not be entirely ignored in the CEA's work. The staff cooperated with the French armed forces from the beginning, chiefly by giving lectures to their personnel on basic nuclear physics and the effects of nuclear weapons. The army could scarcely dream of possessing nuclear bombs of its own, but the generals knew that others might well drop them on France, and they wanted to prepare for civil defense. The War Ministry meanwhile detached a few bright junior officers to get specialized training in various laboratories, such as at the CEA and the Collège de France. "These young people," one of their eventual leaders recalled, "were later to constitute a group of 'atomician specialists' whom we would recover when we began to study the possibility of making nuclear arms in France, and who would then render immense services to the development of our program."[22]

In all these external relations—cooperation with the armed forces, discussions of disarmament policy, negotiations with private suppliers—the top scientists of the CEA had many chances to meet their counterparts in the military, political, and industrial leadership. Unlike the ideological ties that had bound the scientists of the L'Arcouest circle to certain other groups before the war, the new contacts sprang from the practical needs of the different groups. In other countries, where the same forces were at work, the process went still farther. In the United States the new National Science Foundation and Atomic Energy Commission were run by elite scientists along roughly the same lines as the CNRS and CEA, but this was only the most visible part of a wider activity. Important parts of the faculties of leading American universities were devoted by the 1950s to far-ranging studies of such matters as the construction of nuclear weapons and their use in battle.[23] American scientists were constantly on the move, meeting one another in committees which interlocked with groups of industrialists, military commanders, government executives, and statesmen on the highest levels. In comparison with this activity and similar activity in Great Britain and the Soviet Union, Joliot and his colleagues had only limited contacts with military, industrial, and political leaders. While the Scientific Committee of the CEA enjoyed independence of action, they lacked a highly developed system of external support.

Most of the Scientific Committee's time was taken up with details internal to the CEA. They made arrangements for buildings, personnel, and equipment. They engaged in negotiations with Belgians and

Portuguese on the remote chance that some uranium could be won. They made continual changes in the CEA's structure, such as creating an external council of prestigious scientists to advise and, particularly, to support them. And they probed all sorts of technical ideas, some of them futile, such as a prolonged and expensive attempt to produce pure beryllium for use as a moderator. Every detail of France's nuclear program interested the Scientific Committee. For example, at a typical meeting on March 25, 1947, they discussed among other matters the position that France should take in the nuclear disarmament talks at the United Nations, a French law embargoing fissionable materials, the investigation of a site where traces of uranium ore had been found, admittance of visitors to Châtillon, the hiring of three engineers, relations with the press, and the lack of space at Châtillon.[24]

The increasing inadequacy of Châtillon was one of the Committee's major worries. At the beginning of 1947 there were 140 employees in the old fortress complex; at the end of 1949 there were 530; and at the end of 1951 there were 778. Workshops of all possible types—electronic, mechanical, and chemical—were crammed into every corner, and makeshift wooden buildings were thrown up as the existing barracks and underground chambers became overcrowded. Naked light bulbs dangled from the vaulted stone ceilings of tunnels where youthful workers at tables lined the walls. "The whole is as discordant as possible," a manager of the installation admitted in 1950; "buildings have frequently been improvised, almost without recourse to outside building firms."[25]

Long before this state was reached, the CEA had determined to build a new research center worthy of nuclear energy. They chose a site at Saclay on the southern outskirts of Paris in a region suggested by Irène Curie, who had often taken walks on the rural plateau. The scientists planned to build an entire city of modern laboratories, employing thousands of people working in teams on the most advanced devices, not only reactors but also instruments of basic research. The CEA hoped not only that basic studies of the nucleus would benefit the reactor engineers but also that a pure research center would attract the best scientists, who could then be asked to give a part of their time to technical problems. But as Perrin recalled:

These reasons were not the only ones nor even the strongest ones for developing important sectors of fundamental research in the heart of the Commissariat à l'Energie Atomique. Its founders, and particularly Frédéric Joliot, clearly understood that when the great material and human resources required for the development of nuclear energy were put to work, this and only this would provide an opportunity to give fundamental scientific research the rapid expansion that was required if France were to retrieve her position.[26]

In short, the French scientists, as convinced as ever that basic research should have a share of the funds resulting from the applications of science, seized this opportunity when nothing was being denied nuclear physicists. A student of French science policy has called Saclay the most significant accomplishment of the CEA's early years: "In place of individualistic professors there were now teams of researchers and supporting technicians. The industrialization of basic research was begun."[27]

To the research scientist, fission was only one of the various types of nuclear behavior, and seldom the most interesting. In the postwar years and up to the present, anyone who wanted to know more about what the world is made of sought to uncover the general laws that govern how particles come together to build a nucleus. The Scientific Committee took it for granted that France must win a strong position in such research or someday suffer for her backwardness. The most promising line of attack was through particle accelerators, for it had become possible to build instruments with far greater power and sophistication than any cyclotron built before the war. In 1947 the CEA created an Accelerator Division which, aided by Surdin's electronics service, quickly built an excellent electrostatic research machine and began work on a giant cyclotron.

Because Joliot, Curie, Auger, and Perrin were not only commissioners of the CEA but also professors at the leading schools of Paris, they linked nuclear energy work with the traditional community of academic scientists. When the CEA established laboratories, a few of these outside scientists requested permission to work in them. But the "Canadians" held to their promises to retain strict control over knowledge of nuclear fission, and the CEA laboratories remained a world apart from the academic community.

In contrast, some CEA staff moved into university laboratories in order to work on problems of interest to the CEA. By 1949 there were, for example, five such people at the Radium Institute and four in Joliot's Collège de France laboratories. This was a reasonable expedient during the days when the CEA was overcrowded in Châtillon, even though the system created, as Joliot remarked, a confusion of authority. Between 1946 and 1949 the problem was gradually straightened out by putting the external laboratories on subventions. Almost from the beginning these funds were also used to support pure research outside the CEA. The grants grew rapidly, until by the end of 1950 they totaled over 140 million francs. More than half of this total was spent by only four laboratories, those of Joliot, Perrin, Auger, and Irène Curie, although a number of other institutions also benefited.[28]

The postwar revival of French science was not due to Joliot and his collaborators alone, for conservative physicists had also survived the war and were organizing notable laboratories. These scientists on several occasions gave vital support to the CEA, which in turn made grants to their laboratories. But the chief new French center for basic research, Saclay, was the personal creation of Joliot and his colleagues.

The inhabitants of the Saclay plateau were not pleased when they learned of the CEA's plans. They were afraid of reactors, thinking that the devices might somehow blow up like a bomb, or attract nuclear bombardment in the next war. The local people circulated petitions and held up construction for months. Joliot fought to overcome this opposition, not only in official circles but directly. In February 1947 he came to the Saclay schoolhouse to speak to the assembled farmers. He had continued to develop a pronounced taste and formidable talent for public speaking, which he now brought to bear with all his personal force. In a few hours he persuaded the crowd to have confidence in him. He drove away exhausted, but exhilarated by his rapport with the villagers.[29]

Since the war, according to Francis Perrin, Joliot "had often seemed a tormented spirit, plagued by deep self-doubt despite his brilliant successes and seeking in the adulation of crowds compensation for the reserve that he sometimes perceived among his peers." Adulation was easy to find, for most of the French still saw Joliot as the son-in-law of Marie Curie, Resistance hero, nuclear wonder-worker, and victor of the "Battle of Heavy Water." A romanticized film of that name, widely distributed, included Joliot and others playing themselves in a reenactment of the events of 1940.[30] Joliot remained as much as ever an attractive personality, a spellbinder both in personal contacts and before crowds.

Yet Joliot had never entirely shared the life of the French elite; it was only among simple workmen and village fishermen that he felt truly at ease. Moreover his prestige among the common people was not appreciated among the conservatives, who regarded him and his leftist colleagues with increasing suspicion. As Biquard saw it, "the more progress the Commissariat made, the more numerous became those who were astonished to see it headed by a scientist who was also an engineer, who was not part of the grand family of Polytechniciens, and who, moreover, permitted himself to be a Communist."[31] To the conservatives in France it made little difference that the Communist party had participated in the government since the Liberation and was urging workers to devote themselves dutifully to the task of rebuilding industry. Tensions were building month by month, and Communists were increasingly isolated.

Communism was still attractive in the early postwar years to many scientists and other intellectuals. The party's coherent ideology and its call to selfless struggle continued to contrast with the ambiguous programs of most other parties. Moreover, with the working class joining up in great numbers and with the Red Army only a few hundred miles away, victory seemed imminent. The American writer Saul Bellow recalled that when he lived in Paris in the late forties, banalities about revolution "had reached all levels of French society. Anticipating the coming victory of Communism, many French intellectuals prepared themselves opportunistically for careers in the new regime." Whatever their reasons for joining, the party was glad to have such people, as a leader wrote to his comrades in a circular letter of 1947: "The presence of a great number of intellectuals in the ranks of our Party extends its sway among wide strata of our country's population. We have considered making practical use of this precious capital for the coming electoral campaign."[32] Joliot and many others responded by speaking and writing in support of Communism.

While electioneering vigorously, the Communists found it difficult to function as an ordinary electoral party seeking power only within parliament. France's economy was persistently weak; production remained well below the level of 1938, which in turn fell below the 1929 peak. The working classes, kept in penury, in 1947 erupted in wildcat strikes. Meanwhile sordid colonial warfare burst out in Indochina and Madagascar. Worst of all, the wartime alliance between the Western powers and the Soviet Union was breaking up with glacial slowness and finality. The division ran right through France, for the Communist party remained loyal to Stalin, while most other parties leaned farther and farther toward cooperation with the United States. For all these reasons and more, in May 1947 the Communist party's ministers were dismissed from the French government.

The next winter the Communists led a general strike which seemed to put the Republic in danger. But the government broke the strike, and the party was left in isolated, bitter opposition. Its isolation also reflected international developments: the descent of the Iron Curtain across Eastern Europe, the growth of the Indochinese war, the fall of Czechoslovakia to the Communists, and the Berlin blockade. It began to seem surprising that any Communist, even Joliot, should be allowed to hold a high position of power and trust within the government.

The situation could persist because Joliot was still seen as an honored scientist rather than a politician, still less a typical Communist politician. With his comfortable house in a well-to-do suburb, his summers at L'Arcouest in the house inherited from Marie Curie, and his winter ski vacations with Irène Curie at a chalet which they owned in the Alps, he did not quite fit into the aggressively proletarian party leadership, even though as a Communist he did not quite fit into L'Arcouest either. He did not claim to be an original political thinker, and in speeches he simply set forth those parts of Communist policy with which he could agree strongly, such as the refusal to accept American suzerainty over Europe and the opposition to nuclear armament as a misuse of science.

His views on the role of science were perhaps most deeply expressed in a piece he wrote in 1947 for the World Federation of Scientific Workers. Using words which could almost have been uttered by the anticlerical Berthelot fifty years earlier, Joliot declared, "The spread of scientific knowledge among the people has always encountered strong resistance from the devotees of mysticism and superstition. And," he continued, in the rhetoric of Marxism, "from those whose profits and privileges depended upon the timid ignorance of the beings they dominated." The nuclear bomb, Joliot noted, had increased the anxieties about science. But he rose to the defense of basic research, placing it above any use or misuse. Whereas Berthelot had claimed a high position for science because it provided beliefs that everyone in France could share, Joliot extended the argument to all humanity: "Pure scientific knowledge brings peace to our spirits and a firm faith in the ascent of man by casting out superstition . . . *Science is besides, and it is one of her proudest titles, a fundamental element in the unity of the thoughts of men all over the world.* For all these reasons science has in itself an undeniable moral value and an exalted part in social affairs."[33]

Phrases like this, however, did not attract political leaders of postwar France as strongly as they had attracted earlier generations. The Communist party was not interested in giving science an exalted part in politics, and the old Radical and Socialist parties, which had long made use of the ideology of science triumphant, were in decay and beginning a search for new ideologies. Although in funding and organization the alliance between science and the French state had never been stronger, this alliance was for practical benefits rather than for ideological ends.

In a world where international relations were coming under the dominance of scientific weapons and processes, the universality of scientific thought looked more like a hazard than a benefit. With the arrest in 1946 of the Communist physicist Alan Nunn May for having passed some of the Montreal group's nuclear secrets to the Soviet embassy in Canada, a new image emerged: the scientist as traitor and spy. It was the Americans, sure of their superiority to everyone else, who were most unwilling to see knowledge shared around and most suspicious of Communists. In 1947 and 1948 the British, still trying to negotiate cooperation with the United States, noticed a great distrust there of the French Communist scientists. An official in the American State Department wondered whether Joliot could be dislodged, perhaps by a direct approach to the French prime minister. This idea came to nothing, since it was agreed that Joliot's prestige and his indispensability at the CEA made him untouchable.[34]

The Americans nevertheless found ways to express their distrust of the French scientists. For example, in March 1948 Irène Curie arrived in the United States to lecture for a group which was on the attorney general's list of supposedly subversive organizations. On her arrival she was seized without warning by immigration authorities and detained for a day. *Time* magazine suggested that Curie had been treated better than she deserved, since "all Communists in a democracy are potential spies and traitors." Curie, who was not a party member but often spoke like one, replied in kind. "Americans," she said after her release, "look with much more favor on fascism than on communism . . . Americans think fascism has more respect for money."[35]

As the United States government became more and more worried about Joliot and his circle, the French government came to agree more and more with the United States on many matters. This change was partly because of the onset of the Cold War and the coming of Marshall Plan aid and partly because of the departure of the Communist ministers, which had shifted the government's center of balance rightward. Bidault's attempt to mediate between East and West was a failure, anti-Communism grew stronger and louder, and Joliot emerged as a target.

Already in March 1948 he was attacked indirectly in the National Assembly. A deputy, Henri Monnet, announced his disapproval that high government positions were held by people "whose thought shows a disturbing synchronism with that of Stalinist Russia." He proposed that the CEA's budget be cut by a small but symbolic one million francs as an expression of displeasure. It was a concern for national defense, explained Monnet, that had led him to raise the issue. "You have not put a Communist general staff at the head of your army," he told the government. "I ask you not to put Communist operatives at the head of scientific research, which concerns the national defense in the highest degree." Monnet was frequently interrupted by sarcastic shouts from the Communist benches, and a reply by a Communist representative was also jeered, until the session was suspended in tumult, an example of the parliamentary theater of the times. Monnet's proposal was finally turned down by only a few votes, with the majority abstaining.[36]

Since these political attacks had no effect on the technical progress of the CEA, it might seem that nuclear engineering and French politics were effectively sealed off from one another. Yet it was the political views of Joliot and his colleagues, such as their wish for rapid changes in the life of the masses and their hope for state science agencies under the control of scientists, that had brought the CEA into being and continued to shape it. Joliot's distance from the conservatives and his drive to build a reactor were two different expressions of a single set of beliefs. If the connection now seems remote, it did not seem so to Monnet when he protested the presence of an independent leftist in the CEA precisely because he did not want that sort of person in command of that particular technology. As the first French reactor moved toward concrete existence, it was becoming more and more a political as well as a technological object.

17

The First French Reactor

To have the idea of a nuclear reactor is a long way from having a reactor itself. The CEA leaders had to struggle continually with problems of a political nature in order to get the authority, personnel, supplies, industrial support, and funding they needed to bring a reactor into being. Equally important, they had to overcome serious technical obstacles. As much as the political difficulties, with which they were intimately connected, the technical ones helped to determine both the timing and the direction of the CEA's growth.

The largest single division of the CEA was the Direction des Recherches et Exploitations Minières (DREM), whose mission was to break the Anglo-American uranium monopoly. Approaches to the Union Minière had yielded no uranium at all, only the loan of a few grams of radium. Negotiations with the Portuguese government dragged on for years, winning only a few dozen tons of stony ores mined in Mozambique, and that not until late 1949. The CEA was so desperate for uranium that Guéron was asked to look into methods of working ores which had less than one part in a thousand of the element.[1] The 16 tonnes of uranium compounds that were found in France and Morocco at the end of the war would provide a start in building reactors, but France would eventually have to locate far more or fall hopelessly behind the English-speaking countries. Since only a few deposits of uranium ores were known in French territory, all of them too small to be of use, the CEA set out to discover more.

Young geology students were snatched straight from the Sorbonne. But as there were too few of these, the CEA advertised for young untrained men willing to work as prospectors. The news spread by word of mouth among soldiers and guerillas fresh from the Free French and the Maquis; accustomed to an outdoor life and eager for adventure, they were easily recruited. Orcel and Barrabé put them through special courses and in four years trained 120 prospectors. The shop at Châtillon provided them with geiger counters and the like. Surdin was soon designing and producing all the devices the CEA needed, up to 3500 geiger counters of various types in 1948 alone, and even selling instruments to outside laboratories. Not only electronics but every sort of mining and prospecting tool was in short supply in France. For example, jeeps and motorcycles were difficult to find and tires for them were scarcer still, so on Curie's suggestion the young prospectors made use of bicycles wherever possible. By the end of 1946 prospecting teams were already in the field in France, Madagascar, the Congo, and the Ivory Coast. They adopted the Maquis fashion of beards and affected bush hats brought back from Africa. The teams roamed everywhere, geiger counters in hand, astonishing French peasants who came upon them in the fields.[2]

The prospectors' first task was to range about in search of traces of radioactivity in the soil. Any uranium deposit will give rise to mildly radioactive compounds that dissolve in water and diffuse widely through the neighboring countryside. As a leader of the work pointed out, "It is in fact practically impossible not to detect . . . the presence of uranium in a region," even when no visible sign of uranium-bearing minerals appears on the surface. Once a likely region was found, more precise instruments were brought in to map out the radioactivity so that its source could be located. The final step was to drill test holes and lower radioactivity detectors into them in hopes of finding a vein.[3]

A number of possible deposits were turned up, but the young prospectors' hopes always proved to be higher than the uranium content of their finds. At the start of 1948 the search was concentrated on France, particularly on a promising region in the center of the country where student prospectors on a training exercise had turned up interesting minerals. A new mission was organized with experienced personnel, which in November 1948 hit a thick vein of the tar-black uranium ore pitchblende. The prospectors were enraptured, and the farmers in the neighborhood caught their enthusiasm, for no doubt the uranium under their fields would fetch a fine price. And besides, a café orator in the nearest village declared, "It's thanks to us that France will be able to make her atomic bomb!" The DREM redoubled its efforts, drilled thousands of test holes, and in 1950 began mining ore from a shaft at La Crouzille.[4]

Meanwhile prospecting continued elsewhere, and in the next few years much larger deposits were found in several parts of France, in Madagascar, and in Gabon. It turned out that the CEA's early fears were not justified: uranium is a widely distributed and easily located mineral which almost any country can find. But the DREM's energetic campaign had assured France a large stock early enough so that her reactor program was never delayed for lack of uranium.

Even with uranium in hand, the CEA's future would not be secure until it had in hand the other essential ingredient of reactors, a moderator. Guéron made a survey of the possibilities for producing heavy water on

French soil, while the CEA bought hundreds of tons of beryllium compounds in case the metal should prove suitable. But in the end the most
accessible moderator appeared to be graphite. Private industry seemed
likely to be of help here, so the CEA negotiated with several corporations. In May 1946 they settled on Pechiney, a giant science-oriented
chemicals combine, and soon signed a contract for 15 tonnes of pure
graphite. The CEA planned to work unusually closely with industry,
setting up a shop in one of Pechiney's factories where engineers from the
two organizations could work as a single team. Neither the firm nor the
CEA was fully at ease with such a close arrangement. Dautry's help was
needed to persuade Pechiney, and Guéron recalled that "to some extent I
had to push it through the teeth of the Commissariat."[5]

The reactor demanded unprecedented purity. A few parts per million
of an element like cadmium or boron which readily absorbs neutrons
would be enough to prevent the reactor from working. To avoid contamination, even the materials used to handle the graphite had to be
extremely pure. By 1948 the engineers had processed a great deal of
graphite, but tests carried out laboriously in large piles, like Halban's and
Fermi's experiments of 1940, showed that the graphite was still not pure
enough to serve as a moderator. In mid-1949 the CEA and Pechiney
signed a contract for 2000 tonnes of extremely pure graphite, to be made
in a completely new factory in which the graphite would be processed
entirely by remote control so that the engineers could maintain strict
cleanliness. Finally in 1950 graphite of "nuclear purity" began to come
out in quantity.[6]

Besides producing graphite in collaboration with the Pechiney engineers, Guéron's division had the complex task of monitoring the purity
of the graphite, heavy water, uranium, and numerous other materials
that the CEA needed. Since the accuracy that they required went beyond
anything previously attempted in France, the CEA had a special need for
expert analytical chemists and their equipment. It proved impossible to
find enough experienced people, so Guéron had to train young scientists;
for several years recruitment and teaching were his most acute problems.[7]

Chemistry proved to be a bottleneck for the CEA's program in other
ways, above all when it came to getting uranium metal. This would be
the job of a group working under Goldschmidt. Their process had two
steps: first the raw yellow or black compounds had to be purified and
transformed to UO_2, a chocolate-brown oxide; then the latter had to be
reduced to pure metal. The process was simple in principle and easily
done in the laboratory, but the CEA needed pure materials in greater
amounts than any scientific laboratory could turn out. Goldschmidt still
felt more like an academic chemist than an engineer, but his work had

reached the pilot-plant stage. In the spring of 1946 members of the Scientific Committee met with representatives of the Société des Terres Rares, a company closely tied with Pechiney. Terres Rares had long furnished special chemicals to nuclear physicists and had helped the Collège de France team to treat their uranium oxide in 1940. The company now signed a contract to build a factory for converting the CEA's stock of uranium compounds to brown oxide.

The factory went up in buildings leased from the government explosives service, the Direction des Poudres, in a corner of a vast gunpowder works south of Paris at Le Bouchet. To get complete freedom from contaminants, the reduction was done entirely in stainless steel containers in rooms air-conditioned to be free of dust; the workers were obliged to take precautions that were less typical of a factory than of a surgical operating room. The plant started up in January 1948 and in a few months produced several tonnes of pure brown oxide.[8]

The relationship with the Terres Rares was difficult and led to ill feelings, in part because the CEA scientists were inexperienced in dealing with industry. The contract did not contain adequate controls over costs and delays, which rapidly mounted. There were also problems of secrecy, for the scientists, thinking it likely that nuclear energy would turn out to be enormously profitable, wanted to prevent any particular private company from gaining an unfair advantage through the use of their knowledge. In mid-1948 the CEA finally canceled its contract with Terres Rares and took over the factory whole, its own engineers replacing the company's.

Relations between the CEA and private firms were supposed to be overseen by Denivelle, the secretary-general midway between the scientific and administrative divisions, but this arrangement only caused more problems. Denivelle came under attack from several of the chief scientists, who felt that he was altogether too close to the industrialists to represent the CEA's interests. In early 1948 he resigned.[9]

Still, the main problem that the CEA had with industry was the simple lack of capabilities. To produce uranium metal, the standard technique that the French had learned from the British was to convert uranium oxide into uranium tetrafluoride, a green salt, and then mix this with calcium inside a strong vessel. The combination was ignited, producing a violent reaction which left a gray puddle of pure uranium metal at the bottom of the vessel. Unfortunately this process demanded tons of very pure calcium and Guéron could see no way for French industry to produce such a quantity soon. He tried to develop an alternative method, substituting magnesium, as the Americans had done, since magnesium metal was easily bought, but his method did not yield satisfactory uranium ingots. The problem was serious enough to hold up the entire French effort to build a reactor.

Goldschmidt, aware of the troubles in the uranium program, began to feel that the delay was something the young CEA could ill afford. As he recalled: "In June 1947 I met Bruno Pontecorvo again in Canada and he explained to me the value of building a first pile as swiftly as possible; it happened that the official existence of the little pile Kowarski had built at Chalk River . . . with the help of secrecy had satisfied the Canadian Parliament and, among other things, had forestalled general discussion of serious delays that had appeared in the construction of the larger pile, which in reality was incomparably more important." In short, while a small reactor might have limited technical value, its impact on government ministers and the public would give it high political value. Since political gains could be as important for the CEA's progress as strictly technical ones, Goldschmidt decided that the CEA should rush to put up a reactor even if the problems of producing metallic uranium meant the reactor had to be much cruder than orginally planned. Instead of waiting indefinitely for metallic uranium, he proposed that a first pile be made with the brown uranium oxide itself. Unfortunately nothing was known about the properties of this compound—its thermal conductivity, its resistance to corrosion under heat, and so forth. The pile therefore could not be operated at a high temperature. Except for using oxide rather than metal, it would almost be a copy of Kowarski's ZEEP reactor at Chalk River. The French would have to defer pilot studies of cooling systems, the production of substantial quantities of plutonium, and other things that could be done only with a bigger reactor; but they would still learn much from building the smaller reactor, and they would gain all the psychological benefit of lighting a nuclear fire in France. Goldschmidt presented this proposal to the Scientific Committee in the summer of 1947 at L'Arcouest, and they immediately agreed.[10]

Kowarski naturally was put in charge of the project, a partial repeat of his work in Canada, although he would have preferred to do something more original. He facetiously called the proposed reactor the "French Low Output Pile" until someone told Joliot what FLOP meant in English. Kowarski eventually named it Zoé, using the initials of zéro énergie, oxide, and eau lourde (heavy water). The Scientific Committee gave Kowarski's work an absolute priority over all other CEA activities, for as Joliot told the Atomic Energy Commission in November 1947, "the present aim of the Commissariat is essentially the realization of a pile in 1948," that is, within roughly a year.[11]

There was still a problem to be solved before the uranium oxide could be used. It came out of Le Bouchet as a powder which was not nearly dense enough to serve as a reactor fuel. The metallurgist Jacques Stohr proposed to sinter it, or compress it at high temperature, in a hydrogen atmosphere. His group first tried to do this in commercial furnaces, but none of those available in France could reach the necessary 1500 degrees

Celsius. It was a difficulty typical of the CEA's problems in a country whose industry was not always up-to-date. Stohr's team finally built their own battery of furnaces in a tunnel at Châtillon, working at high speed with a minimum of tools, and soon were stamping out fat cylinders resembling brown ceramic.[12]

The workers pushed on in haste, for failure to complete Zoé on schedule by the end of 1948 would cost the CEA prestige, threatening loss of the funds they needed to go on: the CEA might be smothered in its cradle. A theoretical group worked on the sizes and spacings of the reactor components, but for lack of information on the properties of uranium oxide they could not get precise results. Rough estimates had to suffice, and given the necessary five tonnes of heavy water and three and a half tonnes of uranium oxide, the group had considerable latitude in design. A hangar was swiftly thrown up at Châtillon to house Zoé, which would be a cylindrical vat surrounded by a graphite neutron reflector and concrete shields, containing an arrangement of aluminum tubes filled with the cylinders of uranium oxide. The engineers' main problem was cleanliness, since even the natural oils of a fingerprint or the normal humidity of the atmosphere were unacceptable impurities.

Even when completed, the hall had the look of a place under construction, like a factory in which the machines had yet to be installed. On the bare floor stood a concrete cube some five meters tall, furnished here and there with pipes and electrical switches, looking more like a large storage shed than a nuclear device. Toward dawn on December 15, 1948, the leaders of the CEA gathered in the hall to watch as heavy water was added, slowly and carefully in case anything went wrong. All were tense, assuring themselves of success yet wondering whether something might have been overlooked. About nine in the morning the heavy water passed a level at which the pile was predicted to begin showing some activity, but nothing happened. A little later, Kowarski recalled, a needle on a dial suddenly quivered and stopped: "A moment of intense emotion." As more heavy water was added, geiger counters about the hangar started a steady clatter. Shortly after noon the doors were opened on a crowd of anxious CEA personnel, and Joliot appeared to announce victory. The ovation he received touched him deeply.[13]

The French public were equally delighted. Newspaper headlines trumpeted the peaceful nature of the French program and its conquest of the Anglo-Saxon monopoly on nuclear energy. Since the reactor near Moscow was still secret, Zoé was, as far as most of the world knew, the first reactor built outside the English-speaking countries. The pile at Châtillon, one newspaper announced, was "a great achievement, French and peaceful, which strengthens our role in the defense of civilization."[14] The Communist press was particularly pleased with the new reactor, both as a setback to American dominance and as a triumph of one of the party's most famous members.

The press in the United States and Britain was less enthusiastic. Seeing the French success and not knowing that the Soviets were much farther along, anti-Communists feared that the CEA knew secrets which could be important to the Soviet Union. According to the London *Economist* of December 25, 1948, "Atomic research in France with Communist participation is hardly compatible in the long run with French military commitments in Western or Atlantic Union, and sooner or later it seems that France will have to face the problem of purging Communists from posts affecting national security." Such pronouncements, which were believed to reflect governmental views, were taken seriously in France.

Joliot replied to his enemies at a luncheon of the Anglo-American Press Association in Paris on January 5, 1949. He flatly stated that no honest French citizen, Communist or otherwise, would ever deliver secrets to any foreign power.[15]

This statement, which did little to please the Anglo-Saxons, was also unlikely to please the Communist party. Although they were fond of featuring Joliot's activities in their press, the Communists neglected to report his patriotic promise. Before the end of the month the party secretary, Jacques Duclos, declared, "Every progressive man has two fatherlands, his own and the Soviet Union." Joliot's statement clearly deviated from the party line. One French newspaper purportedly quoted from the Communist political bureau that "M. Joliot-Curie has committed the error of including the USSR, the fatherland of all workers, in the phrase 'any foreign power.' A French Communist should have no secrets from the USSR."[16]

Joliot was caught in one of the touchiest issues of his times. The French Communist party was ill at ease with many of the intellectuals who had joined it during and after the war, finding them excessively independent and given to doubting the universal wisdom of Stalinist thought. Such "Titoist" independence was beginning to become a serious issue in 1948, and by February 1949 the French Communist party's Central Committee was insisting that Communist intellectuals emphatically defend every one of the party's positions. For a time anyone who did not unhesitatingly put the Soviet Union first or who, as a renegade party member recalled, "made the slightest remark, expressed the simplest doubt or showed the slightest intellectual independence, immediately found himself lumped with the worst enemies."[17] A number of intellectuals abandoned the party or were driven out.

To maintain discipline among the rest, the party viciously reviled the dissidents in its press and kept a close eye on its cells, whose members were supposed to meet every fortnight with their fellows to receive directives from the leadership and discuss politics (Joliot's cell was probably number 546, at the Collège de France).[18] These methods were usually effective in keeping people inside "the family." Many party members, as one observer noted, "in time found it hard to imagine an existence outside the ranks in which their friendships had been made and their life organized."[19]

According to Goldschmidt, the weeks following Joliot's promise to keep his nuclear secrets were lacerating: "he had been publicly disavowed by his party, and we wondered whether he was not about to break with it. I recall one evening when I had an appointment with him at the Collège de France; he made me wait more than three hours while he was in conversation with a leader of the Communist party. When the latter left, Joliot, white with fatigue and nervous tension, did not conceal the anguish he was going through."[20]

The fact that he was a scientist did not spare Joliot from these conflicts, for Communists refused to allow science a higher sphere of its own. Rather the party insisted that its "proletarian science," infused with political ideology, directly contradicted inferior theories of "bourgeois science."[21] This attitude sometimes led to excesses where commissars thought themselves fit to make declarations, on the sole basis of Marxism, about physical or biological facts; but it also represented a turn in modern thought.

From the beginning of organized science in the seventeenth century down to the time of Pierre and Marie Curie, most scientists had insisted that they could improve humanity's material condition most swiftly by sticking to research, far from political involvements. This program had seemed to be successful, for through research people learned how to stem epidemics, grow better crops, replace human labor with electrical motors, and much more. However, the First World War and the Great Depression showed that the advance of knowledge can lead as easily to destruction and unemployment as to universal prosperity. Jean Perrin, Bernal, and others concluded that science, like a worthy machine that can either be used well or be misused, requires a proper social system to employ it for the benefit of all. A different conclusion was reached by some Communists under Stalin as well as by some Nazis under Hitler. Science, they believed, takes on the color of the society that nurtures it, so that when nurtured by the wrong people not only the applications of science but even its theories are perverted. The Communists were seldom entirely comfortable with this doctrine, but it served to strengthen their demand that intellectuals conform to party dogma. Scientists themselves usually refused to agree that political ideology could play a role in legitimate scientific theory, and Joliot never showed any sympathy for "proletarian" scientific ideas as opposed to other kinds.

Fortunately for Joliot, the party had other work for intellectuals. The leadership had recently decided that there were "great possibilities for rallying the masses" in a campaign linking economic and political issues to propaganda against war.[22] Among other measures, a movement in defense of peace was started among intellectuals. This movement, designed to attract the left's traditional pacifists, was tied to attacks on the rearmament of Germany and the military alliance with the United States, government policies that were widely disliked. Not only the Communists but many others believed that the United States was preparing to launch a preventive attack against the Soviet Union, making opposition to nuclear warfare urgent.

Joliot was already prominent in groups, like the World Federation of Scientific Workers, whose members opposed military uses of science, and in February 1948 he had been one of some thirty people who formed the Combattants de la Liberté, later renamed the Mouvement de la Paix, the Peace Movement. Later that year a Communist-led World Congress for Peace was held in Poland where the French delegation, which included Irène Curie, was the strongest of the Western contingents. The language used in such meetings proved too extreme for many anti-Communists, who began to find it difficult to distinguish between genuine pacifism and mere subservience to the Moscow line.[23]

Early in 1949 Joliot was approached by Aimé Cotton's wife Eugénie, a survivor of the group of bright young people who had gathered around Pierre and Marie Curie at the turn of the century and a fellow-traveler very active in Communist organizations. She persuaded Joliot to accept the presidency of a new World Congress of Partisans of Peace which was being hastily organized. The Congress met in Paris that April in unseasonable heat. Impressive delegations from many nations crowded the hall, enduring "a flood of words in a miasma of perspiration," according to an American journalist, who also reported that the best speech was Joliot's opening address.[24]

In this speech, as in others, Joliot mingled Communist rhetoric with his vision of the possibilities of science. The capitalists, he warned, were rearming Germany and preparing an anti-Soviet war in hopes of temporarily staving off "the economic crisis congenital to capitalist regimes." But as Joliot had been arguing ever since the 1930s, if only a fraction of the money spent on war could be diverted to peaceful research, scientists could produce unimaginable benefits.[25] Nuclear bombs were therefore a shameful misappropriation of funds; if the effort spent on them were put to peaceful work, science could advance to a point where all of France's electrical power would be supplied by only twenty nuclear centers—which was his long-range program for the CEA. Joliot was thus transforming a technical statement into a political one. He went even further, declaring that nuclear weapons must be placed outside the law. This was not only a political but a moral statement, claiming for scientists the right to be heard above nationalistic quarrels. For, he asked, "is not science the fundamental factor of unity between all men dispersed upon the globe?" To tremendous applause he concluded by calling on people of goodwill to avoid the scourge of war.[26]

Joliot's internationalism and pacifism not only fitted the Communist party's political program but also agreed with the feelings of many people around the world. Other scientists were throwing themselves into campaigns for international cooperation and the control of nuclear bombs. Although the effects of their work cannot be measured, it is certain that scientists in various countries were the first and most persistent to insist that nuclear weapons are so unlike any other weapons that they can never morally be used, that this message was listened to, and that eventually so many people came to share this view that it became a force in world affairs. Joliot and his colleagues may have been right to feel that their work, however harshly criticized and however laborious, was worthwhile.

A few days after this speech, Joliot announced to the Congress that although the French government was presently directing its nuclear program strictly toward peaceful applications, *"if tomorrow they ask us to do war work, to make an atomic bomb, we will reply: No!"* As the Congress closed, scientists and technicians made the same pledge, signing on bands of gold paper. The declaration was widely reported, and soon postcards appeared bearing Joliot's portrait and the statement, sealed with his elegant signature.[27]

These declarations, and particularly Joliot's presence at the head of a Congress largely devoted to anti-American propaganda, angered many in France. One newspaper demanded that he be fired from the CEA. Later that year it repeated its criticism, warning that there were over a hundred Communists at Châtillon and a disproportionate number in the Scientific Committee.[28]

Joliot's correspondence files yield indirect evidence bearing on such claims. In March 1948 he got a letter from "the communists of the three cells of Châtillon" sympathizing over the detention of his wife in the United States, and in April 1949 he received birthday greetings signed by representatives of four cells there. Cells were supposed to redivide once they had more than thirty to fifty members, so a figure of around a hundred party members at Châtillon in early 1949, or about a quarter of the workers there, would not be excessive. This percentage does not imply that Joliot tried to pack the CEA with Communists, but only that his leadership was congenial to them, and that there were many leftists in the pool of young engineers and scientists from which the CEA necessarily recruited. At one point a quarter of the students of the Ecole Normale were also reportedly party members.[29] Convinced anti-Communists, both in France and abroad, thus had reason to feel that the CEA was not their territory.

Such a conviction was now represented in the heart of the CEA itself. After Denivelle had resigned as secretary-general, both Joliot and Dautry had suggested candidates for this sensitive post midway between their divisions. Eventually a compromise was struck with a man proposed by the military representative on the Atomic Energy Commission. The new secretary-general, René Lescop, held a minor position in the Radical party, but his career was chiefly associated with ordinary work in military engineering. He was an alumnus of the Ecole Polytechnique. The scientists in the CEA soon had reason to believe that Lescop was an anti-Communist who meant to have Joliot and his collaborators replaced, if possible with Polytechniciens, so that the nuclear energy program would come under the control of industrial and military circles.[30] At first the new secretary-general kept in the background, but he found opportunities for action when, starting in 1949, the CEA ran into administrative difficulties.

With the completion of Zoé the early, heroic period of the nuclear energy program had come to a satisfactory end. The CEA, no longer pursuing a simple, immediate goal but still growing with startling speed, needed to get itself better organized. At the start of 1949 Lescop lectured the Scientific Committee on the need for better administration, and the Committee agreed on three points: "(1) While accounting in the legal sense is done well, there does not exist at the CEA accounting in the industrial and technical sense . . . (2) The Personnel Services do not exist. (3) The Purchasing and Sales Services remain entirely to be organized.[31] The leaders had to devote their attention to straightening out these matters and establishing a clear table of organization for the scientific services.

The CEA's budget also was a problem in 1949, for its approval ran into unexpected delays. Shabby lies were being whispered about Joliot: that he had diverted CEA funds, or that he was liable to defect to the Soviet Union. The scientists had to mobilize all their forces, including their advisory Scientific Council, who wrote the prime minister that proposed reductions "would compromise the development of Atomic Energy in France." Dautry too fought to win over the government, as when he visited the president to offer reassurances about the political situation within the CEA. Even so the CEA did not get quite all it had requested. Later that year Joliot advised that "from now on the members of the Scientific Committee should forewarn their parliamentary friends in order to arouse them to intervene again when there is a vote."[32]

The Scientific Committee of the Commissariat à l'Energie Atomique, 1947: (standing) Bertrand Goldschmidt, Pierre Biquard, Léon Denivelle, Jean Langevin; (seated) Pierre Auger, Irène Curie, Frédéric Joliot, Francis Perrin, Lew Kowarski

The neutron studies room
with graphite pile
at Châtillon, c. 1948

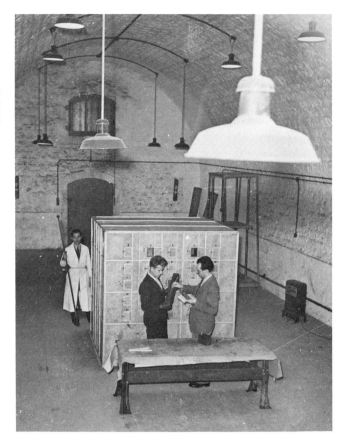

Journalists surrounding Joliot and Francis Perrin after startup of Zoé, 1948

Marcoule, an industrial complex of the CEA, in the 1970s (large buildings on right house reactors)

The CEA leaders took the trouble to develop such friends. For example, during the budgetary fight Goldschmidt had looked over the National Assembly and noticed a rising young deputy who was expert in matters of finance, Félix Gaillard. Goldschmidt invited him to visit Châtillon and see Zoé. At the end of the morning's visit, Gaillard was brought to the Collège de France to meet Joliot, with whom he stayed, caught by the scientist's spell, until three in the afternoon. "He had given up lunch," Goldschmidt recalled, "but had been definitively conquered by atomic energy." From then on Gaillard seconded the scientists' efforts.[33]

This sort of politicking, building on widespread agreement that nuclear energy was essential for France, assured the CEA a smooth growth of funding throughout its early years. From its origins until about mid-1948 the CEA's rate of spending doubled every six months or less, and from then through 1950 the expenditures, now much larger, doubled every twelve months. By the end of 1950 a total of nearly eight billion francs had been spent, about half of it on buildings, permanent equipment, and primary materials such as ores and graphite, a third on salaries, and the remainder on consumable materials, other overhead, and miscellaneous expenses. Dautry prided himself on the smoothness of the funding operation.[34]

Although the main concern of the Scientific Committee was technical progress, the bulk of the time in their frequent meetings was spent on other matters—administration, salary levels, press relations, ceremonial visits to Zoé (including a trip by Halban), and a hundred other arrangements. In 1949 perhaps the most persistent item on their meeting agendas was the construction at Saclay, and they spent many hours on such matters as the architecture of the buildings and arrangements for electricity.[35]

Zoé was brought to its top power of ten kilowatts in the spring of 1949. The scientists and technicians monitored it closely in order to understand how the power output went up and down and how this output was affected by temperature. The CEA was thus developing a group of reactor experts. Perhaps the greatest technical advantage of owning a reactor came when the French had to measure the purity of materials like graphite and uranium oxide. Where once the scientists had needed a ten-tonne pile of graphite and three months of work to determine the quality of a sample, they now could simply put a piece inside the reactor tank and calculate its neutron absorption precisely by seeing how it affected the reactor's functioning.[36]

Because it had a working reactor, the CEA had to give much closer attention than before to problems of health and radioactivity. A biological and medical division had been established in early 1947, but it had been devoted more to ordinary industrial safety and biomedical research with radioactive isotopes than to protecting people from radioactivity. Zoé posed a real hazard to the workers, so at the beginning of 1949 Kowarski paid a friendly visit to the new British nuclear center at Harwell and on his return advised revision and enlargement of the health services. His chief concern was worker safety; the problem of polluting the countryside did not worry the French scientists at this stage, since they scarcely had the means yet to create much of a hazard.[37]

A few weeks after Zoé went critical, Kowarski set his staff to work designing their next reactor, to run at over a megawatt. As with the first reactor, the chief purpose of the second and larger one was political: it would maintain France's international prestige and also her bargaining position with other countries when seeking nuclear information and materials. A larger reactor with cooling would also be a step toward producing industrial power. But with their increasing experience of the difficulties in building reactors, the scientists were coming to feel, as Goldschmidt predicted rather accurately, that it would take at least thirty years before nuclear energy could start seriously to supplement coal in commercial power production. In August 1949 the Scientific Committee, meeting in the Curie house at L'Arcouest to review their programs, approved the plans for a second reactor. By the end of the year construction had begun at Saclay.[38]

Zoé, built in a hurry with makeshifts and guesswork, was essentially a scientific research tool; the second reactor would be unequivocally a pilot plant. It would be built less from physicists' simplified calculations based on first principles than from engineers' detailed estimates and experience. The basic shape was fixed by the materials that the CEA could lay hands on: uranium metal, assuming that it could be produced, and heavy water, if necessary reusing Zoé's. As usual with the early reactors, cooling posed the most painful choices. An engineering group studied a number of schemes in detail, finally committing the CEA to compressed gas as a coolant.[39]

The Scientific Committee had asked Goldschmidt to take over the job of getting pure uranium metal by the tonne. Sensible and reliable, but enthusiastic once he was embarked on a program, Goldschmidt took on more and more responsibilities. As he shuttled between Paris and Le Bouchet, he was also shuttling between his roles as administrator and laboratory chemist. He had chosen the method that Kowarski and Guéron had learned from the British, in which uranium oxide was reduced to metal using calcium of "nuclear" purity. At the end of 1947 the Scientific Committee had arranged a contract with the Société Electrométallurgique du Planet to build a new factory for distilling calcium metal to this unprecedented purity, and through the next two years the company's work proceeded under the control of CEA engineers. As in other areas of uranium production, the CEA reserved ownership of the installation for itself. By the end of 1949 satisfactory calcium was coming forth, and uranium ingots soon followed.[40]

Meanwhile, in the fall of 1949 a rod of uranium oxide was removed from Zoé and set aside for six weeks while its intense radioactivity fell off. Then Goldschmidt's chemists set to work handling the rod by remote control behind thick steel radiation shields, attempting to extract the plutonium that had been created within it. Goldschmidt had been working out improved methods of plutonium extraction, but for this first sample he relied on a process similar to the one he and his colleagues had developed in Canada.

Finally one Friday in late November the team obtained a liquid which they expected would contain about one milligram of plutonium. They planned to precipitate out a solid plutonium compound the next week, but on Saturday Goldschmidt sent messages to his helpers: "I can't wait until Monday, you absolutely must come Sunday." His colleagues gathered to watch as Goldschmidt personally combined the chemicals in glass vessels. Nothing happened. Someone exclaimed that the method had failed, and then a quantity of plutonium iodate appeared like snow. The jubilant group immediately carried the vessel off to Joliot's home. It was an unforgettable experience, he later said, to see this element he had heard so much about.[41]

Joliot had made it a practice about twice a year to address the workers of the CEA, from scientists to cleaning women, in order to inspire them with a common drive and purpose. He called a meeting that Tuesday for the entire staff, by now nearly a thousand people. After reviewing the CEA's general progress, he held up the tube with its tiny quantity of whitish powder and announced their first plutonium. "That day," one of the administrators wrote, "enthusiasm was total and all the staff of the Commissariat, from the directors to the lowest employees, shared the same élan."[42]

But this community of spirit was fragile. When the plutonium precipitate appeared in Goldschmidt's liquid, it signaled a turning point for the CEA. The early years of struggle against technical problems were over; now virulent political problems, inherent in the CEA from the start, were beginning to appear in the open. One cause was the very success of the scientists. For it was now obvious that the CEA was a thriving organization which would play a decisive role in the future of France.

18

Politics and Plutonium

As the CEA moved toward its goal of big reactors, capable of producing both industrial power and plutonium, it was thereby moving month by month deeper into political territory. Joliot too, his destiny intertwined with his agency's, was getting deeper into politics. But he made a distinction that reactors cannot make: he wanted to produce abundant power but not plutonium for bombs. Accordingly he devoted more and more of his time to public action against nuclear weapons, believing this might help save the world from catastrophe. During 1949 he gave frequent speeches on the subject around France and in foreign capitals.

Goldschmidt, Guéron, and Kowarski were disturbed by Joliot's increasing commitment to propaganda, if only because it meant that he was seldom around to lend them his advice and authority in their daily administrative problems. Shortly after the first extraction of plutonium, Joliot complained to the "Canadians" that it had been done without his knowledge. Kowarski brusquely replied that such things would not happen if Joliot were not away so often, and Joliot walked out in a bad mood. The next day there was a blowup in the CEA offices, which had moved to the Rue de Varenne several months earlier. Joliot reproached his colleagues for lacking confidence in him and declared that he in return had lost confidence in them, particularly in Guéron. Thereupon, according to Goldschmidt:

Guéron got up in a huff and said, "If it's like that I resign," and left the room and went upstairs to his office. So there was a great silence. Joliot said, "I believe I have exaggerated a bit." And Kowarski and I said, "Yes." Then he said, "Perhaps I should apologize to Guéron." And Kowarski and I said, "Yes." And then he said, "I really would like to go apologize to Guéron but I don't even know where his office is!"

So we told him and he went and looked for Guéron. And then we had an extraordinary session, everybody speaking his heart out, saying what he thought. By four o'clock we went to lunch, a spaghetti lunch in one of those small restaurants where Joliot probably had been during the Resistance, and by then we had made peace. It was understood that every Wednesday he would give us the whole of the afternoon to discuss all the things of managing the Commissariat . . .

He was an extraordinary man for human relations. I would say that those last five or six months we worked with him were the ones where I felt the closest to him and where I admired him the most. I'm not speaking of science but of his human qualities.[1]

While Joliot agreed to give them the necessary attention, his colleagues agreed not to criticize his politics publicly. Yet despite their mutual respect, despite their wholehearted cooperation, something was wrong beyond repair. As Kowarski recalled, the leader of the CEA seemed as if "stranded on a boat which was driven further and further from shore on a strong current. Joliot more and more identified with the Communist party, and the role of the Communist party was dwindling. Joliot was carried away from us before our eyes."[2]

There would have been little problem if Joliot, like scientists before him, had used his political connections simply to raise support for disinterested science; but in fact he brought pressure on governments to change their general policies. This action subjected him to pressure in return. The counterpressure was the more intense because of his decision, in his bid for political power, to cast his lot with the French Communists.

While the Communists were mimicking A. A. Zhdanov, who vilified independent intellectuals in the Soviet Union, the French conservatives were imitating Senator Joseph McCarthy, who vilified independent intellectuals in the United States. With the strength of the right in the French government growing, anti-Communist pressure increased. Although the sometime Resistance hero Bidault was prime minister again, his government was weak and fragmented, and he went along with the anti-Communism of his political allies.

One sore point was the Communist head of the CNRS, Georges Teissier, Joliot's friend since the days of the National Front. Teissier proclaimed that France must discard the "imperialist civilization of Truman" and embrace the way of Joseph Stalin; under capitalism, charged Teissier, science would only be starved, disorganized, and misused. In January 1950 Bidault's government fired Teissier. The Communists claimed that a purge was underway, with Joliot liable to go next. They pointed to attacks in the American press which lumped him with the recently discovered British traitor Klaus Fuchs and other scientist-spies, as proof that the Americans hated Joliot, and charged that the ambassador to Paris had been instructed to push for the scientist's removal. According to the party press, if Bidault dared to touch Joliot, it would confirm that the French government was Washington's slave.[3]

The United States State Department did regard Joliot as a dangerous enemy, while the United States Central Intelligence Agency secretly listed half of the members of the CEA's Scientific Committee as "Communist sympathizers" and judged only Goldschmidt and Guéron to be "politically reliable." The Americans, however, wanted to avoid any appearance of directly pressuring France to remove Joliot; seeing the CEA as a nest of Communists with or without him, the State Department simply did whatever it could to impede the CEA's progress. As the official in Washington responsible for such matters remarked, "Joliot-Curie's presence . . . made it easier to say 'No' to any French request for cooperation in the atomic field." But the Americans, in common with Bidault's government, were finding it harder and harder to tolerate Communists.[4]

The Communists were doing everything possible to discredit and impede the French government. To obstruct the war in Indochina, the rearmament of Germany, and the formation of the North Atlantic Treaty Organization, Communist-led dockers refused to load war materials, railwaymen sabotaged arms shipments, and workers hit industries and services with wildcat strikes. The party linked every political and economic issue to the central theme of peace and nuclear disarmament. The strategy was effective, for people across the political spectrum felt naked before the threat of yet another invasion from the east, and feared that American belligerency would lead to the annihilation of France.

In the midst of this agitation Joliot stood out. In March 1950 he attended a meeting in Stockholm of the Committee of the Congress of Partisans of Peace, and he became the first to sign its Stockholm Appeal, which demanded the interdiction of nuclear weapons. Around the world millions followed him with their signatures, placing beyond narrow partisan boundaries the belief that the use of nuclear weapons would be a crime against humanity. The French Communist party made the Stockholm Appeal a centerpiece of its propaganda and set quotas for gathering signatures by each of its cadres.[5]

The campaign reached a crescendo at the party's twelfth National Congress, held in the working-class suburb of Gennevilliers in April 1950. Some nine hundred delegates, a majority of them under thirty years old, crowded into a hall for twelve hours a day to listen to speeches, applaud, and sing. Observers remarked that the Congress resembled partly a mass demonstration and partly a religious retreat. A huge red banner above the podium declared, "The French people do not and never shall make war against the Soviet Union." When Joliot appeared on April 5, there was an enthusiastic uproar. He spoke of the danger of war fomented by imperialists and of the scientists' struggle for liberty and peace. The delegates stood and applauded wildly when he reached his peroration: "progressive scientists and communist scientists shall not give a jot of their science to make war against the Soviet Union . . . we shall hold firm, sustained by our conviction that in so doing we serve France and all of humanity."[6]

Joliot was again declaring his independence of government orders: if they ordered him to build nuclear weapons, he would refuse. In part this was a personal statement of a deep moral feeling. He had made it clear as early as 1944, when Groves's men interrogated him, that he saw nuclear energy as something special which should never be used for destruction. But because he was the head of the CEA, his statement was not only personal but also political. His closest colleagues did not doubt that he had been impelled by his party to defy the government. The day after this speech, Perrin recalled, Joliot told his friends, "If the government doesn't fire me after what I've said, I don't know what more they need."[7]

Yet at heart he did not expect the irresolute government to act against him. His statutory five-year term as high commissioner was due to run out at the end of the year; the government, rather than cashier France's foremost scientist and raise a storm, could simply fail to renew his appointment. No doubt they did not intend to renew him in any case, so Joliot must have felt that there was less to risk in speaking out than in keeping silent.

The situation at the CEA increasingly embarrassed the government. The proportion of Communists in the upper levels was believed by some to be as high as 65 percent, although this estimate actually exceeded even the proportion of fellow travelers. CEA scientists and technicians, following Joliot's lead, were also circulating a declaration that if they were ordered to work on nuclear weapons, they would refuse.[8] Although the government had not seriously considered building such weapons, nor had yet the means to do so, its authority was being deliberately undermined.

On the day of Joliot's Gennevilliers speech a large group of moderate and rightist deputies petitioned the government to take "all legitimate measures of active defense against the anti-French activities of a foreign national party known as the Communist party." They asked the government to pay special attention to the "elimination of any risk of espionage and Communist penetration in the various centers of scientific research." Joliot's speech helped confirm him as a symbolic target for the anti-Communists, and personal attacks against him in parliament and the press grew virulent.[9]

On April 26 Joliot was scheduled to be far from partisan struggle, among friends at the regular Wednesday conference with his colleagues at the CEA, then in the evening at a delayed celebration of his fiftieth birthday. That afternoon in the midst of discussion with his fellow-scientists at the CEA headquarters he got a telephone call, summoning him to see the prime minister. Bidault, Joliot's old comrade from the Resistance, greeted him with visible emotion. He told Joliot that he was to be removed from his post at the CEA immediately.

Although Joliot had known this could happen, he was stricken. The blow was not announced at the party with his friends that evening, and the atmosphere remained merry while the food and wine, which he always savored, were good. Then he rose and began a rambling speech; for hours, it seemed he could not stop himself. Only the few who had heard the news understood that his wide-ranging observations were a sort of testament.[10]

For a long time afterward, Joliot seemed a broken man. He was cut off from the organization and the work which meant more to him than any other. During the next years he was further wounded not only by the predictable scorn of his enemies but by the unexpected silence and even disregard of some people, including Communists, he had thought his friends. He had lost not only his power but even the admiration that power may confer.

Joliot's dismissal marked the end of an era in the history of the French nuclear physicists. It became clear before long just how great a change his passing symbolized. Of the whole complex, controversial, and obscure history of the CEA during the next years, a few characteristic events suggest what Joliot's fall meant.

Bidault justified his dismissal of Joliot to the National Assembly on the grounds that Joliot had encouraged defiance of the government, not only by promising to refuse to build weapons but also by supporting the workers who were on strike or sabotaging arms shipments. The Communists retorted that the government had acted as a puppet of the United States. They ignored Bidault's ambiguous statement that France was not at present building nuclear weapons, for according to the Communists, France's American masters would never allow her to own a nuclear bomb. The Assembly approved the government's position by a strong majority, with the Communists alone opposing. In this debate, as in most public discussions at the time, nobody touched on the issue that would eventually be the most important—the fact that the CEA was swiftly building up a capability to produce plutonium and the government had never specified just what would be done with this plutonium once it became available.[11]

It is true that Joliot was fired because he was a Communist who embarrassed the government. But beyond transitory political maneuvers, Joliot's Communism laid bare tensions that had always been present but might otherwise never have come to light. He was only the most vehement of many who believed that the voices of "progressive" scientists must be heard when science is put to work. The government would not brook such independence, being inclined to bind scientific work more closely to its own military and industrial interests.

With Joliot gone, Dautry was in sole command of the CEA. The remaining scientists thought it best to put off naming a new high commissioner until things settled down. "Perrin, Kowarski, Guéron, and myself," Goldschmidt recalled, "knocked on numerous official doors to explain this point of view, which was understood and adopted." But the continuing vacancy left the administrative division in a position to extend its authority. Lescop, the minor Radical party functionary who openly advocated military and industrial interests from his post as the CEA's secretary-general, had already begun to make changes with Dautry's backing. The result, Kowarski noted in December 1950, was "guerrilla warfare" between his people and the administrators. The administrative division by gradual steps seized control of hiring and promotions, contracts with outside firms, medical services, and many other matters which the Scientific Committee had once run to suit themselves. Protests to Dautry brought only bland assurances that problems would be straightened out.[12]

Meanwhile a group of deputies demanded a systematic purge of the Communists and foreigners who were thought to "honeycomb" certain agencies, including the CEA. While Dautry and Lescop did not undertake a wholesale purge, they tried not to hire any new Communists and dismissed a few, some from the upper ranks and some technicians. As a sign of its demand for strong measures at the CEA, the National Assembly cut the 1951 nuclear energy budget back to four-fifths of the 1950 level, which gave the administrators an opportunity summarily to dismiss more people.[13]

The reduction of personnel and the uncertainty over who would go next, combined with the sudden loss of funds, demoralized many in the CEA. Almost the only substantial project left was the experimental gas-cooled reactor going up at Saclay under Kowarski's supervision. Kowarski himself felt more and more insecure in his position and was on the watch for other possible jobs, if necessary outside France.[14] The CEA reached its nadir in the winter of 1950-1951. Irène Curie's term on the Atomic Energy Commission was allowed to run out, which left two places to fill. It seemed to the scientists that the CEA was about to be captured, following Lescop's plans, by a politically rightist group interested in aiding private business and building nuclear weapons.

Francis Perrin emerged at this point as the pivot of the scientists' resistance. In the thirties people had seen him as a man held back by the overpowering personality of a famous father, but he now revealed a political drive and maturity of his own.[15] Affable, modest, and less inclined than his father to push visionary schemes, he was no less adept at

administrative affairs and political arrangements. He found allies among
the chiefs of the Socialist party and won Dautry's support as well. With
Goldschmidt, Guéron, and Kowarski he again made contact with politicians and officials and, equally important, with eminent scientists and
technocrats. After months of office visits, dinners, and telephone calls,
a compromise emerged.

In April 1951 Perrin was confirmed as the new high commissioner.
But in the meantime the government adjusted the CEA to make the high
commissioner and his scientists clearly subordinate; from now on there
would be only one head, the administrator-general. The Atomic Energy
Commission was at the same time enlarged to include officials from other
government agencies and representatives of private industry, in order
to tie the CEA more closely to these groups. Perrin and his colleagues
retained considerable autonomy, particularly in the furtherance of pure
science. But the lesson of Joliot's dismissal was driven home: the government, not the scientists, would have the final say over the uses of nuclear
fission.[16]

In August 1951 Dautry suddenly died. A long-time believer in the
value of scientific research and cooperation with engineers, as well as
a born administrator, Dautry had probably suffered from the political
turmoil that beset the CEA.[17] A few months later a new administrator-
general was named, Pierre Guillaumat, a forceful and capable personality as well as an engineer and Polytechnicien. Guillaumat eliminated
Lescop's post of secretary-general and strengthened the administrative
division, and also committed himself along the same policy lines that
Lescop had promoted: close cooperation with French industry and the
manufacture of plutonium.

From 1946 when Joliot presented the CEA's first plan for reactor development, the scientists had oriented their program toward the eventual
construction of one or more megawatt-class graphite-moderated reactors. At meetings of the Scientific Committee in 1948 and 1949 Perrin
and Kowarski had given estimates of how much pure uranium metal
and graphite these would require, and by 1950 the CEA's factories were
turning out both these materials by the tonne.[18] The projected reactors
were to serve as a step toward electrical power plants, but everyone
understood that the laws of physics required the reactors to produce not
electricity alone but also plutonium.

The scientists had a number of reasons to extract the plutonium. Since
in all countries reactor technology was embryonic, nobody could guess
which of many possible forms the future nuclear industry would take,
and some of the possibilities involved the use of plutonium in reactors.
Moreover, the extent of both France and the world's uranium deposits
was still unknown, and many scientists, fearing that the deposits would

be meager, hoped to extend them by "breeding" plutonium in reactors. These reasons are less compelling today, now that plutonium breeders are known to be more hazardous than other reactors and uranium-235 has become widely available as a substitute for plutonium, but in 1951 it seemed unlikely that a country could stay economically strong over the long run without producing plutonium.[19]

Enough plutonium for a substantial reactor is also enough for a good many nuclear weapons. Leading members of the government understood this simple fact, and it must have been in the fore during their deliberations over the future of the CEA. Ever since Hiroshima some French political and military leaders from de Gaulle down had felt that a country without nuclear weapons would not be taken seriously, and they had waited impatiently for the day when France could return to her former role as a great power.[20] They had not exerted pressure on the CEA while nuclear bombs had been well beyond France's reach, but the CEA's progress was changing the situation.

Some of the scientists, including Goldschmidt, Guéron, and Kowarski, were willing to produce plutonium even if it might be usable for weapons in some distant future; others, such as leftists who still survived in the lower ranks of the CEA, hoped to stay as far away from weaponry as possible. Perrin was in between, but he felt that the few bombs that France might eventually build would be more dangerous than useful to possess. It was not clear when the issue would have to be faced, for as Perrin noted, "a decision to fabricate atomic bombs cannot be taken until after the success of the first development stages, stages almost independent of the final end, be it productive or military." Nevertheless, when confronted in mid-1951 with plans for the next stage—multimegawatt reactors—Perrin had misgivings. Might these not create a temptation to produce bombs? John Cockcroft, head of nuclear energy work in Britain, warned Perrin: "You shouldn't build these reactors. Then you will have the military on your back from the start . . . It will be unbearable for you if you go on with such reactors." Perrin tended to agree, so in August 1951 he proposed that the CEA develop reactors in a leisurely fashion with emphasis on research. He argued that the funds and uranium at the CEA's disposal were too slight to allow substantial production of plutonium.[21]

That same month the government named a new secretary of state responsible for nuclear energy: Félix Gaillard, the youthful deputy who two years earlier had visited Châtillon with Goldschmidt and been won over by the promise of nuclear power. With support from other political figures he now offered the CEA an impressive sum of money to be spent on a greatly expanded program over the next five years; the offer was contingent on the expectation that the CEA would use a good part of the money to make plutonium. This offer of funds nullified Perrin's argument that the CEA's program was too small to produce much plutonium. From then on he and his colleagues had little choice, for they saw no sensible way to spend the money except on plutonium-producing reactors. With the scientists' cooperation or without it, these were going to be built.

Kowarski and Goldschmidt now drew up plans for two different graphite-moderated reactors and a plutonium extraction factory. In a few years these could begin producing roughly 20 kilograms of plutonium a year. Reluctantly Perrin endorsed the plan. Soon after, a third reactor was added; since it was identical to the second one, it could be of no use for either experiment or development, but only for making more plutonium. As Guéron recalled, at this point he found the military implications inescapable: "To my knowledge, there was no formal decision to make a bomb, but it was clear that lots of people wanted it, and Guillaumat certainly was not averse—he went at it in a somewhat slow, underhand way, but it was clear that he was preparing for it. We could still hope at that time that the government would abide by its U.N. pledge. But it became less and less probable."[22]

When the five-year plan for the CEA was presented to the National Assembly, for the first time France's nuclear program was subjected to extensive public debate. Few deputies objected to the production of plutonium, and when the Communists proposed that France promise not to use it for weapons, the motion was soundly defeated. "It would be strange indeed," said a government spokesman, "at a time when so many countries on both sides of the iron curtain are making weapons of mass destruction, for France to refuse on principle the possibility of doing the same to assure her own defense." The government repeated what its predecessors had maintained, that France had no plans to build nuclear bombs. This was strictly true, for the CEA had not yet begun technical studies of their construction. Nevertheless from its birth the CEA had followed a program which, whether the scientists willed it or not, logically culminated in the production of enough plutonium for bombs. The government would not have given the CEA so much money had this not been true.[23]

While many politicians, like many scientists, hoped bombs could be staved off for a long time, other hopes were held by some of the people who had fought for an enlarged CEA. In a secret survey for American officials, the Central Intelligence Agency reported that General Paul Bergeron, who since 1948 had been the military representative on the CEA's Atomic Energy Commission, and others in the French military knew by 1952 that enough plutonium for several bombs could be produced within a few years: "If and when this happens, it seems certain that the Military will try to take over the plutonium for use in atomic weapons . . . Plans are already being laid to accomplish this purpose which is directly contrary to the stated aims of the CEA."[24]

From 1952 to 1960 the CEA doubled its personnel and its budget every two years. Massive gas-cooled graphite reactors appeared at Marcoule on the banks of the Rhone. The first reactor, G-1, was built to yield 40 megawatts of thermal energy, and it was designed explicitly to optimize production not of power but of plutonium.[25] It started up in 1956. A full-sized plutonium extraction factory began operating soon afterward.

Unlike previous projects, these plants were in the hands not of the CEA itself but of industrial contractors. This was deliberate government policy, for the new leaders of the CEA felt that their tasks were now so large and technical that private industry must be brought in as a full partner. Perrin recalled that the change of control led to one of his few serious conflicts with the administrator-general: "When a factory was to be erected near Narbonne to purify uranium from the ore to uranium oxide, this factory was to be paid for by the Commissariat. And the question arose, What should be its status? Should it be a private company, in which the Commissariat would have maybe a fifty percent share [or] . . . fully under the Commissariat? I said it should be under the Commissariat, like the factory at Le Bouchet . . . Guillaumat said, No, it should be under the control and responsibility of a private industry." The two could not reach an agreement and finally had to take the matter to the responsible minister, who ruled in favor of private industry. To foster such ties, the CEA formed in 1952 an Industrial Equipment Committee which included eminent representatives of large private and industrial enterprises. From this point forward the CEA depended upon a number of important French corporations.[26]

In time nuclear energy became a part of France's industrial growth. By the 1970s most reactors were no longer in the CEA's hands, for Electricité de France, a government-controlled electrical power corporation, had taken over their operation. The electricity combine vigorously pushed a program of nuclear power stations, seeing no alternative except to mortgage France to foreign fuel suppliers, or turn off the lights; under the circumstances they were glad to have nuclear energy at hand. By the late 1970s reactors were providing more than 10 percent of the national electricity production. It was planned that by the mid 1980s two-thirds of France's electricity would come from nuclear energy, assuring the nation some independence.[27]

Through the early 1950s French governments continued to avoid an explicit decision about bombs. Meanwhile work on them proceeded with increasing vigor, for no definite decision was necessary either in public or in secret. For example, on December 26, 1954, the then prime minister,

Pierre Mendès-France, called an interministerial meeting to thrash out the question of making bombs. At the United Nations earlier that year the prime minister had, as Goldschmidt put it, "tried to convince the Russian and American delegates to stop nuclear testing. They laughed in his face." On the way back from New York Mendès-France had explained his defeat to the nuclear expert accompanying him, Francis Perrin: "You see, when there are discussions among the four permanent members of the Security Council, America, the Soviet Union, England and myself, we are like gangsters gathering around a table. Each one takes his knife and thumps it under the table—'Now we can talk!'—with their bombs. And I have no bomb to talk with them." Mendès-France had presided over other humiliating defeats, including the surrender of France's colonial empire in Indochina and the resurrection of the dreaded German army. In the December meeting, therefore, after hearing the pros and cons of building a bomb, Mendès-France announced that he was in favor: "A country is nothing without nuclear armament."[28] Although this decision alone was not enough to commit the government, and in due course Mendès-France fell from power, such incidents left Guillaumat, his group within the CEA, and their military allies free to press ahead with studies of nuclear weapons. The scientists, whatever their feelings, did not mount strong opposition.

When de Gaulle returned to power in 1958, he had only to endorse the work that had been done by the organization he had founded. In 1959 the French undertook the greatest of all types of nuclear engineering projects, a uranium-235 isotope separation plant which would help them to build hydrogen bombs.[29]. In February 1960 the first French plutonium bomb exploded above the Sahara. By the 1970s France had a full-scale nuclear strike force.

Meanwhile the French exported their reactor technology to a number of countries. Noticeable among these was Israel, whose tightly guarded Dimona reactor, a 24-megawatt plant supplied by France, has produced plutonium equivalent to at least one bomb a year since it began operation in 1965, although there is no proof yet that the plutonium has been put into the form of weapons.

Reactors using heavy water and natural uranium, like the Dimona plant, efficiently produce power as well as plutonium of military grade. The consequences of the work begun at the Collège de France in 1939 are perhaps most visible in the worldwide spread of heavy water reactors, all of which can be traced back to the French experiments in Paris and Cambridge. The American heavy water reactor and its descendents owed much to the French impetus, while the team that Halban had established in Canada produced even more direct results. The large NRX reactor at Chalk River, along with the pilot ZEEP, proved uncommonly successful. The Canadians have since continued to develop heavy water-natural uranium reactors. Their initial lead has allowed Canada to become the only country of small population among the serious competitors in the world nuclear energy industry. The highly successful Canadian Deuterium-Uranium (CANDU) reactor is of all reactors, aside from the controversial plutonium breeder, the most efficient in use of the world's finite stock of uranium fuel.[30]

The Canadians exported their technology to several other countries. Notorious among these was India, an early customer for a Canadian heavy water reactor. To the Canadians' dismay, in 1974 the Indians used plutonium produced in this reactor to construct a nuclear explosive. India has also pressed the use of heavy water reactors for peaceful power production.

The British program also had its roots at Chalk River, and for years many British personnel studied or visited there. Several of the British leaders had been with Kowarski when he built ZEEP, and the first substantial reactor in Great Britain used both designs and graphite originating in Canada. Britain set out to create an extensive reactor program, comparable with those in the United States and the Soviet Union. Engineering experiments for the next generation of reactors were carried out not only in Britain but also in Canada using ZEEP and NRX; the laboratory and pilot plant work for a plutonium extraction plant, begun by Goldschmidt, was completed largely at Chalk River. By the end of the 1950s British reactors were already producing several hundred megawatts of electricity along with several hundred kilograms of plutonium per year. As planned from the start, this plutonium was used to stock an arsenal of weapons.[31] By the late 1970s the British, like the French, were drawing over 10 percent of their electrical energy from uranium and possessed a dangerous nuclear strike force.

In short, the Collège de France team of 1939 was ancestor to an important portion of global nuclear developments, peaceful and otherwise. No one can yet say whether, as the French scientists intended, their work has been a blessing for humanity. Nuclear weapons may prove to deter great wars, although the weight of evidence indicates that they will only make the destruction which comes in such wars more swift and lasting, unless such weapons are recognized as useless for any positive purpose and are put aside. Nuclear reactors may prove to be too hazardous, expensive, or unpopular for widespread service, although the weight of evidence indicates that in the long run they may be less harmful than other available energy sources, whose hazards we are only beginning to recognize, and still less harmful than no energy at all. All that can now be said for certain is that the scientists have far extended humanity's range of choice.

The members of the little group that took up fission in France in 1939 devoted their later years to building instruments of pure research. The CEA continued to support fundamental research, and Francis Perrin, who remained high commissioner until his retirement in 1970, was particularly proud of his work in promoting the construction of a great particle accelerator at Saclay. He was gratified that this facility was built even though it would admittedly lead the CEA away from the applications of nuclear energy and indeed from any foreseeable practical application whatsoever.[32]

Halban left Britain in 1955 to take command of a nuclear physics group at the Ecole Normale, and over the next few years he directed the construction of an important French linear electron accelerator. But his heart, never strong, was failing. He retired a few years before his premature death in 1964.

Kowarski had thought of leaving the CEA since before Joliot's dismissal. Beginning in 1952, his activity was transferred gradually to Geneva, where he helped found the European Council for Nuclear Research (CERN), a major center for particle accelerators in which many other CEA scientists participated. He subsequently led CERN's computer division and continued to take a hand in nuclear policy questions.

Auger, who left the CEA in 1948 to become scientific director at UNESCO, concluded his career as Director of the French and European space programs. Guéron left the CEA in 1958 to become scientific director at Euratom, a Europe-wide nuclear energy agency, and in 1968 returned to the French university system as a professor. Goldschmidt was the only one of the wartime team to remain in the CEA; he stayed on as director of chemistry for some years and took charge of international relations until his retirement in 1977.

Joliot, after the shock of his dismissal from the CEA, slowly recovered his balance and got back to work with his usual energy. He continued to dedicate himself to nuclear disarmament, engaging without hesitation in endless correspondence and meetings. Meanwhile he encouraged the construction of an extensive research center outside Paris at Orsay, where Halban's accelerator and a new institute founded by Irène Curie were joined by other important facilities.

The couple continued to spend their summer vacations at L'Arcouest, which has retained some of its special atmosphere down to the present. The Joliot-Curie children, Pierre and Hélène (the latter married to Michel Langevin, grandson of Marie Curie's old friend), and the grandchildren are pursuing careers in science or other intellectual fields. The L'Arcouest circle no longer dominates French science as it once did, if only because of its success, for in the sprawling establishment resulting from the work of the founders of the CNRS and CEA, the influence of a few Normaliens and children of L'Arcouest is diluted by great numbers of Polytechniciens and university graduates of diverse origins. Nor does any one group of professors have the intimate relationship with politicians enjoyed by the L'Arcouest circle in the 1930s. Under de Gaulle and his successors new and stronger organizations at the highest levels have linked the entire community of academic science more tightly than ever before to government, industry, and the armed forces. But government planners still have not been able to solve all their problems, and were Jean Perrin alive, he would surely set to work against the various rigidities and inequities that have grown up.[33]

Joliot and Curie's personal influence was ended prematurely by death. She died in 1956 after the last of a number of difficult illnesses, and Joliot was hard hit by the loss. He too was ill and outlived her by only two years, just long enough to complete the institute she had begun at Orsay. He was a scientist to the end, following the processes of his own sickness—like Fermi and Szilard, who also died untimely in these years—with precision and a certain intellectual detachment, as if he were following the transformations of a radioactive solution.

Even while fostering ever larger teams of scientists, Joliot had worried about the prospects of research performed on an assembly line. Such research, he felt, was liable to be less exciting and productive than work done by trial and error, in close contact with the apparatus, alone or with one or two companions. Toward the end he looked back fondly to the days when a scientist could work like an artisan or an artist, personally summoning forth knowledge to reshape the world.[34]

Afterword

This book has been about the relationship within society between knowledge and power. By drawing together both technical and social facts, I have presented as much·as I reliably know about that relationship. I would now like to give, rather than a conclusion to this unfinished history, a few tentative and personal reflections on it.

In the early years of our century Pierre Curie, detecting with delicate instruments the warmth of radium, recognized that a new source of energy was at hand; before the deaths of his famous daughter and son-in-law his nation and others had committed their future to nuclear reactors and nuclear bombs. This history raises a troubling question: What forces brought such an outcome from such a beginning?

In the narrative I did not aim so much at a causal explanation in terms of preceding happenstance as at a primitive sort of structuralist explanation. I assembled types of events that kept recurring under different forms, such as visions of the power of science in peace and war; quests for money, authority, and personnel; transitions from laboratory puzzle to industrial project; and attempts to control consequences of research through patents, secrecy, or direct administration. All these events may well be outcroppings of a deeper structure of thought and social action. What can be said about this structure?

Claims have lately been made that humanity is helpless before science and technology—that discoveries come out of nature in a logically predetermined series, instigating technologies which have little relation to what human beings truly want. A few months after France exploded her first nuclear bomb, one Frenchman declared, "Since it was possible, it was necessary." Others have used the French case to illustrate a supposed principal law of our age, that technique dominates human considerations.[1] I believe this is false; we cannot excuse ourselves of responsibility so easily. Things become "possible" not just because of nature's dictates but also because of human imperatives every child understands. A nuclear plant producing electrical power is a tangible expression of the primitive desire for material goods and services; a nuclear weapon is an expression of equally familiar, darker wishes. Since France was a country poor in coal and oil, and subject to cruel invasions, she worked to provide herself with reactors and bombs. We must not underestimate the power of our wills, conscious and unconscious, to shape our world.

The nuclear physicists, from Pierre Curie's time down to Joliot's, made their choices deliberately. They chose to study the atomic nucleus, partly in hope of discovering a way to use its energy; they chose to force the discovery on the attention of government and industry; and they chose to develop its practical consequences. Generations of scientists before them had chosen to build up science's funding, prestige, and organization so that precisely this sort of discovery could be made, appreciated, and exploited. These early decisions are to this day transforming the world's politics and economy. In this sense the scientists, whether they sat with government ministers or not, were in power.

The laws of physics nevertheless strictly limit human power. Many of us have wished that cheap, perfectly safe reactors could be made, fulfilling H. G. Wells' dream of a nuclear engine in every car, and many have wished that nuclear bombs were impossible to build. Such wishes have no effect on the capture cross-section of uranium or the number of neutrons emitted per fission. We must pay careful attention to scientific facts, for our prosperity, or even our survival, may depend on the precise value of a constant like v. We may not even know what the physical laws are which define our possible actions, and we may not be able to make use of what we do know. Nuclear fission is a simple phenomenon, yet scientists struggled for years to see through it, and then large teams of engineers got into a thousand difficulties turning it to practical use. We are not magicians to make objects obey our commands at the wave of a hand. Rather, we must painfully negotiate with nature.

In this dialog between nature and human will, the scientists and engineers are the interpreters. Such intermediaries have their own characteristics which impose a form on whatever results. To understand what shaped the history of French nuclear physics therefore requires a close look at the principal actors.

The motives of Joliot, Halban, Kowarski, and their colleagues were complex: they yearned for respect as scientists, they took pleasure in discovering the ways of nature, they hoped to improve humanity's lot, and of course they harbored personal ambitions. Those ambitions might seem to have been particularly important in their history, for the scientists certainly sought and reached positions of high authority as they developed nuclear energy. But this outcome was a surprise to them, for in choosing scientific careers, they originally hoped only to win a good position in life, not to wield personal authority over grand events. Although like most people they struggled to rise, the scientists' chief ambition went far beyond their careers. They meant to set in motion forces that would transform society for all time to come, and they did.

They did not, in fact, keep control once they had set events in motion. Scientists like Halban and Kowarski in Cambridge and Montreal, Szilard and his colleagues in Chicago, or Joliot and the Scientific Committee of the CEA in Paris, for all their efforts, personally controlled events only as long as matters moved in the direction desired by other, more powerful groups. Scientists who failed to attract and keep outside support found themselves shunted aside. But the officials who then took command were not themselves the masters of history. Scientists held their sway not as individuals issuing decrees but as a body of people organizing discoveries and their application.

While scientists seem unable to keep power by their separate wills, events may be shaped by the scientific community, that is, by general characteristics shared by all scientists. Although little is known about such characteristics, psychological studies suggest that the sort of people who succeed in science are often better at reasoning situations through by logic than at feeling out their human meaning; they are more comfortable with things than with people.[2] This trait seems to be at least as true for engineers. Thus the history of nuclear physics would not be expected to result in social innovations; it would result quite naturally in physical devices. Among the hundreds of thousands of documents written by scientists and engineers in various countries over the first five years of nuclear fission work, not a single memorandum or letter has been found that attempts to analyze at length the consequences for society of developing reactors. Even the general implications of nuclear bombs were almost never discussed with care until around the time they were exploded upon cities.

The tendency to rush nuclear energy into use without examination was shared by many people, all of whom filled the same role in society; therefore it is less the character of any one person than the characteristics of the role itself that must be examined. People who fill the role of scientist are rarely pressed to discuss the social implications of discoveries; usually, as in the case of fission, it has been left to the scientists eventually to force such discussions on others. But pressures do bear on anyone who is a scientist, pressures exerted by every other member of society. Scientists, in order to get the money, authority, and personnel they need to make and exploit discoveries, must continually negotiate with other people. These negotiations give form to the scientist's role, for people expect something in return.

For generations France called upon scientists to render a number of services. The scientists best known to the public were medical doctors, and in a field as far removed from medicine as physics Marie Curie and Joliot were praised because their work might help cure disease. Scientists were also expected to advance the material wealth of their nation in everything from crops to chemicals. By the prestige of their work scientists could enhance the status of those who supported them, bolstering claims to leadership by either a group within the nation or the nation in the world. They were called upon for vital aid in times of war. As society changed, scientists like Jean Perrin and Joliot might also try to give a sense of direction toward an idealized future of prosperity and goodwill. And they could seek out and teach fundamental knowledge of the natural universe. All these are functions that human societies have needed since the distant past—these are precisely those functions once exercised by tribal shamans.[3]

While the functions have evolved very slowly, the structure in which they are realized has changed considerably since Berthelot's day. Then science was practiced mostly by individuals in personal relationships with each other; now it is practiced mostly by groups within organizations. Among the various functions of scientists, the modern structure particularly encourages the orderly and the practical. While scientists have personal and professional urges toward intellectual ferment, abstract visions, and sweeping change, these are not the urges of engineers, or of the administrators whose careers are linked with engineering. Such managers, as they labor to plan the most rational use of resources, serve and shape the managerial goals: organizational survival, stability, and measured expansion.[4] People like Sengier, McGowan, Groves, and Dautry had no burning desire that nuclear energy should revolutionize world society. They wished mainly to protect established interests, such as corporations and nations, and to see that these interests got a share in any advance. Such goals became dominant once the CNRS and CEA took shape as large, centrally managed bodies, interlocking with industrial corporations and government. The new structure necessarily had more to do with plans for practical applications than with ideals of truth.

From these organizations there issued nuclear reactors and weapons, technological devices which in turn demand centralized bureaucracies and industrial and governmental backing for their employment. The results reflected the means used to produce them. At first glance this congruence may seem puzzling, for ordinances cannot affect the nature of fission, but the explanation is simple. The physicists early recognized the nucleus as something not to be quickly understood and exploited without an organized, large-scale effort. Had Jean Perrin's circle and their colleagues elsewhere failed to seek and win the support of government and industry, places like Joliot's Collège de France laboratory complex would never have come into being, and without such institutions the studies leading to nuclear energy would have been done much later or not at all. Afterward, when the physicists decided to build working reactors and bombs, they had to call on the largest available governmental and industrial organizations, for otherwise little could have been done. The results reflected the means used to produce them because, out of the whole range of possible solutions to social problems, reactors and bombs were what the organization of the work made practical.

In short, the overall movement of these events was determined less by physics and technique, less by society at large, than by what lay in between: the particular structure within which modern scientists fulfill their role, a structure molded on one side by what can be done with poorly known laws of nature, on the other by an industrialized and bureaucratized society of competing national governments.

The question arises whether the French could have found better solutions to their problems of material welfare and defense. I think that the story could have turned out otherwise only in a very different society, one in which the people who chose to contemplate nature and aid humanity lacked either the desire to study and exploit nuclear fission or the means. Whether those hypothetical people would have put their effort into other approaches, and whether in this way they would have come up with better solutions to the dilemmas of energy and war, I do not know. But I do believe that there is no energy or other benefit of any kind that can be had without long work and care, and which may not bring with it sad weapons.

On the one hand, this means that we should be cautious about building up scientific organizations only to direct them urgently toward material needs—even when scientists encourage this development in hope of getting more money for their research. When we cherish research only to push each scrap of knowledge into use without reflection the moment it is won, when we press science to respond to wishes that are incompletely thought through or unconscious, we must not be surprised if the pressure returns upon society with frightening force. Then we may find ourselves, as now, staking our future on a peculiar instability of a rare element. As has often been said in recent years, science is a dangerously narrow base on which to rest our future.[5]

On the other hand, the solution can hardly be to abandon such a powerful tool. As French scientists often pointed out, knowledge springing from basic research has freed millions of people from grinding servitude and can take us much farther still. It would be stupid to try to improve our lot without using the powers of technology and science, which are our only powers in the natural world. Pure knowledge is itself the most wonderful and durable product of our civilization. Those who love the quest for truth will support basic research; in any case it is too late to choose to be ignorant.

So our best hope is to use science, but with care. We must understand what we want from it and make judicious decisions at an early point, before the scientific-technological process has gone too far. How can we be wise so soon? It may be that if we hold within our view not only the scientific facts of a matter but also the social conditions that bring the facts into human affairs, we will find ourselves wiser than we thought. We may be able to discern in the structure of research and its application a silhouette of what will result. And this structure is something we can all understand and, if need be, influence.

In particular, we must make some improvements in whatever influences the scientist's role. Each of us can help develop this role—and all the other roles it interacts with, including our own—as open rather than limited, as wide-ranging rather than fixed within bureaucratic constraints. Each of us can become more responsive to both technical logic and human feeling, to the laws of both our physical world and our social structures. Furthermore, we can step outside the boundaries of our jobs in order to act publicly on the basis of what we understand. Among such public acts, the unfolding of a social or even a moral vision may well be the most powerful. These prescriptions are well-known, even banal; yet they may run counter to the way society has been moving. Scientists in particular seem increasingly remote from what may be their natural role as unorganized, constantly thoughtful, and sometimes visionary seekers and helpers.

Fortunately there are examples to lead us. In France and elsewhere, from before Berthelot's time to Joliot's and after, many people understood that technical experts can and must take a hand in social and moral problems, while nonspecialists can and must grasp scientific and administrative issues. It is not easy; when Joliot's vision of equality and disarmament proved incompatible with the organizational and technical circumstances of nuclear energy, he was almost torn apart. Nevertheless he was aiming in the right direction, attempting to recombine within himself two ways of seeing and acting upon the world which have grown far apart. If we also work to escape the constrained thinking and activities of narrowly defined roles, we might look with more confidence to the future.

BIBLIOGRAPHY

NOTES

INDEX

Bibliography

Only works cited in footnotes are included. Omitted are all strictly scientific and technical articles, newspaper accounts, and mimeographed, microfilmed, or unpublished materials, as well as published works of scant relevance. For some hard-to-find works the call number in the Bibliothèque Nationale (BN), Paris, is appended.

Abelson, Philip H. "A Sport Played by Graduate Students." *Bulletin of the Atomic Scientists* 30, no. 5 (May 1974): 48-52.

Ailleret, Charles. *L'Aventure atomique française: Souvenirs et réflexions.* Paris: Grasset, 1968.

Alfaric, Prosper. "Pour la science dirigée." *Revue Socialiste* n.s. no. 30 (1949): 129-141.

Allier, Jacques. "Les Premières piles atomiques et l'effort français." *Revue Scientifique* 89 (1951): 331-350.

———. "Souvenirs sur la collaboration franco-britannique au début de la science atomique." *Energie Nucléaire* 2 (1960): 355-359.

Amaldi, Edoardo. "The Production and Slowing Down of Neutrons." In *Handbuch der Physik*, vol. 38, pt. 2. Berlin: Springer-Verlag, 1959.

Anderson, Herbert L. "Early Days of the Chain Reaction." *Bulletin of the Atomic Scientists* 29, no. 4 (April 1973): 8-12.

———. "Fermi, Szilard and Trinity." *Bulletin of the Atomic Scientists* 30, no. 8 (October 1974): 40-47.

———. "The Legacy of Fermi and Szilard." *Bulletin of the Atomic Scientists* 30, no. 7 (1974): 56-62.

Appell, Paul. "L'Institut Edmond de Rothschild." *Revue de France* 7, no. 12 (June 15, 1927): 715-723.

———. "Pour la science." *Revue Scientifique* 59 (1921): 245.

———. "La Recherche scientifique." *Revue Scientifique* 61 (1923): 161-164.

Audiat, Pierre. "Au Collège de France." *Revue de Paris* 53, no. 2 (July 1946): 99-109.

Auger, Pierre. "Hommage à Henry Laugier." *Les Cahiers Rationalistes* no. 300 (1973): 308-319.

L'Avenir de la culture. Addresses by George Cogniot, Paul Vaillant-Couturier, and Jean-Richard Bloch. Paris: Parti Communiste Français, Comité Populaire de Propagande, n.d. BN 8°S.Pièce. 18452.

Bagge, Erich; Kurt Diebner; and Kenneth Jay. *Von der Uranspaltung bis Calder Hall.* Rowohlts deutsch Enzyklopädie. Hamburg: Rowohlt, 1957.

Barrabé, Louis. "Quelques souvenirs sur Frédéric Joliot-Curie." *La Pensée* n.s. no. 87 (September-October 1959): 58-62.

Barrès, Maurice. *Chronique de la Grande Guerre*. 14 vols. Paris: Plon, 1924.

———. *Pour la haute intelligence française*. Paris: Plon-Nourrit, 1925.

———. "Que fait l'université pour la recherche scientifique? . . ." *Revue des Deux Mondes* 7th ser. 55 (1920): 241-282.

Bataillon, Marcel. "Une Matinée avec Joliot." *La Pensée* n.s. no. 87 (September-October 1959): 69-70.

Berget, Alphonse. "La Science." In *L'Avenir de la France: Réformes nécessaires*. Paris: Alcan, 1918, pp. 490-508.

Bernal, J. D. Address. In Association of Scientific Workers, *In Memory Paul Langevin*. London: Porteous, 1947.

———. "Langevin et l'Angleterre." *La Pensée* n.s. no. 12 (May-June 1947): 17-20.

———. *The Social Function of Science*. London: Routledge & Kegan Paul, 1939.

Berthelot, Marcelin. In *Commémoration du banquet Berthelot . . .* Paris: Imprimerie Nouvelle (association ouvrière), 1895. BN Ln^{27}44127.

Bichelonne, Jean. *Conférence . . . sur la réorganisation industrielle et commerciale de la France . . .* Paris: Imprimerie Municipale, 1943. BN 8°Lk^{18}4003(4).

———. *L'Etat actuel de l'organisation économique française*. Ecole Libre des Sciences Politiques, Conférences d'Information, 1941-1942, no. 3. Mesnil: Firmin-Didot, 1942. BN 8°R.47399(3).

Bidault, Georges. *Resistance: The Political Autobiography of Georges Bidault*. Trans. Marianne Sinclair. London: Weidenfeld & Nicolson, 1967.

Binion, Rudolph. *Defeated Leaders: The Political Fate of Caillaux, Jouvenal, and Tardieu*. New York: Columbia University Press, 1960.

Biquard, Pierre. *Frédéric Joliot-Curie et l'énergie atomique*. Savants du Monde Entier, vol. 3. Paris: Seghers, 1961.

———. "Mon ami Frédéric Joliot-Curie." *La Pensée* n.s. no. 87 (September-October 1959): 71-79.

———. *Paul Langevin: Scientifique, éducateur, citoyen*. Savants du Monde Entier, vol. 38. Paris: Seghers, 1969.

Blackett, P.M.S. "Jean Frédéric Joliot, 1900-1958." *Biographical Memoirs of the Fellows of the Royal Society (London)* 6 (1960): 87-105.

Blondel, F. "Le Ravitaillement de la France en matières premières minérales." *Revue des Questions de Défense Nationale* 1 (1939): 429-441.

Blum, Léon. *L'Oeuvre du Léon Blum*. Ed. Robert Blum. Vol. 4, pt. 1, *Du 6 février 1934 au Front Populaire*. Paris: A. Michel, 1964.

———. *Pour être socialiste*. 1919. 12th ed. Paris: Librairie Populaire, 1936.

Booth, Eugene T., et al. *Beginnings of the Nuclear Age*. New York: Newcomen Society in North America, 1969.

Borel, Emile. "La Crise économique et la science." *Revue de Paris* 38, no. 2 (March-April 1931): 756-767.

———. *Emile Borel, philosophe et homme d'action*. Ed. Maurice Fréchet. Les Grands problèmes des sciences, vol. 19. Paris: Gauthiers-Villars, 1967.

———. "L'Organisation de la recherche scientifique." *Revue Scientifique* 63 (1925): 545-548.

Borel, Marguerite [Camille Marbo]. *A travers deux siècles: Souvenirs et rencontres (1883-1967)*. Paris: B. Grasset, 1967.

Bothe, Walther, and S. Flügge, eds. *Nuclear Physics and Cosmic Rays*. Office of Military Government for Germany, Field Information Agencies Technical, FIAT Review of German Science. Wiesbaden: Dietrich, 1948.

Boucherie. "La Forêt française au secours de la Défense nationale: Les problèmes des carburants en France." *Revue des Questions de Défense Nationale* 1 (1939): 521-542.

Bourgin, Hubert. *De Juarès à Léon Blum: L'Ecole Normale et la politique.* 1938. Reprint. Paris: Gordon & Breach, 1970.

Boyer, Jacques. "Une Visite à M. et Mme. Joliot-Curie." *La Nature* 63, pt. 2 (Dec. 15, 1935): 585-586.

Brayance, Alain. *Anatomie du Parti Communiste Français.* Paris: Denöel, 1952.

Bromberg, Joan. "The Impact of the Neutron: Bohr and Heisenberg." *Historical Studies in the Physical Sciences* 3 (1971): 307-341.

Brower, Daniel D. *The New Jacobins: The French Communist Party and the Popular Front.* Ithaca, N.Y.: Cornell University Press, 1968.

Bupp, Irvin C., and Jean-Claude Derian. *Light Water: How the Nuclear Dream Dissolved.* New York: Basic, 1978.

Bush, Vannevar. *Pieces of the Action.* New York: William Morrow, 1970.

Cambridge University, Cavendish Laboratory. *A History of the Cavendish Laboratory.* London: Longmans, Green, 1910.

Canac, François. "Un Nouveau corps d'état: Les Chercheurs." *Revue des Deux Mondes* 8th ser. 18 (1933): 685-693.

────. "L'Organisation de la recherche scientifique." *Revue des Deux Mondes* 8th ser. 55 (1940): 354-365.

Casanova, Laurent. *Responsabilités de l'intellectuel communiste.* Paris: Nouvelle Critique, 1949. BN 16°R.Pièce.866.

Caute, David. *Communism and the French Intellectuals, 1914-1960.* London: Deutsch, 1964.

Chadwick, James. "Some Personal Notes on the Search for the Neutron." In *Proceedings*, 10th International Congress of the History of Science, N.Y., 1962, I, 159-162. Paris: Hermann, 1964.

Chastenet, Jacques. *Histoire de la Troisième République.* Vol. 6, *Déclin de la Troisième, 1931-1938.* Paris: Hachette, 1962.

Chatelet, Albert. "La France devant les problèmes de la recherche." La Documentation Française, *Notes et Etudes Documentaires* no. 2552 (June 20, 1959), no. 2580 (October 20, 1959), no. 2671 (May 28, 1960).

Childs, Herbert. *An American Genius: The Life of Ernest Orlando Lawrence.* New York: E.P. Dutton, 1968.

Clark, Ronald W. *The Birth of the Bomb.* New York: Horizon Press, 1961.

────. *Tizard.* Cambridge: Massachusetts Institute of Technology Press, 1965.

Clarke, I.F. *Voices Prophesying War, 1763-1984.* London: Oxford University Press, 1966.

Cogniot, Georges. "Un Homme véritable." *La Pensée* n.s. no. 87 (September-October 1959): 63-64.

Cohen, Francis; Jean Desanti; Raymond Guyot; and Gérard Vassails. *Science bourgeoise et science prolétarienne.* Paris: La Nouvelle Critique, 1950. BN 16°R.Pièce. 922.

Collège de France (Paris). *Annuaire.*

Collingwood, E.F. "Emile Borel." *Journal of the London Mathematical Society* 34 (1959): 488-512, 35 (1960): 384.

Compton, Arthur. *Atomic Quest: A Personal Narrative.* New York: Oxford University Press, 1956.

Conant, James Bryant. *My Several Lives: Memoirs of a Social Inventor.* New York: Harper & Row, 1970.

Cotton, Eugénie. *Aimé Cotton: L'Optique et la magnéto-optique.* Savants du Monde Entier, vol. 34. Paris: Seghers, 1967.

——. *Les Curie et la radioactivité.* Savants du Monde Entier, vol. 14. Paris: Seghers, 1971.

Coulomb, Jean. "Charles Maurain." In Association Amicale de Secours des Anciens Elèves de l'Ecole Normale Supérieure, *Annuaire,* 1969, pp. 27-28.

——. "Jean Perrin, Fondateur du Centre National de la Recherche Scientifique." In Institut de France, Académie des Sciences, *Notices et Discours* 4 (1962): 675-683.

Crowther, J.G. *The Cavendish Laboratory, 1874-1974.* New York: Science History, 1974.

——. *Fifty Years with Science.* London: Barrie & Jenkins, 1970.

——. "M. Louis Rapkine." *Nature* 163 (1949): 162-163.

——. *Science in Liberated Europe.* London: Pilot, 1949.

——. *The Social Relations of Science.* New York: Macmillan, 1941.

Curie, Eve. *Madame Curie.* Trans. Victor Sheean. Garden City, N.Y.: Doubleday, Doran, 1937.

Curie, Irène. *See* Joliot-Curie.

Curie, Marie. *Pierre Curie.* Trans. C. and V. Kellogg. New York: Macmillan, 1923.

——. *Radioactivité.* Paris: Hermann, 1935.

Curie, Marie, and Irène Curie. *Correspondance: Choix de lettres, 1905-1934.* Ed Gilette Ziegler. Paris: Editeurs Français Réunis, 1974.

Dalimier, Dr., and Louis Gallie. *La Propriété scientifique: Le Projet de la C.T.I. . .* Paris: A. Rousseau, 1923. BN 8°F.30511.

Dautry, Raoul. "L'Energie atomique." *Cahiers du Monde Nouveau* 3 (1947): 65-76.

——. *L'Organisation de la vie sociale.* Cahiers du Redressement Français, no. 24. Paris: Editions de la S.A.P.E. [Société Anonyme de Propagande et d'Editions], 1927.

De Broglie, Louis, "Notice sur la vie et l'oeuvre de Emile Borel." In Institut de France, Académie des Sciences, *Notices et Discours* 4 (1957): 1-24.

——. "Notice sur la vie et l'oeuvre de Frédéric Joliot." In Institut de France, Académie des Sciences, *Notices et Discours* 4 (1959): 222-248.

Documentation Française, La. "Le Centre National de la Recherche Scientifique." *Notes Documentaires et Etudes* no. 608 (April 25, 1947).

——. "Le Développement nucléaire français depuis 1945." *Notes et Etudes Documentaires* no. 3,246 (Dec. 18, 1965).

Duclos, Jacques. *Communism, Science and Culture.* Trans. Herbert Rosen. New York: International, 1939.

——. *Mémoires.* Vol. 3, *Dans la Bataille Clandestine.* Part 1, *1940-1942: De la drôle de guerre à la ruée vers Stalingrad.* Paris: Fayard, 1970.

Duclos, Jacques; Victor Joannès; et al., eds. *Le Parti Communiste Français dans la Résistance.* Paris: Editions Sociales, 1967.

Dupouy, Georges. "Raoul Dautry." *Revue Générale des Sciences* 62 (1955): 139-145.

Earle, Edward Mead, ed. *Modern France: Problems of the Third and Fourth Republics.* Princeton: Princeton University Press, 1951.

Eggleston, Wilfrid. *Canada's Nuclear Story.* Toronto: Clarke, Irwin, 1965.

Eisenmann, Mme. Zadoc-Kahn. "Les Savants français et l'énergie atomique." La Documentation Française, *Cahiers Français d'Information,* special no., November-December 1948.

Ellul, Jacques. *The Technological Society.* Trans. John Wilkinson. 2nd ed., rev. New York: Knopf, 1964.

Elsasser, Walter M. *Memoirs of a Physicist in the Atomic Age.* New York: Science

History (Neale Watson), 1978.

Fabre, Robert. "La France manque de Charbon." *Revue de Paris* 52, no. 5 (August 1945): 63-75.

"Une Famille de savants." *Annales Politiques et Littéraires* 106, no. 2544 (December 1935): 609.

Fauré-Fremiet, E. "Le Mouvement actuel pour la réorganisation des recherches scientifiques en France." *Revue Scientifique* 59 (1921): 1-10.

Fauvet, Jacques, with Alain Duhamel. *Histoire du Parti Communiste Français.* Vol. 2, *Vingt-cinq ans de drames, 1939-1965.* Paris: Fayard, 1964.

Feather, Norman. "The Discovery of the Neutron." In *Proceedings,* 10th International Congress of the History of Science, Ithaca, N.Y., 1962, I, 135-147. Paris: Hermann, 1964.

Fermi, Enrico. *Collected Papers (Note e memorie).* Ed. Emilio Segrè et al. 2 vols. Chicago: University of Chicago Press, 1962-1965.

———. "Physics at Columbia University." *Physics Today* 8, no. 11 (November 1955): 12-16.

Fermi, Laura. *Atoms in the Family: My Life with Enrico Fermi.* Chicago: University of Chicago Press, 1954.

Flanner, Janet [Genêt]. *Paris Journal, 1944-1965.* Ed. William Shawn. New York: Harcourt Brace Jovanovich, 1965.

Forman, Paul. "The Financial Support and Political Alignment of Physicists in Welmar Germany." *Minerva* 12 (1974): 39-66.

Forman, Paul; John L. Heilbron; and Spencer Weart. *Physics circa 1900: Personnel, Funding, and Productivity of the Academic Establishments.* Historical Studies in the Physical Sciences, vol. 5. Princeton: Princeton University Press, 1975.

France, Caisse des Recherches Scientifiques. *Rapport annuel* (1922). By M. C. Colson. Melun: Imprimerie Administrative, 1923. BN 8°Lf^{242}.175.

France, Commission Chargée d'Enquêter. *Les Evènements survenus en France de 1933 à 1945.* Assemblée Nationale no. 2344, Session of 1947.

France, Présidence du Conseil, Commissariat à l'Energie Atomique. *Commissariat à l'Energie Atomique, 1945-1956.* Paris: Commissariat à l'Energie Atomique, 1956?

———. *Rapport d'activité du Commissariat à l'Energie Atomique de 1er Janvier 1946 au 31 Décembre 1950.* Paris: Imprimerie Nationale, 1952. BN 4°Lf^{275}.7 (1946-50).

"French Atomic Scientists Report on Their Work in 1949." *Bulletin of the Atomic Scientists* 6 (1950): 299-302.

Friedmann, Georges. *La Crise du progrès: Esquisse d'histoire des idées, 1895-1935.* Paris: Gallimard, 1936.

Frisch, Otto R. "How It All Began." *Physics Today* 20, no. 11 (November 1967): 43-48.

———. "Lise Meitner." *Biographical Memoirs of the Fellows of the Royal Society (London)* 16 (1970): 405-420.

———. "Somebody Turned the Sun on with a Switch." *Bulletin of the Atomic Scientists* 4, no. 4 (April 1974): 13-18.

Gilpin, Robert. *France in the Age of the Scientific State.* Princeton: Princeton University Press, 1968.

Goldschmidt, Bertrand. "Hans Halban, 1908-1964." *Nuclear Physics* 79 (1966): 1-11.

———. *Les Perspectives actuelles de l'énergie atomique . . .* Paris: Les Echos, 1949. BN 8°Z.29625(77).

———. "Les Premiers pas." *Echos du CEA,* special no. (October 1965): 9-14.

———. "La Purification de l'uranium." *Atomes* no. 35 (February 1949): 52-53.

————. *Les Rivalités atomiques, 1939-1966*. Paris: Fayard, 1967.

Goldsmith, Maurice. *Frédéric Joliot-Curie: A Biography*. London: Lawrence & Wishart, 1976.

Golovin, Igor Nikolaevich. *I.V. Kurchatov: A Socialist-Realist Biography of the Soviet Nuclear Scientist*. Trans. William H. Dougherty. Bloomington, Ind.: Selbstverlag Press, 1968.

Goodman, Clark, ed. *The Science and Engineering of Nuclear Power*. Cambridge: Addison-Wesley, 1947.

Goudsmit, Samuel A. *Alsos*. New York: Henry Schuman, 1947.

Gowing, Margaret M. *Britain and Atomic Energy, 1939-1945*. New York: St. Martin's, 1964.

————. *Dossier secret des relations atomiques entre Alliés, 1939 / 1945*. Trans. and ed. Bertrand Goldschmidt. Paris: Plon, 1965.

Gowing, Margaret M., with Lorna Arnold. *Independence and Deterrence: Britain and Atomic Energy, 1945-1952*. 2 vols. New York: St. Martin's, 1974.

Graetzer, Hans G., and David L. Anderson, eds. *The Discovery of Nuclear Fission: A Documentary History*. New York: Van Nostrand Reinhold, 1971.

Graham, Loren R. "The Formation of Soviet Research Institutes: A Combination of Revolutionary Innovation and International Borrowing." *Social Studies of Science* 5 (1975): 303-329.

Greene, Nathanael. *Crisis and Decline: The French Socialist Party in the Popular Front Era*. Ithaca, N.Y.: Cornell University Press, 1969.

Griffith, George. *The Lord of Labour*. London: F.V. White, 1911.

Groueff, Stephane. *Manhattan Project: The Untold Story of the Making of the Atomic Bomb*. Boston: Little, Brown, 1967.

Groves, Leslie R. *Now It Can Be Told: The Story of the Manhattan Project*. New York: Harper, 1962.

Guéron, Jules. "Chimie analytique et pureté nucléaire." *Atomes* no. 35 (February 1949): 53-56.

————. "L'Eau lourde et le graphite." *Atomes* no. 85 (April 1953): 135-137.

————. "Energie nucléaire et politique générale." *Atomes* no. 5 (July 1946): 3-6.

Hahn, Otto. *My Life: The Autobiography of a Scientist*. Trans. Ernst Kaiser and Eithne Wilkins. New York: Herder & Herder, 1970.

————. *A Scientific Autobiography*. Trans. and ed. Willy Ley. New York: Scribner's, 1966.

Haïssinsky, Moïse. *Le Laboratoire Curie et son apport aux sciences nucléaires*. Report CEA-R-4201. Paris: Service Central de Documentation du Commissariat à l'Energie Atomique, 1971.

Halban, Hans. "L'Oeuvre scientifique de Frédéric Joliot." *Journal de Physique et le Radium* 8th ser. 20 (October 1959): 38S-45S.

————. *Travaux et titres scientifiques*. Paris?: n.p., 1958.

Herriot, Edouard. *Créer*. 2 vols. 1919. Reprint. Paris: Payot, 1925.

————. *Esquisses*. Paris: Hachette, 1928.

Hewlett, Richard G., and Oscar E. Anderson, Jr. *The New World, 1939 / 1946*. A History of the United States Atomic Energy Commission, vol. 1. University Park: Pennsylvania State University Press, 1962.

Histoire du Parti Communiste Français. By "Anciens membres du P.C.F." 3 vols. Paris: n.p., 1960-1964?

Hoffmann, Stanley, et al. *In Search of France*. Cambridge: Harvard University Press, 1963.

Holton, Gerald. *The Scientific Imagination: Case Studies*. Cambridge, Eng: Cambridge University Press, 1978.

————. "Striking Gold in Science: Fermi's Group and the Recapture of Italy's Place in Physics." *Minerva* 12 (1974): 159-198.

Hommage à Raoul Dautry 1880-1951 . . . Paris: Cité Universitaire de Paris, 1952.

Hostache, René. *Le Conseil National de la Résistance: Les Institutions de la clandestinité*. Paris: Presses Universitaires, 1958.

Hunebelle, Danielle. "Une Epopée de l'après-guerre: L'Epanouissement de notre industrie atomique." *Réalités* no. 155 (December 1958): 110ff.

Imperial Chemical Industries. *Annual Report*.

L'Institut International de Coopération Intellectuelle, 1925-1946. Paris: Institut International de Coopération Intellectuelle, 1946?

Irving, David. *The Virus House: Germany's Atomic Research and Allied Counter-Measures*. London: William Kimber, 1967.

Irving, R.E.M. *Christian Democracy in France*. London: Allen & Unwin, 1973.

Jaubert, Jean. "Enquête sur les industries chimiques françaises." *Revue Scientifique* 53 (1905): 97-128.

Joliot-Curie, Frédéric. "Le Centre National de la Recherche Scientifique." *La Pensée* n.s. no. 5 (October-December 1945): 3-7.

————. "L'Energie atomique en France." *Atomes* no. 15 (June 1947): 187-191.

————. "Les Grandes découvertes de la radioactivité." *La Pensée* n.s. no. 74 (July-August 1957): 3-15.

————. "Note sur la science soviétique." *La Pensée* n.s. no. 7 (April-June 1946): 29-32.

————. "Le Programme du Commissariat." *Echos du CEA*, special no. (October 1965): 15-16.

————. "La Recherche scientifique est-elle menacée?" *La Pensée* n.s. no. 25 (July-August 1949): 11-18.

————. *Textes Choisis*. Paris: Editions Sociales, 1959.

Joliot-Curie, Frédéric and Irène. *Oeuvres scientifiques complètes*. Paris: Presses Universitaires, 1961.

Joliot-Curie, Irène. *Souvenirs et Documents*. Ed. Association Frédéric et Irène Joliot-Curie. Paris: J. London, n.d.

————. "La Vie et l'oeuvre de Marie Sklodowska-Curie." *La Pensée* n.s. no. 58 (November-December 1954): 19-30.

Joliot-Curie, Irène and Frédéric. "La Découverte de la radioactivité artificielle." *Atomes* no. 85 (January 1951): 9-12.

Kevles, Daniel J. *The Physicists: The History of a Scientific Community in Modern America*. New York: Knopf, 1977.

Kilian, W. "Sur le recrutement du personnel des laboratoires scientifiques." *Revue Scientifique* 60 (1922): 717-719.

Kohl, Wilfrid L. *French Nuclear Diplomacy*. Princeton: Princeton University Press, 1971.

Kopelmanars, L. "Le Centre National de la Recherche Scientifique et les problèmes actuels de la recherche française." *Revue Socialiste* n.s. no. 22/23 (June-July 1948): 120-134.

Kowarski, Lew. "Après Zoé la deuxième étape." *Atomes* no. 85 (April 1953): 113-115.

————. "Atomic Energy Developments in France. *Nucleonics* 2 (May 1948): 59-65.

————. "Atomic Energy Developments in France During 1946-1950." *Nature* 165 (1950): 382-383.

————. *Les Avant-projets de distribution du gaz: Transport à distance, Distribution locale, Gazification rurale*. Paris: Dunod, 1938.

————. "Psychology and Structure of Large-Scale Physical Research." *Bulletin of the Atomic Scientists* 5, no. 6/7 (June-July 1949): 186-191.

————. *Réflexions sur la science.* Geneva: Institut Universitaire de Hautes Etudes Internationales, 1978.

————. "Zoé: Le Départ des piles françaises." *Echos du CEA,* special no. (October 1965): 21-23.

Kramish, Arnold. *Atomic Energy in the Soviet Union.* Stanford, Cal.: Stanford University Press, 1959.

Kwal, Bernard, and Marc Lesage. "La Physique au Palais de Découverte." *La Nature* 65, pt. 2 (1937): 194-253.

Labeyrie, Jacques. "Au Fort de Châtillon." *Atomes* no. 22 (January 1948): 18-21.

Laborde, Albert. *Pierre Curie dans son laboratoire.* Conférences du Palais de la Découverte, ser. A, no. 220, Mar. 24, 1956. Paris: Palais de la Découverte, n.d.

Langevin, André. *Paul Langevin, mon père: L'Homme et l'oeuvre.* Paris: Editeurs Français Réunis, 1971.

Larmour, Peter J. *The French Radical Party in the 1930's.* Stanford, Cal.: Stanford University Press, 1964.

Laugier, Henri. "Le Centre National de la Recherche Scientifique en France." *Revue d'Alger* no. 1 (1944): 6-20.

————. *Combat de l'exil.* Montreal: Editions de l'Arbre, 1944.

————. "How Science Can Win the War." *Free World* 1 (October 1941): 55-63.

————. "The National Center of Scientific Research in France." In *Quartier Latin,* trans. Sybil Kane Walker. Montreal: University of Montreal, 1941?

————. "Reconstruction in France: Educational Equality." *Free World* 6 (1943): 161-165.

Le Chatelier, Henry. *Science et industrie.* Paris: Flammarion, 1925.

Legendre, R. "L'Office National des Recherches Scientifiques et Industrielles et des Inventions." *Revue Scientifique* 61 (1923): 165-172.

Le Meur, Eugène. "La Mécanique de la pile." *Atomes* no. 35 (February 1949): 43-45.

Leprince-Ringuet, Louis. *Des atomes et des hommes.* 2nd ed. Paris: Fayard, 1966.

Lévy-Bruhl, Henry. "Pas de science dirigée." *Revue Socialiste* n.s. no. 27 (1949): 249-255.

"Lew Kowarski." *Les Archives Internationales* no. 250, Document no. 971 (November 1950).

Long, T. Dixon, and Christopher Wright, eds. *Science Policies of Industrial Nations: Case Studies of the United States, Soviet Union, United Kingdom, France, Japan and Sweden.* New York: Praeger, 1975.

Lot, Fernand. *Jean Perrin et les atomes.* Savants du Monde Entier, vol. 16. Paris: Seghers, 1963.

Louis-Antériou, Jacques, and Jean-Jacques Baron. *Edouard Herriot au service de la République.* Paris: Dauphin, 1957.

Luchaire, Julien. *La Crise de la science pure.* Enquête sur la situation du travail intellectuel, 2ᵉ série; La Vie intellectuelle dans les divers pays: France, Brochure no. 14. Geneva: Société des Nations, Commission de Coopération Intellectuelle, 1923. BN 8°R.32309(14).

Lumière, Auguste. "A propos de la recherche scientifique libre ou dirigée." *Revue Générale des Sciences* 54 (1947): 3-6.

McGowan, Harry Duncan. *Speech to the Glasgow Chamber of Commerce . . .* A Newcomen Society Publication. Princeton: Princeton University Press, 1944.

Maillart, Henri. *See* Mathieu, Henri.

Marbo, Camille. *See* Borel, Marguerite.

Mathieu, Henri, [Henri Maillart]. *L'Enseignement supérieure: Enquête sur la situation de l'enseignement supérieur scientifique et de l'enseignement technique.* Paris: Editions de La Bonne Idée, 1925.

Matignon, Camille. "Souvenirs sur Marcellin Berthelot." *Revue de Paris* 34, no. 6 (November-December 1927): 362-380.

Maublanc, René. "L'Université Française et la Résistance." *La Pensée* n.s. no. 15 (November-December 1947): 51-65.

Maurois, André. "Raoul Dautry." *Revue de Paris* 47, no. 3 (May 1940): 20-31.

Mayer, André. *Sur quelques questions relatives à l'organisation des recherches scientifiques* . . . Paris: Davy, 1920. BN 8°R.Pièce.26844.

Mendl, Wolf. *Deterrence and Persuasion: French Nuclear Armament in the Context of National Policy, 1945-1969.* London: Faber & Faber, 1970.

Michel, Henri. *Histoire de la Résistance en France (1940-1944).* Que sais-je? no. 429. 3rd ed. Paris: Presses Universitaires, 1962.

Montel, Paul. "Emile Borel." *Revue générale des sciences* 63 (1956): 137-141.

Moureu, Charles. *La Chimie et la guerre: Science et l'avenir.* Les Leçons de la guerre. Paris: Masson, 1924.

———. "Les Gaz de combat." *La Nature* 48/2 (1920): 294-296, 316-319, 332-335.

———. "La Science dans la vie moderne et les conditions générales de la recherche scientifique en France." In *1914-1924: Dix ans d'efforts scientifiques et industriels.* Paris: Chimie et Industrie, 1926. BN 4°R.Pièce.5135.

Noble, David F. *America by Design: Science, Technology and the Rise of Corporate Capitalism.* New York: Knopf, 1977.

Noguères, Henri. *Histoire de la Résistance en France de 1940 à 1945.* Vol. 1, *La Première Année, Juin 1940-Juin 1941,* with M. Degliame-Fouché and J. -L. Vigier. Vol. 3, *Et du Nord au Midi* . . . *Novembre 1942-Septembre 1943,* with M. Degliame-Fouché. Paris: Robert Laffont, 1967, 1972.

Nye, Mary Jo. *Molecular Reality: A Perspective on the Scientific Work of Jean Perrin.* London: Macdonald, 1972.

———. "Science and Socialism: The Case of Jean Perrin in the Third Republic." *French Historical Studies* 9 (1975): 141-169.

———. "The Scientific Periphery in France: The Faculty of Sciences at Toulouse (1880-1930)." *Minerva* 13 (1975): 374-403.

Oliphant, Mark. *Rutherford: Recollections of the Cambridge Days.* Amsterdam: Elsevier, 1972.

Orcel, Jean. "Frédéric Joliot-Curie." *La Pensée* n.s. no. 81 (September-October 1958): 51-60.

———. "Irène Joliot-Curie (1897-1956)." *La Pensée* n.s. no. 80 (July-August 1958): 79-102.

"Organisation et financement de la recherche scientifique." *Réalités Françaises* no. 8 (May 15, 1938): 1-2.

Palmer, Archie M. *University Research and Patent Policies, Practices and Procedures.* National Research Council Publication no. 999. Washington, D.C.: National Academy of Sciences, 1962.

Paris, University, Institut de Biologie Physico-Chimique. *Cérémonie en l'honneur du 25e anniversaire* . . . Paris: Hermann, 1953. BN 8°R. Pièce.25481.

Le Parti Communiste et l'avenir du pays. L'Ecole Elémentaire du Parti Communiste Français, IV.4ème leçon. Paris: Section Nationale d'Education du Parti Communiste Français, 1938. BN 8°Lb^{57}.18977(II,4)A.

Pash, Boris T. *The Alsos Mission.* New York: Award House, 1969.

Paul, Harry W. "The Crucifix and the Crucible: Catholic Scientists in the Third Republic." *Catholic Historical Review* 58 (1972): 195-219.

———. "The Debate over the Bankruptcy of Science in 1895." *French Historical Studies* 5 (1966): 299-327.

———. *The Sorcerer's Apprentice: The French Scientist's Image of German Science, 1840-1919.* Gainesville: University of Florida Press, 1972.

Paxton, Robert O. *Vichy France: Old Guard and New Order, 1940-1944.* New York: Norton, 1975.

Perrin, Francis. "Allocution." In *Comptes rendus du Congrès International de Physique Nucléaire . . .*, ed. P. Gugenberger, I, 21-28. Paris: Centre National de la Recherche Scientifique, 1964.

———. "Allocution." In Institut de France, Académie des Sciences, *Funérailles nationales de Frédéric Joliot . . .* Paris: Institut de France, 1958. BN 4°Z.1617(1958, XIVbis).

———. "Frédéric Joliot." In *Dictionary of Scientific Biography*, VII, 151-157. New York: Scribner's, 1973.

———. *Notice sur les travaux scientifiques de Francis Perrin.* Paris: Hemmerlé, Petit, 1951.

———. "Vingt ans après." *Echos du CEA*, special no. (October 1965): 42-48.

Perrin, Jean. *Les Atomes.* 2nd rev. ed. Paris: Alcan, 1924.

———. *Libération de l'humanité par la science.* Paris: Cahiers Rationalistes, 1936. BN 8°R.41165(11).

———. *L'Organisation de la recherche scientifique en France: Discours prononcé au conseil supérieure de la recherche scientifique.* Paris: Hermann, 1938.

———. *La Recherche scientifique.* Actualités scientifiques et industrielles no. 58. Paris: Hermann, 1933. BN 4°V.12012(58).

———. *La Science et l'espérance.* Paris: Presses Universitaires, 1948.

Peyret, Henry. *La Bataille de l'énergie.* Que sais-je? no. 863. Paris: Presses Universitaires, 1964.

Pickles, Dorothy. *French Politics: The First Years of the Fourth Republic.* London: Royal Institute of International Affairs, 1953.

Prost, Antoine. *Histoire de l'enseignement en France, 1800-1967.* Paris: Colin, 1968.

Puget, Henry. "Les Problèmes administratifs en France relatifs à l'utilisation pacifique de l'énergie atomique." La Documentation Française, *Notes et Etudes Documentaires* no. 2,856 (Feb. 1, 1962).

Purcell, Edward M. "Nuclear Physics Without the Neutron: Clues and Contradictions." In *Proceedings*, 10th International Congress of the History of Science, Ithaca, N.Y., 1962, I, 121-137. Paris: Hermann, 1964.

Ranc, Albert. *Le Budget du personnel des recherches scientifiques en France.* Paris: Chimie et Industrie, 1926. BN 8°R.34654.

———. *Jean Perrin, un grand savant au service du socialisme.* Paris: Editions de la Liberté, 1945.

"Raoul Dautry." *Etudes* 242 (January-March 1940): 58-62.

Reader, William J. *Imperial Chemical Industries: A History.* Vol. 1, *The Forerunners, 1870-1926.* Vol. 2, *The First Quarter-Century, 1926-1952.* London: Oxford University Press, 1970-1975.

Reid, Robert. *Marie Curie.* New York: E.P. Dutton, 1974.

Richet, Ch., Jr. "La Situation matérielle des savants." *Revue Scientifique* 63 (1925): 97-100.

Roubault, Marcel. "La Découverte de l'uranium français." *Echos du CEA*, special no. (October 1965): 17-19.

———. "La Recherche et l'exploitation des minerais d'uranium." *Atomes* no. 85

(April 1953): 128-133.

Rousseau, Pierre. "L'Avenir de l'énergie atomique." *Revue de Paris* 56, no. 8 (July 1949): 126-136.

Rouzé, Michel. *Frédéric Joliot-Curie.* Paris: Editeurs Français Réunis, 1950.

Rutherford, Ernest; James Chadwick; and C.D. Ellis. *Radiations from Radioactive Substances.* Cambridge: Cambridge University Press, 1930.

Savel, Pierre. "Radioactivité et radioéléments artificiels." *Revue de Chimie Industrielle* 45 (1936): 122-126, 154-158.

———. "Vingt-sept années de collaboration scientifique avec Frédéric Joliot-Curie." *La Pensée* n.s. no. 87 (September-October 1959): 80-83.

Scheinman, Lawrence. *Atomic Energy Policy in France under the Fourth Republic.* Princeton: Princeton University Press, 1965.

Schroeder-Gudehus, Brigitte. "The Argument for the Self-government and Public Support of Science in Weimar Germany." *Minerva* 10 (1972): 537-570.

Seaborg, Glenn T. *History of Met Lab Section C-I.* Vol. 1, *April, 1942 to April, 1943.* Vol. 2, *May, 1943 to April, 1944.* Lawrence Berkeley Laboratories PUB 112. Berkeley, Cal.: Lawrence Berkeley Laboratories, 1977-1978.

Segrè, Emilio. *Enrico Fermi, Physicist.* Chicago: University of Chicago Press, 1970.

Sherwin, Martin J. *A World Destroyed: The Atomic Bomb and the Grand Alliance.* New York: Knopf, 1975.

Shirer, William L. *The Collapse of the Third Republic: An Inquiry into the Fall of France in 1940.* New York: Simon & Schuster, 1969.

Siegfried, André. "Le Collège de France." *Revue de Paris* 45, no. 6 (1938): 510-521.

Silva, Raymond. "In Memoriam Raoul Dautry." *Monde Nouveau* 7, no. 51-52 (1951): 3-6.

Smith, Alice Kimball. *A Peril and a Hope: The Scientists' Movement in America, 1945-1947.* Chicago: University of Chicago Press, 1965.

Smyth, Henry D. *Atomic Energy for Military Purposes: The Official Report on the Development of the Atomic Bomb under the Auspices of the United States Government, 1940-1945.* Princeton: Princeton University Press, 1946.

Snow, C.P. *Science and Government.* Cambridge: Harvard University Press, 1962.

Soddy, Frederick. *The Interpretation of Radium: Being the Substance of Six Free Popular Experimental Lectures Delivered at the University of Glasgow.* 3rd ed., rev. and enl. London: John Murray, 1912.

———. *Science and Life: Aberdeen Addresses.* London: John Murray, 1920.

Stohr, Jacques A. "La Fabrication des comprimés d'oxyde d'uranium." *Atomes* no. 35 (February 1949): 57-58.

Strauss, Lewis L. *Men and Decisions.* Garden City, N.Y.: Doubleday, 1962.

Surdin, Maurice. *L'Electronique au Commissariat à l'Energie Atomique.* Commissariat à l'Energie Atomique, Conférence no. 1. Châtillon, Fontenay-aux-roses: C.E.A., 1950. BN 4°V.20825(1-40).

Szilard, Leo. *The Collected Works of Leo Szilard.* Vol. 1, *Scientific Papers.* Ed. Bernard T. Feld and Gertrud Weiss Szilard. Cambridge: Massachusetts Institute of Technology Press, 1972.

———. *Leo Szilard: His Version of the Facts. Selected Recollections and Correspondence.* Ed. Spencer R. Weart and Gertrud Weiss Szilard. The Collected Works of Leo Szilard, vol. 2. Cambridge: Massachusetts Institute of Technology Press, 1978.

Taranger, Pierre. *Les Rapports de l'industrie avec le C.E.A.* Commissariat à l'Energie Atomique, Conférence no. 21. Gif sur Yvette: Centre d'Etudes Nucléaires de Saclay, 1956. BN 4°V.20825(1-40).

Thérive, André. "Le Collège de France." *Revue de Paris* 45, no. 9 (May 1938): 91-105.

Thoma, J. André. "Le Martyre de Fernand Holweck." *La Pensée* n.s. no. 27 (November-December 1949): 21-28.

Thomson, George P. "Anglo-U.S. Cooperation on Atomic Energy." *Bulletin of the Atomic Scientists* 9, no. 2 (March 1953): 46-48.

———. *Nuclear Energy in Britain During the Last War.* Oxford: Clarendon, 1962.

Thorez, Maurice. *Oeuvres.* 20 vols. Paris: Editions Sociales, 1950-1955.

Tiersky, Ronald. *French Communism, 1920-1972.* New York: Columbia University Press, 1974.

Toutée, Jean. "Naissance du Commissariat." *Echos du CEA*, special no. (October 1965): 5-8.

United States Department of State. *Foreign Relations of the United States: 1949.* (Washington, D.C.: U.S. Government Printing Office, 1976).

Vaillant-Couturier, Paul. *Au service de l'esprit: Pour la convocation des états généraux de l'intelligence française* . . . Paris: Editions Sociales Internationales, 1937. BN 8°R.Pièce.21044.

Vilan, Michel. *La Politique de l'énergie en France, de la seconde guerre mondiale à l'horizon 1985.* Paris: Cujas, 1969.

Vinet, E. "La Guerre des gaz et les travaux des services chimiques françaises." *Chimie et Industrie* 2 (1919): 1377-1415.

Voguet, André. "Souvenirs de l'année universitaire 1940-1941." *La Pensée* n.s. no. 89 (January-February 1960): 23-30.

Weart, Spencer R. "The Physics Business in America: A Statistical Reconnaissance." In *The Sciences in the American Context*, ed. Nathan Reingold. Washington, D.C.: Smithsonian Institution Press, in press.

———. "Scientists with a Secret." *Physics Today* 29, no. 2 (February 1976): 23-30.

———. "Secrecy, Simultaneous Discovery, and the Theory of Nuclear Reactors." *American Journal of Physics* 45 (1977): 1049-1060.

Weil, André. "Science française?" *La Nouvelle N[ouvelle] R[evue] F[rançaise]* 5 (January 1955): 97-109.

Weiner, Charles. "Institutional Settings for Scientific Change: Episodes from the History of Nuclear Physics." In *Science and Values*, ed. Arnold Thackray and Everett Mendelsohn, pp. 187-212. New York: Humanities Press, 1974.

Weiner, Charles, ed., with Elspeth Hart. *Exploring the History of Nuclear Physics.* American Institute of Physics Conference Proceedings no. 7. New York: American Institute of Physics, 1972.

Weisskopf, Victor F. "Physics in France." *Physics Today* 4, no. 12 (December 1951): 6-11.

Wells, H. G. *The World Set Free: A Story of Mankind.* New York: E. P. Dutton, 1914.

Werth, Alexander. *France, 1940-1955.* 2nd ed., enl. Boston: Beacon, 1966.

Wheeler, John. "The Mechanism of Fission." *Physics Today* 20, no. 11 (November 1967): 49-52.

———. "Niels Bohr and Nuclear Physics." *Physics Today* 16, no. 10 (October 1963): 36-45.

Wheeler-Bennett, John. *John Anderson, Viscount Waverley.* New York: St. Martin's, 1962.

Williams, Philip M. *Crisis and Compromise: Politics in the Fourth Republic.* 3rd rev. ed. Hamden, Conn.: Anchor, 1964.

World Federation of Scientific Workers. London: World Federation of Scientific

Workers, 1947.

Ydewalle, Charles d'. *L'Union Minière du Haut-Katanga: De l'âge colonial à l'indépendence.* Paris: Plon, 1960.

Zay, Jean. *Souvenirs et solitude.* Paris: R. Julliard, 1945.

Zola, Emile. *Paris.* Paris: E. Fasquelle, 1898.

Notes

This book could be written only because the scientists who evacuated the Collège de France in 1940 took along a suitcase stuffed with their most important papers. These papers went to Britain, on to Montreal, back to Britain, and thence to the Commissariat à l'Energie Atomique in Paris. They now rest as file CEA.51 in the Curie Laboratory of the Radium Institute. The day I first looked through these papers—letters, memoranda, laboratory notes, and scraps—was the day I began to think about a book. As the study grew, I paged through many other stacks of paper. Copies of wartime British and American documents often turned up in several locations, but I cite only one repository for each item, while trying in the sum of the notes to give an idea of where such materials were found.

Those who participated in the events will report that the most important things were not always written down, and when there were documents, these were often lost or destroyed. Therefore, while relying above all on primary documents, I also made extensive use of reminiscences, written and oral. Some were tape-recorded interviews and others were informal conversations or printed recollections. Where not stated otherwise, an interview was by myself. A less commonly used sort of testimony was given by Halban and Kowarski during patent hearings in 1957, subject to cross-examination under oath. Unfortunately few reminiscences on tape and in print are so controlled, and in every case the fallability of memory is notorious. In a conversation as later recalled by someone, for example, the words may well not be the exact ones used. People try to summarize what an incident meant to them and rarely give a flawless report. But such reminiscences can recapture from a situation essential matters which would otherwise be lost. Nowhere in history is there a single fact with the solidity of a fact of physics, and historians must labor to weave a pattern which will hang together however weak each individual thread. For the more important points I used reminiscences only where, as was often the case, they were confirmed by independent evidence; in less important matters I used reminiscences as long as they were consistent with other evidence. In a few cases my judgments were shaped (but never entirely based) on documents or confidences I was asked not to cite.

It is not customary for historians to point out information they lack, but the reader deserves fair warning. To my knowledge Joliot never told his version of this story in detail, although the gap may be partly made up from his extensive papers. Halban, on the contrary, told his version on several occasions but kept few of his papers. Worse, of the great mass of French government papers created in connection with such agencies as the CNRS and CEA, all but an almost random sample are lost, or in private hands but out of sight, or still in the hands of government agencies and rigorously sealed by law and custom for at least fifty years. So this book is restricted and cannot pretend, for example, to give the full story of the administrations of the CNRS and CEA or of events in higher governmental circles. The object of my study was to see what the scientists themselves did inside their laboratories and outside, and what affected this, and what it affected. If all the documents I hunted for unsuccessfully were suddenly to appear on my desk, I believe they would enrich but not greatly alter this history.

All translations are my own unless otherwise noted. Beginning in the 1930s, Joliot and his wife often signed their last name Joliot-Curie. Halban's earlier publications are signed Hans von Halban, Jr.

ABBREVIATIONS

AHQP	Archive for the History of Quantum Physics. Copies deposited at AIP; American Philosophical Society, Philadelphia; Niels Bohr Institute, Copenhagen; University of California, Berkeley; University of Minnesota, Minneapolis. The Bohr Scientific Correspondence is also located at these repositories.
AIP	Center for History of Physics, American Institute of Physics, New York.
AN	Archives Nationales, Paris.
ANL	Metallurgical Laboratory files, temporarily deposited at Argonne National Laboratory, Argonne, Illinois; to be transferred to NA.
CEA	Archives of Commissariat à l'Energie Atomique, Paris.
CF	Archives of Collège de France, Paris.
CNRS	Archives of Centre National de la Recherche Scientifique, Paris.
CR	Comptes Rendus of the Académie des Sciences, Paris.
CUL	Columbia University Library, New York.
DOE	Historian's Office of U.S. Department of Energy (formerly of U.S. Atomic Energy Commission), Germantown, Maryland.
EOL	Ernest O. Lawrence Papers, Bancroft Library, University of California, Berkeley, cited by carton number: folder number.
G-	Captured German report; copies available from National Technical Information Service, Oak Ridge, Tennessee.
HalbT	Testimony of Hans Halban, June 20, 1957, PCB.
JA	Jacques Allier Papers, Paris.
J. Phys.	Journal de Physique et le Radium, 7th series unless otherwise noted.
J.O.	Journal Officiel of French Parliament.
KoT	Testimony of Lew Kowarski, March 16, 1967, PCB.
LK	Lew Kowarski Papers, Geneva.
LS	Leo Szilard Papers, La Jolla, California.
MED	Manhattan Engineer District Papers, RG 77, NA, which include Leslie Groves correspondence, entry 1, cited as MED (Groves), and George Harrison-Harvey Bundy files, entry 20, cited as MED (Harrison-Bundy).
NA	National Archives building, Washington, D.C.
OHI	Oral History Interview, tape-recorded, on deposit (usually with edited

transcript) at AIP.

OSRD   Atomic Energy Papers of Office of Scientific Research and Develop-
             ment, RG 227, NA, which include Vannevar Bush-James B. Conant files,
             cited as OSRD(BC), with preliminary box number: folder title.

P   Paris

PCB   Docket 18 of Patent Compensation Board, U.S. Atomic Energy Com-
             mission, at Department of Energy patent offices, Germantown, Mary-
             land.

Phys. Rev. *Physical Review.*

PRO   Public Record Office, London.

RI   Papers of Marie and Irène Curie and Frédéric Joliot at Curie Laboratory,
             Radium Institute, Paris.

1. PROFESSORS AND POLITICIANS

1. F. Perrin, "Allocution," p. 2. See also Biquard, *Joliot-Curie*; Goldsmith, *Jo-liot-Curie*; Rouzé, *Joliot-Curie*, p. 33; OHI Lew Kowarski by Charles Weiner, March, October 1969.

2. See A. Langevin, *Paul Langevin*, pp. 52-54; Cotton, *Aimé Cotton*; Lot, *Perrin*; M. Curie, *Pierre Curie*; E. Curie, *Madame Curie*.

3. Marbo [M. Borel], *Souvenirs*, pp. 67-68.

4. Berthelot in *Commémoration du banquet Berthelot*, pp. 466-467; "Bert," *Dictionnaire de biographie française.* See also Matignon, "Souvenirs sur Marcellin Berthelot"; Paul, "The Crucifix and The Crucible"; Paul, "The Debate over the Bankruptcy of Science"; Prost, *Histoire de l'enseignement.*

5. Henry E. Guerlac, "Science and French National Strength," in Earle, ed., *Modern France*, pp. 81-105; OHI Wolfgang Gentner by Weiner, November 1971.

6. Lucien Herr to Director, Ecole Normale, Dec. 11, 1887, 61AJ.82(Perrot), AN; Bourgin, *De Juarès à Léon Blum*, ch. 14; "Renseignements concernant la demande d'une bourse," 1894, F^{17}24,822 (Perrin), AN.

7. G. Darboux, "Eloge," in *Oeuvres de Henri Poincaré* (P: Gauthier-Villars, 1916), II, lx; Zola, *Paris*, p. 143; J. Perrin, *La Recherche scientifique*, p. 16.

8. J. Perrin, address to Conseil Supérieure de la Recherche Scientifique, Mar. 2, 1938, in Ranc, *Perrin*, p. 42.

9. Laborde, *Pierre Curie*, p. 11.

10. P. Curie to Berthelot, Dec. 11, 1901, MS 3962 (Berthelot), Bibliothèque de l'Institut de France, Paris. See also P. Curie to Ch. Ed. Guillaume, Dec. 30, 1898, 30.D3, Curie Papers, Bibliothèque Nationale, Paris.

11. Biquard, *Joliot-Curie*, pp. 14-15. Curie may have been thinking of radium more as a poison than as a source of energy, but the parallel with explosives is suggestive.

12. Griffith, *The Lord of Labour*, p. 270. See also Clarke, *Voices Prophesying War*, ch. 3.

13. Reid, *Marie Curie*, p. 235.

14. Daniel P. Jones, *The Role of Chemists in Research on Gases in the U.S. During World War I*, Ph.D. diss., University of Wisconsin, 1969 (Ann Arbor, Mich.: University Microfilms no. 66-22,406), pp. 52-54; Vinet, "Guerre des gaz," pp. 1384, 1390.

15. "André Mayer 1875-1956," in Paris, Université, Institut de Biologie Physico-Chimique, *Cérémonie*, p. 24; Moureu, *La Chimie et la guerre*, ch. 3.

16. Vinet, "Guerre des gaz" (draws heavily on Moureu, "Les Gaz de combat"); Augustin M. Prentiss, *Chemicals in War* (New York: McGraw-Hill, 1937), p. 179.

17. "Rapport sur l'activité de l'Office National," 1925, in folder, "Office National . . . Renouvellement," F^{22}317, AN; Moureu, *La Chimie et la guerre*, ch. 3; OHI F. Perrin, November 1977; Cotton, *Aimé Cotton*, ch. 6; A. Langevin, *Paul Langevin*, ch. 3.

18. E. Curie, *Mme. Curie*, pp. 312-321; Marbo, *Souvenirs*, ch. 13.

19. M. Curie to I. Curie, Oct. 11, 1921, in M. and I. Curie, *Correspondance*, p. 231.

20. Rouzé, *Joliot-Curie*, pp. 22-24; Biquard, *Joliot-Curie*, pp. 24-25.

21. Mathieu [Maillart], *L'Enseignement supérieure*, p. 61. Membership counts from the *Procès-Verbaux*, later *Bulletin*, of the Societé Française de Physique. Membership rose from about 900 in 1900 to 1600 in 1913-1918, dropped to 1100 in 1923, and remained below 1300 through 1945. The decline may have involved changed attitudes or simple economic factors, such as a raise in fees. *Procès-Verbaux*, 1921, p. 3; *Bulletin*, Jan. 15, 1926, pp. 2-6. Publication counts: from *J. Phys.*, where number of main articles was 80-90 per year after 1900 but rose after 1931; and from *CR*, where number of pages, which was in constant proportion to number of articles, was around 2100 per year before the war, 1000 after, and rose after 1931.

22. Barrès, "Que fait l'université," pp. 245-246. See also OHI F. Perrin, November 1977.

23. Georges Lemoine, Presidential address, *CR* 173 (1921): 1217, reprinted in *Revue Scientifique* 60 (1922): 105-109; Richet, "Situation matérielle," p. 98. See also Guillaume Bigourdan, Presidential address, *CR* 179 (1924): 1475; Luchaire, *La Crise de la science pure*.

24. E. Borel, *Supplément (1921) à la Notice (1912) sur les travaux scientifiques de M. Emile Borel* (Toulouse: Edouard Privat, 1921); Mayer, *Sur quelques questions*, pp. 5, 11-13.

2. A CAMPAIGN FOR SCIENCE

1. Eugène Thébault in *La Petite République*, Sept. 8, 1904; Richet, "Situation matérielle." See also Forman, Heilbron, and Weart, *Physics circa 1900*; M. Curie, *Pierre Curie*, ch. 6.

2. E. Roux in *Le Temps*, July 25, 1919, p. 4; L. Mangin in *Le Figaro*, Apr. 28, 1921, p. 1.

3. J. Perrin, Nov. 12, 1933, in Lot, *Perrin*, p. 66. See also Appell, "La Recherche scientifique"; Reid, *Marie Curie*, pp. 53-54, 297-298, 313; Nye, *Molecular Reality*, pp. 55-62.

4. Moureu, *La Chimie et la guerre*, pp. 270-277, 370. See also Moureu, "La Science dans la vie moderne."

5. Moureu, preface to Barrès, *Pour la haute intelligence française*, p. i; Barrès, *Chronique de la Grande Guerre*, XIII, 466; Barrès in *J.O., Débats, Chambre* (Dec. 10, 1921): 4875-4876.

6. E. Curie, *Mme. Curie*, ch. 23; *Le Figaro*, Apr. 29, 1921, p. 1; Reid, *Marie Curie*, chs. 20-21; Ranc, *Budget*, pp. 74-75.

7. Aimé Cotton, Presidential address, *CR* 207 (1938): 1269-1279; Emile Bertin, Presidential address, *CR* 175 (1922): 1265-1266; Fauré-Fremiet, "Le Mouvement actuel"; Georges Lemoine, Presidential address, *CR* 173 (1921):1211-1220; Montel, "Emile Borel," p. 140.

8. Forman, Heilbron, and Weart, *Physics circa 1900*, pp. 75-81; France, Caisse des Recherches Scientifiques, *Rapport annuel (1922)* (Melun: Imprimerie Administrative, 1923), p. 40.

9. *Le Temps*, July 18, 20, 22, 25, 27, Aug. 1, 8, 15, 21, 1919; Barrès, "Que fait l'université," p. 252.

10. E. Borel, Presidential address, *CR* 199 (1934):1472; de Broglie, "Notice sur . . . Borel," pp. 16-18; Pierre Girard in Paris, Université, Institut de Biologie Physico-Chimique, *Cérémonie*, pp. 18-19; Appell, "L'Institut Edmond de Rothschild."

11. J. Perrin in Lot, *Perrin*, p. 165.

12. Barrès, "Que fait l'université," pp. 248, 256-258; Kilian, "Sur le recrutement"; Weil, "Science française?"

13. Lemoine, Presidential address, *CR* 173 (1921):1211-1220.

14. Letters to de Broglie, Joliot Dossier, Archives de l'Académie des Sciences, Paris; Biquard, *Joliot-Curie*, p. 26; Crowther, *Fifty Years with Science*, p. 134.

15. Elsasser, *Memoirs of a Physicist*, p. 175.

16. J. Perrin, Presidential address, *CR* 203 (1936): 1412.

17. France, Caisse des Recherches Scientifiques, "Rapport Annuel, 1923," in *J.O., Annexes* (1925): 778.

18. André François-Poncet in *J.O., Débats, Chambre* (Nov. 19, 1924):3649. See also Nye, "The Scientific Periphery in France," p. 399.

19. Weart, "The Physics Business in America"; Kevles, *The Physicists*, chs. 12-18; Forman, "Financial Support"; Schroeder-Gudehus, "Argument for . . . Science in Weimar Germany." See also Graham, "The Formation of Soviet Research Institutes"; Graham, "The Development of Science Policy in the Soviet Union," in Long and Wright, eds., *Science Policies of Industrial Nations*, pp. 12-58.

20. Jaubert, "Enquête sur les industries chimiques françaises"; Berget, "La Science"; Le Chatelier, *Science et industrie*; Urbain in Luchaire, *La Crise de la science pure*, p. 13.

21. E. Borel, "La Propriété scientifique," *Revue de Paris* 29 (1922): 850-860, reprinted in *Emile Borel, philosophe et homme d'action*, p. 394.

22. National Academy of Sciences, National Research Council, *Consolidated Report . . . 1919-1932* (Washington, D.C.: National Research Council, 1932), p. 34; Raymond Weiss in *L'Institut International de Coopération Intellectuelle*, pp. 452-469; Dalimier and Gallie, *La Propriété scientifique*, pp. 62-64.

23. Larmour, *The French Radical Party*, p. 60; Herriot, *Créer*, I, 14-29, 434-447. See also Louis-Antériou and Baron, *Edouard Herriot*, pp. 53-55, 112-115; Herriot, "Berthelot ou la beauté de l'intelligence," in Herriot, *Esquisses*, p. 34.

24. Collingwood, "Emile Borel," p. 496; Marbo, *Souvenirs*, p. 212; E. Borel, "La Crise economique et la science," pp. 756, 757, 759.

25. At the same time, a one-time appropriation of a million francs was voted for the Radium Institute and Perrin's institute. *J.O., Débats, Chambre* (1924): 3648-3650. The tax provided science 3.8 million francs in 1926, 8.8 million in 1932, and 6.7 million in 1936, Chatelet, "La France devant les problèmes de la recherche," no. 2671, p. 27.

26. Curie's was the largest sum given at this session; Perrin got an equal amount. France, Caisse des Recherches Scientifiques, "Procès-Verbaux de la Commission Technique, Registre no. 4, 1925-1931," 2e section, Apr. 24, 1928, F^{17}17,432, AN; Maurice Hamy, Presidential address, *CR* 187 (1928): 1192.

27. J. Perrin, *La Recherche scientifique*, p. 21; Reid, *M. Curie*, p. 302. See also Paul, *The Sorceror's Apprentice*.

28. J. Perrin, *La Science et l'espérance*, pp. 167-168. See also E. Borel, "L'Organi-

sation de la recherche scientifique"; Gilpin, *France in the Age of the Scientific State*; Mathieu, *L'Enseignement supérieur*.

29. France, Haut-Comité de Coordination des Recherches Scientifiques, *Rapport Général*, 1939, p. 102, $F^{17}17$, 490, AN; Prost, *Histoire de l'enseignement*, pp. 329-331; Hoffmann, "Paradoxes of the French Political Community," in Hoffmann et al., *In Search of France*, pp. 1-117.

30. F. Perrin, "Frédéric Joliot," p. 152. See also I. Curie to Marie Meloney, Jan. 28, 1929, Meloney Collection, CUL.

31. J. Perrin, "Faciliter la recherche," in Lot, *Perrin*, p. 166; J. Perrin, *L'Organisation de la recherche scientifique en France*, pp. 11-12.

32. A. Langevin, *Paul Langevin*, pp. 106-113; John D. Bernal, intro. to Biquard, *Langevin*, p. 177.

33. J. Perrin in *CR* 190 (1930): 1533-1534; Marbo, *Souvenirs*, p. 68.

34. Coulomb, "Charles Maurain."

35. J. Perrin, *La Recherche scientifique*, pp. 20-21; J. Perrin, *Libération de l'humanité par la science*, p. 17.

36. E. Cotton, *Les Curie*, p. 95; OHI Aristid V. Grosse by C. Weiner with Lillian Hoddeson, April 1974.

37. Lot, *Perrin*, pp. 81-82; Canac, "Un Nouveau corps d'état."

38. France, Caisse Nationale des Sciences, *Rapport Annuel . . . 1932*, $F^{17}17,458$, AN, p. 16.

39. M. and I. Curie, *Correspondance*, p. 346. In 1932 Joliot was named half-time chargé de recherches (receiving 18,000 francs a year) and also took an equivalent half-time position as assistant at the Radium Institute. In 1933 he was promoted to maître de recherches and in 1935 added the equivalent teaching title, maître de conférences. In 1936 he became a directeur de recherches, the highest CNRS grade, equivalent to full professor. Irène Curie became a chef de travaux (laboratory staff position) in 1932, a maître de recherches in 1935, and a maître de conférences in 1937.

40. Emilio Segrè in Weiner, ed., *Exploring the History of Nuclear Physics*, p. 134; I. Curie to Marie Meloney, Dec. 5, 1935, Meloney Collection, CUL.

41. There was precedent for this: the Ecole Pratique des Hautes Etudes, set up with high hopes under the Second Empire, was soon taken over by the gerontocracy and became a mere conduit for funds to their laboratories.

42. OHI Edmond Bauer by Thomas Kuhn and Théo Kahan, January 1963, AHQP, sess. II, pp. 1-2; Zay, *Souvenirs*, p. 183.

43. [André Mayer?], "Rapport au Président de la République Française," [1932], $F^{17}17,463$, AN; "Note sur la création d'un Conseil Supérieur de la Recherche Scientifique," January 1933, $F^{17}17,463$, AN.

44. I. Curie to M. Curie, May 12, 1932, in M. and I. Curie, *Correspondance*, p. 344; notes to Cavalier, December 1932, $F^{17}17,463$, AN.

45. Coulomb, "Jean Perrin," p. 679.

46. "Organisation et financement"; "Projet de rapport sur l'activité de la Caisse des Sciences en 1933" and "Rapport de Secrétaire au Conseil d'Administration de la Caisse Nationale des Sciences . . . 1934," $F^{17}17,458$, AN; Jean Zay in *J.O., Débats, Sénat* (Dec. 29, 1936): 1864.

47. Bernal, "Langevin et l'Angleterre," p. 19.

48. OHI Perrin, November 1977; Chastenet, *Histoire de la Troisième République*, VI, 88, 144-145; A. Langevin, *Paul Langevin*, pp. 121-139.

49. Greene, *Crisis and Decline*; Blum, *Pour être socialiste*, pp. 5, 21-26; Blum, *L'Oeuvre du Léon Blum*, IV: 1, 243; Ranc, *Perrin*, pp. 51-52. See also Léon Blum, preface to Ranc, *Perrin*, pp. 12-15; Nye, "Science and Socialism."

50. Ranc, *Perrin*, p. 39; I. Curie to Marie Meloney, June 19, 1936, Meloney Collection, CUL; Zay in Coulomb, "Perrin," p. 679. Curie held the post from June to October. See her resignation letter, *Le Temps*, Sept. 27, 1936.

51. Zay, *Souvenirs*, pp. 265-268. The minister was Vincent Auriol.

52. "Organisation et financement." The variety of funding mechanisms makes it almost impossible to specify a single figure as the total research budget. According to Vincent Auriol, the 1935 figure was 15 million and the 1936 figure 28.6 million. *J.O., Débats, Sénat* (Dec. 29, 1936): 1864. See also Chatelet, "La France devant les problèmes de la recherche," no. 2671, p. 27.

53. "Note sur le projet de budget de 1939," c. July 1938, folder "1939 (chapitres reservés)," F^{17}13,332, AN; OHI Guéron, April 1978.

54. "Service Central de la Recherche Scientifique," 1936?, folder "1939," F^{17}13,332, AN. Binion, *Defeated Leaders*, ch. 8; Friedmann, *La Crise du Progrès*, pp. 167-169. See also J. Perrin, "Libération de l'humanité," pp. 3-4; E. Borel, "La Crise économique et la science," p. 756.

55. *J.O., Débats, Sénat* (Dec. 29, 1936): 1863-1865; interview of A. Langevin, October 1976.

3. THE URANIUM PUZZLE

1. J. Perrin, "Atomes et lumière," *Revue du Mois* 21 (1920): 113-166; Soddy, *The Interpretation of Radium*, p. 251; Michael I. Freedman, "Frederick Soddy and the Commercial and Social Aspects of Early Radioactivity Research" (talk delivered at 15th International Congress of the History of Science, Edinburgh, August 1977). Perrin appears to share with A. S. Eddington priority for suggesting hydrogen fusion as the source of solar energy.

2. Moureu, *La Chimie et la guerre*, pp. 356-357. Moureu's estimate is a factor of about six below the energy released through nuclear fission in a kg. of uranium. He was referring to the energy slowly released in natural radioactivity.

3. J. Perrin, *Les Atomes*, pp. 318-319. See also J. Perrin, *La Science et l'espérance*, pp. 141-142; M. Curie, *Radioactivité*, pp. 367, 370.

4. E. Curie, *Mme. Curie*, ch. 26; E. Cotton, *Les Curie*, p. 88; I. Joliot-Curie, *Souvenirs et Documents*, pp. 38-39. See also Haïssinsky, *Le Laboratoire Curie*.

5. [M. Curie], Société des Nations, Commission de Cooperation Intellectuelle, Sous-commission des Relations Universitaires, "Memorandum de Mme. Curie . . . sur la question des bourses internationales," mimeo., C.I.O.I./R.I./39 (Geneva: Société des Nations, June 16, 1926), pp. 7-8.

6. Bromberg, "The Impact of the Neutron"; Purcell, "Nuclear Physics Without the Neutron"; Weiner, "Institutional Settings for Scientific Change." See also Weiner, ed., *Exploring the History of Nuclear Physics.*

7. Gian Carlo Wick, oral communication, November 1976, recalling what Joliot told him about keeping polonium concentrated. I. Curie and F. Joliot, *Journal de Chimie Physique* 28 (1931):201-205. For all such articles, see F. and I. Joliot-Curie, *Oeuvres.*

8. W. Bothe and H. Becker, *Zeitschrift für Physik* 66 (1930): 289-306. For polonium, see also Feather, "The Discovery of the Neutron"; OHI Feather by Weiner, February 1971.

9. F. Joliot, "Grandes Découvertes," p. 11; F. and I. Joliot, "Rayonnement pénétrant des atomes sous l'action des rayons α," in Instituts Solvay (Brussels), Institut International de Physique, Conseil de Physique, Rapports et Discussions, 7th, 1933, *Structure et propriétés des noyaux atomiques* (P: Gauthier-Villars, 1934), 121-125;

Biquard, *Joliot-Curie*, pp. 34-40; I. Curie and F. Joliot, *CR* 194 (1932): 273-275.

10. Chadwick, "Some Personal Notes on the Search for the Neutron"; OHI Feather by Weiner, February 1971.

11. Cambridge University, Cavendish Laboratory, *A History of the Cavendish Laboratory*, p. 24; Crowther, *The Cavendish Laboratory*, pp. 21, 61-64, 170-172, 185.

12. Crowther, *The Cavendish Laboratory*, pp. 200-201; Oliphant, *Rutherford*, pp. 139-141; OHI Oliphant by Weiner, November 1971; Rutherford, Chadwick, and Ellis, *Radiations from Radioactive Substances*, p. 532; OHI Chadwick by Weiner, April 1969, pp. 38, 80.

13. I. Curie and F. Joliot, *Projection de noyaux atomiques*, p. 429.

14. I. Curie, "La Recherche scientifique," 1938, in E. Cotton, *Les Curie*, pp. 190-191.

15. In the notation of nuclear reactions, they assumed that they had observed: $^{27}_{13}Al + ^{4}_{2}He \rightarrow ^{30}_{14}Si + (^{1}_{0}n + ^{0}_{1}e^{+})$.

16. Instituts Solvay, *Structure . . . des noyaux*, discussion, pp. 176-177; I. and F. Joliot-Curie, "Découverte de la radioactivité artificielle," p. 10.

17. F. Perrin, "Allocution," p. 23; OHI Perrin, November 1977.

18. J. Perrin, *Les Atomes*, p. 319. $^{27}_{13}Al + ^{4}_{2}He \rightarrow ^{30}_{15}P + ^{1}_{0}n$; $^{30}_{15}P \rightarrow ^{30}_{14}Si + ^{0}_{1}e^{+}$.

19. I. and F. Joliot-Curie, "Découverte de la radioactivité artificielle," p. 11; F. Joliot, interview in *Republica* (Portugal), Jan. 10, 1955, quoted in Goldsmith, *Joliot-Curie*, p. 54; OHI Wolfgang Gentner by C. Weiner, November 1971.

20. F. Joliot, "Grandes découvertes," p. 13; I. Curie and F. Joliot, *CR* 198 (1934): 254-256, 559-561. See also Biquard, *Joliot-Curie*, pp. 40-51.

21. "Une Famille de savants"; Savel, "Radioactivité et radioéléments artificiels," p. 158.

22. Boyer, "Une Visite à M. et Mme. Joliot-Curie"; Rouzé, *Joliot-Curie*, p. 55; *Regard*, 1935, copy among newspaper clippings, RI.

23. G.IV.K.28, CF.

24. Audiat, "Au Collège de France," p. 99. See also Thérive, "Le Collège de France," pp. 93-94; Siegfried, "Le Collège de France."

25. Savel, "Vingt-sept années de collaboration."

26. J. Perrin to F. Joliot, Oct. 9, 1936, F36, RI; G.IV.K.38, CF; Cogniot, "Un Homme véritable."

27. Barrès, *Chronique de la Grande Guerre*, XIV, 75; Crowther, "Science in France"; *Paris-Soir*, Nov. 9, 1938, p. 1. See also Hugh Paxton to E. O. Lawrence, 1938, EOL; F. Joliot, "Project de création d'un laboratoire spécialisée . . . ," 1936?, F36, RI.

28. Kwal and Lesage, "La Physique au Palais de Découverte," pp. 223, 226.

29. A. S. Eddington, "Subatomic Energy," 2nd World Power Conference, Berlin, 1930, *Gesamtbericht zweite Weltkraftkonferenz . . . Transactions . . . Compte rendu . . .* (Berlin: VDI-Verlag, 1930), XIX, 52.

30. Papers in F31, RI; F. Joliot to Administrateur, Collège de France [Faral], Feb. 27, 1939, K.II, CF. Weiner has pointed out that the use of cyclotrons for medical purposes was international.

31. *Le Progrès*, Feb. 4, 1935.

32. Collège de France, *Annuaire*, 1939, pp. 51-52; Elsasser, *Memoirs of a Physicist*, pp. 196-197.

33. Maurice Montabre in *L'Intran*, Dec. 23, 1938.

34. "Note concernant l'acquisition d'un laboratoire à Ivry par le Centre National

de la Recherche Scientifique," Dec. 18, 1936, and other papers, F36, RI.

35. Laugier, "The National Center of Scientific Research in France," p. 8.

36. George Orwell, *Homage to Catalonia* (reprint London: Secker & Warburg, 1959), p. 248; Biquard, *Joliot-Curie*, p. 64; Nye, "Science and Socialism," p. 165.

37. Cogniot in *L'Avenir de la culture*, p. 6. See also Caute, *Communism and the French Intellectuals*; Brower, *The New Jacobins*.

38. Duclos, *Communism, Science and Culture*, pp. 8, 20, 27-28. For Communist views, see also Thorez, *Oeuvres*, Bk. 2, I, 145, V, 206, IX, 79; Vaillant-Couturier, *Au service de l'esprit*, pp. 24-28; *Le Parti communiste et l'avenir*, pp. 2-3.

39. Crowther, *The Social Relations of Science*, pp. 545, 628-629.

40. See e. g. F. Joliot to Gentner, May 3, July 2, 1937, F28, RI.

41. Segrè, *E. Fermi*, chs. 2, 3; OHI Amaldi by Thomas Kuhn, April 1963, AHQP; Holton, "Striking Gold in Science."

42. E. Fermi, *Collected Papers*, I, 639-644, 650, 1037-1043. The Rome group was helped by O. D'Agostino, who had just returned from the Radium Institute. For scientific developments from 1934, see Graetzer and Anderson, eds., *The Discovery of Nuclear Fission*. For Ida Noddack's guess, see OHI Aristid V. Grosse by Weiner, January 1974; Segrè, *Fermi*, p. 76.

43. I. Curie, Halban, P. Preiswerk, *J. Phys.* 6 (1935): 361-364.

44. Goldschmidt, *Rivalités atomiques*, p. 22; I. Joliot, "La Vie et l'oeuvre de Marie Sklodowska-Curie," pp. 26-27; OHI Kowarski by Weiner, March, October 1969, by Weart, July 1973; OHI F. Perrin, November 1973; Marbo, *Souvenirs*, p. 183.

45. Oct. 30, 1936, Rutherford Correspondence, microfilm edition, AIP. Hahn also chided Rutherford for missing the citation.

46. Schroeder-Gudehus, "The Argument for . . . Science in Weimar Germany"; Kaiser Wilhelm-Gesellschaft zur Förderung der Wissenschaften (Berlin), *Handbuch*, ed. Adolf von Harnack (Berlin: Reimar Hobbing, 1928), p. 51; Hahn, *My Life*, ch. 7.

47. Kaiser Wilhelm-Gesellschaft, *Handbuch*, pp. 20, 31, 38-46; Hahn, *My Life*, pp. 83-85, 102, 139-141, 147-148; "Tätigkeitsbericht der Kaiser Wilhelm-Gesellschaft," *Die Naturwissenschaften* 23 (1935): 414; 25 (1937): 371-372; 26 (1938): 324.

48. Hahn, *A Scientific Autobiography*, chs. 6, 7; OHI Gentner by Weiner, November 1971. The scheme involved isomers (different states of a given isotope).

49. I. Curie and P. Savitch, *J. Phys.* 8 (1937): 385-387. The largest known transformation was the loss of an alpha particle, four mass units.

50. Irving, *The Virus House*, pp. 19-30. Meitner and Hahn to I. Curie, Jan. 20, 1938, letters, RI. The actual half-life of the lanthanum isotope is about 3.7 hours.

51. I. Curie and Savitch, *CR* 206 (1938): 906-908.

52. F. Joliot, *Textes choisis*, p. 35; Biquard, *Joliot-Curie*, p. 53.

53. I. Curie and Savitch, *CR* 206 (1938): 1643-1644. They used fractional crystallization. Yttrium is produced abundantly by fission, resembles lanthanum chemically, and has isotopes with 3.1 and 3.5-hour half-lives. Louis A. Turner, *Reviews of Modern Physics* 12 (1940): 5 (this review of early fission work has not yet been superseded). Promethium is also produced abundantly in fission, resembles lanthanum chemically, and has a 2.7-hour half-life. F. Perrin, "Joliot," p. 158.

54. I. Curie and Savitch, *J. Phys.* 9 (1938): 355-359.

55. Edmond Faral, notes, Sept. 20, 1938, IV.K.44, CF.

56. Feb. 15, 1939, in Nye, "Science and Socialism," p. 166. See also A. Langevin, *Paul Langevin*, pp. 163-164; I. Curie to Meloney, Oct. 4, 1938, Meloney Collection, CUL.

57. L. G. Cook in Samuel Glasstone, *Sourcebook on Atomic Energy* (New York: Van Nostrand, 1950), p. 345; Thomas S. Kuhn, *The Structure of Scientific Revolu-*

tions, 2nd rev. ed. (Chicago: University of Chicago Press, 1970) (this book has been frequently criticized but describes something which does happen in at least some cases); Frisch, "Lise Meitner."

58. "Against our will . . . compelled," Hahn to Norman Feather, June 2, 1939, 5:21, Lise Meitner papers, Churchill College, Cambridge.

59. Hahn and Strassmann, *Die Naturwissenschaften* 27 (1939): 11-15. *Gesetzen* to *Erfahrungen:* proof sheets, Hahn papers, Bibliothek und Archiv der Max-Planck-Gesellschaft, Berlin (Dahlem).

4. FORMATION OF A TEAM

1. HalbT, p. 90; I. Curie and Savitch, *CR* 208 (1939): 343-346; OHI F. Perrin, November 1973; OHI Kowarski by Weiner, March, October, 1969.

2. In 1944 Halban told Leo Szilard that Joliot had tried to detect the fission neutrons by their production of radioactive bromine in a solution of ethyl bromide. Szilard, "Conversation with Halban 2-3-44," 3bs.15A, LS. See also Halban, "L'Oeuvre . . . Joliot," p. 44S; OHI Kowarski by Weiner, October 1969.

3. F. Joliot, *CR* 192 (1931): 1105-1107; 208 (1939): 341-343, 647-649. Dr. H. Langevin kindly showed me the radioactivity decay curves and other notes from this experiment.

4. I. Curie and Savitch, *CR* 208 (1939): 343-346 (their notes are in Dossiers 27 and 29, RI); F. Joliot, *CR* 208 (1939): 341-343.

5. KoT, p. 429; OHI Kowarski by Weiner, March, October, 1969; Frisch to Bohr, Jan. 22, Mar. 15, 1939, reel 19, Bohr Scientific Correspondence, microfilm edition, copies at AHQP repositories; Frisch, "How It All Began"; Meitner and Frisch, *Nature* 143 (1939): 239-240; Frisch, *Nature* 143 (1939): 276.

6. Anderson, "Early Days of the Chain Reaction"; E. Fermi, "Physics at Columbia University"; OHI Léon Rosenfeld by Weiner, September 1968.

7. "Lew Kowarski"; OHI Kowarski, March, October, 1969; KoT, pp. 416-423.

8. HalbT, pp. 82-89, 130-131; Halban, *Travaux et titres scientifiques;* Goldschmidt, "Hans Halban"; Halban to Palewski (Directeur du Cabinet, Ministre de la Justice), Oct. 28, 1938, copies in author's files. For Halban senior, see also M. Kofler, "Hans von Halban, 1877-1947," *Helvetica Chimica Acta* 31 (1948): 120-128; OHI Victor Weisskopf, January 1975.

9. Soddy, *Science and Life,* p. 36; OHI Kowarski by Weiner, March 1969; F. Joliot in *Les Prix Nobel en 1935* (Stockholm: Imprimerie Royale, Norstedt & Söner, 1937). See also Joliot's earlier talk reported in *Le Progrès,* Feb. 4, 1935.

10. Halban, "L'Oeuvre . . . Joliot," p. 44S; F. Joliot, *CR* 208 (1939): 341-343.

11. HalbT, pp. 90-91.

12. OHI Kowarski by Weiner, November 1969; HalbT, pp. 91-92. The original transcript gives "discussed" where I put "discussing."

13. Frisch, Halban, and Jørgen Koch, *Kgl. Danske Videnskabernes Selskab, Mathematisk-fysiske Meddelelser* 15, no. 10 (1938); Halban, *CR* 206 (1938): 1170-1172; OHI Frisch by Weiner, May 1967; Edoardo Amaldi and E. Fermi, *Phys. Rev.* 50 (1936): 899-928. For all E. Fermi papers, see his *Collected Papers.* For a general treatment of neutron diffusion measurements and their history, see Amaldi, "The Production and Slowing Down of Neutrons."

14. CEA.51.A(P18), RI.

15. Kowarski, draft dated "Mardi" on back of letter dated Feb. 28, 1939, F28,RI. Halban wrote Joliot on Mar. 3, "Nous étions déjà assez sûrs de notre résultat mercredi matin," i.e., Mar. 1. F28, RI. Kowarski, *J. Phys.* 8th ser., 7 (1946): 255.

16. Halban to F. Joliot, Mar. 3, 1939, F28, RI; F. Joliot to Kowarski, Mar. 3, 1939, CEA.51.C, RI.

17. Halban to F. Joliot, Mar. 3, 1939, F28, RI; Halban, F. Joliot, and Kowarski, *Nature* 143 (1939): 470-471; OHI Kowarski by Weiner, November 1969.

18. The reaction is $^{32}S + {}^0n \rightarrow {}^{32}P + {}^1p$ with threshold energy around 2 meV, considerably higher than the energy of the neutrons from the source.

19. HalbT, p. 121; interview of Dodé, October 1973; CEA.51.A(P18), RI.

20. Halban, Affidavit, June 5, 1957, PCB, para. 10; CEA.51.C,RI; Dodé, Halban, F. Joliot, and Kowarski, *CR* 208 (1939): 995-997.

21. OHI F. Perrin, November 1973.

22. Goldschmidt, *Rivalités atomiques*, p. 20; Cogniot, "Un Homme véritable," p. 63.

5. THE SECRET OF THE CHAIN REACTION

1. F28, RI. This and other letters are printed in *Szilard: His Version*. Part of this chapter was published in a different form as Weart, "Scientists with a Secret."

2. Szilard gives Wells' book as reference no. 0 in his 1940 paper on chain reactions. Szilard, *Collected Works*, p. 226. My account draws heavily on Szilard's reminiscences, LS, most of which appear in *Szilard: His Version*.

3. The problem was a falsely low mass measurement for helium, which made it seem likely that $^1n + {}^9Be \rightarrow {}^4He + {}^4He + 2{}^1n$ would release energy. Hans Bethe, *Phys. Rev.* 47 (1935): 633-634.

4. Szilard to C. S. Wright, Feb. 26, 1936, in Szilard, *Collected Works*, p. 734; Szilard to John Cockcroft, May 21, 1936, R-46, LS; Szilard to Prof. Singer, June 16, 1935, Hist-J, LS; Szilard to Lindemann, June 3, 1935, Hist-J, LS.

5. Szilard to Singer (draft), June 16, 1935, Hist-J, LS; Kenneth T. Bainbridge to James Chadwick, Oct. 19, 1973, Mark Oliphant to Bainbridge, Oct. 31, 1973, Miscellaneous-Bainbridge, AIP; Bernal, *The Social Function of Science*, pp. 150, 182.

6. Mar. 26, 1936, bbs.42, LS.

7. Maurice Goldhaber, R. D. Hill, and Szilard, *Phys. Rev.* 55 (1939): 47-49; Wigner ms. dated Apr. 16, 1941, bbs.41, LS; Szilard to Bohr, Nov. 11, 1938, B.3, LS.

8. George B. Pegram to Edmund Astby Prentis (draft), Nov. 11, 1952, Box 12, Pegram Collection, CUL; Kevles, *The Physicists*, p. 324.

9. *Szilard: His Version*, p. 54.

10. Szilard, "Record 1A," 1956, Hist-G, LS, transcript also in Marshall MacDuffie Collection, CUL; E. Fermi to F. Joliot, Feb. 4, 1939, F28, RI.

11. KoT, p. 436; Paxton to "Maurice" [French cyclotron worker], Feb. 12, 1939, F28, RI.

12. Lawrence to Stanley N. Van Voorhis, Feb. 9, 1939, 17:43, EOL; G. K. Green and Luis W. Alvarez, *Phys. Rev.* 55 (1939): 417; Philip Abelson, *Phys. Rev.* 55 (1939): 418, 670; D. R. Corson and R. L. Thornton, *Phys. Rev.* 55 (1939): 509. Joliot notified Lawrence of his first fission experiment by a letter which I have not found, mentioned by Edwin McMillan, *Phys. Rev.* 55 (1939): 510. See also OHI Robert R. Wilson, May 1977, p. 55; Abelson, "A Sport Played by Graduate Students"; Alvarez, "Adventures in Nuclear Physics" (Faculty Research Lecture, University of California, Berkeley, 1962), Lawrence Radiation Laboratory publication UCRL-10476, p. 22.

13. Frisch, "Somebody Turned the Sun on with a Switch." See also OHI Frisch by Weiner, May 1967; Frisch to Bohr, Mar. 18, 1939, reel 19, Bohr Scientific Correspondence, AHQP.

14. L. Fermi, *Atoms in the Family*, pp. 150, 159; Anderson, E. T. Booth, J. R.

Dunning, E. Fermi, G. N. Glasoe, and F. G. Slack, *Phys. Rev.* 55 (1939): 511-512; Placzek to Halban, Mar. 3, 1939, copy in author's files; Anderson, "Legacy of Fermi and Szilard," pp. 57-58, 60; OHI Grosse by Weiner, April 1974.

15. Michael Pupin, *From Immigrant to Inventor* (New York: Scribner's, 1927); "New York Committee on Submarine Defense: Sub-Committee of National Research Council: Report of Secretary . . . July 1, 1919," Box 21, Pegram Collection, CUL; Pupin, "The Meaning of Scientific Research," *Science* 61 (1925): 30.

16. Pegram?, "Research Problem: Department of Physics," in folder "Pure Science Research Committee, 1927-1928," Box 47, Pegram Collection, CUL; letters in Box 29, Pegram Collection.

17. R. Morton Adams to Szilard, May 19, 1939, B.3, LS; Szilard to Liebowitz, Dec. 4, 1939, M.17, LS; other letters in *Szilard: His Version*, pp. 63-65; Walter Zinn in Booth et al., *Beginnings of the Nuclear Age*, p. 11; Bk.f.2, M.16, M.27, R-1 and O.0, LS.

18. Pegram to Szilard, Apr. 6, 1939, Bk.f.2, LS; Szilard and Zinn, *Phys. Rev.* 55 (1939): 799-800; Anderson, E. Fermi, and H. B. Hanstein, *Phys. Rev.* 55 (1939): 797-798; Anderson, "Early Days of the Chain Reaction"; Anderson, "Legacy of Fermi and Szilard"; Halban, F. Joliot, and Kowarski, *Nature* 143 (1939): 470-471.

19. Anderson et al., *Phys. Rev.* 55 (1939): 511-512.

20. CEA.51.C, RI, on draft of Dodé, Halban, F. Joliot, and Kowarski, *CR* 208 (1939): 995-997.

21. Letters to de Broglie, 1959, in F. Joliot dossier, Archives of the Académie des Sciences, Paris; Halban, *Helvetica Physica Acta* 7 (1935): 856-875; Kowarski, *Avantprojets de distribution du gaz.*

22. CEA.51.A(P15), RI: Halban, Kowarski, and Savitch, *CR* 208 (1939): 1396-1398; HalbT, pp. 121-122. Anderson and Fermi did a similar experiment independently. *Phys. Rev.* 55 (1939): 1106-1107.

23. Weart, "Secrecy, Simultaneous Discovery, and the Theory of Nuclear Reactors," *American Journal of Physics* 45 (1977): 1049-1060; Halban, Kowarski, and Savitch, *CR* 208 (1939): 1396-1398.

24. CEA.51.A(P18), RI.

25. The French knew that the area under the density-distribution curve, S, was proportional to the number of neutrons in their tank. The number of neutrons in the tank was equal to the number of neutrons produced per second, Q, multiplied by the mean time a neutron spent in the tank before being absorbed, which they called $1/A$. Therefore $S \propto Q/A$. By elementary calculus, the small difference ΔS measured between the area of the curve with and without uranium present would be $(\Delta S/S) = (\Delta Q/Q) - (\Delta A/A)$. The team knew the value of A and of ΔA, since A was simply the total probability per unit time that a neutron would be absorbed by the various substances in the tank, a quantity which depended only on known cross-sections, including the values for uranium absorption which they had just measured. Now the French made an error: they said, "Let A_f be the term for the capture leading to fission. Every neutron has a probability A_f/A of causing a fission and, since one individual fission process liberates v neutrons on the average, the total number ΔQ of neutrons thus created is $Q(A_f/A)v$." Halban, F. Joliot, and Kowarski, *Nature* 143 (1939):680. This is correct, except that this ΔQ, which represents the number of neutrons created in fission, is not the same as the ΔQ used in their earlier equation involving the area of the density-distribution curve, and the French proceeded to equate the two. The two would be the same in the traditional sort of experiment where particles are shot through a screen. Adding uranium to the screen would increase the number of neutrons coming out the other side by exactly the number of neutrons created by fission in the

screen. But when neutrons diffuse through a tank, each neutron produced by fission may cause additional fissions, which produce neutrons that go on to create more fissions, so that adding uranium to the tank may cause a great increase in the number of neutrons produced per second. Louis Turner in Princeton recalculated the French data in January 1940 after solving the theoretical problem and got $v = 2.6 \pm 0.6$. *Phys. Rev.* 57 (1940): 334.

26. Clark, *The Birth of the Bomb*, p. 21.

27. OHI Kowarski by Weiner, Oct. 1969.

28. L. Fermi, *Atoms in the Family*, p. 164; Pegram to E. A. Prentis (draft), Nov. 11, 1952, Box 12, Pegram Collection, CUL; Wigner to Szilard, Apr. 17, 1939, Bk.f.2, LS; Strauss, *Men and Decisions*, p. 238; H. W. Graf, "Memorandum for File," Mar. 17, 1939, 13:Compton, OSRD.

29. E. Fermi, "Physics at Columbia"; *Szilard: His Version*, p. 56.

30. Szilard to F. Joliot, Apr. 7, 1939, F28, RI; Zinn to Pegram, copy of draft letter, n.d., in Szilard memorandum to Compton, Nov. 12, 1942, bbs.49, LS.

31. OHI Weisskopf, January 1975; letters and cables in *Szilard: His Version*, pp. 70-73. Blackett's favorable reply to Weisskopf was received Apr. 8, 1939. Copies of letters and cables are in Bk.f.3, LS.

32. Bohr to Frisch (draft), n.d. [March 1939], Bohr Scientific Correspondence, AHQP. John Heilbron helped with the translation. Frisch does not recall receiving this letter. For Bohr's acquiescence, see Wigner to Dirac, Mar. 30, 1939, Bk.f.3, LS; Wigner, "Historical Notes," Apr. 16, 1941, bbs.41, LS, pp. 4-6.

33. R. B. Roberts to E. T. Roberts, Jan. 29, 1939, Dept. of Terrestrial Magnetism, Carnegie Institution of Washington, Washington, D.C., copied for me by R. B. Roberts; Roberts, R. C. Meyer, and P. Wang, *Phys. Rev.* 55 (Mar. 1, 1939): 510; Teller to Szilard, two letters, n.d. [February 1939], B.8, LS, in *Szilard: His Version*, p. 66. See also Roberts, L. R. Hafsted, Meyer, and Wang, *Phys. Rev.* 55 (1939): 664.

34. Roberts to author, May 6, 1976; [Potter], "Exploding Uranium Atoms May Set Free Neutrons That Will in Turn Explode Other Atoms, in 'Cascade' Effect," Science Service, New York, Feb. 24, 1939, printed in *Science News Letter*, Mar. 11, 1939, p. 140. Science Service gave me a copy of the original release.

35. Weisskopf to Blackett, n.d. [c. Mar. 31, 1939], copy in Bk.f.3, LS; Goldhaber to Szilard, Apr. 12, 1939, Bk.f.3, LS.

36. Original is lost; from copy sent by Weisskopf to Szilard, Bk.f.3, LS.

37. Halban to Weisskopf, May 9, 1939, copy in author's files; Clark, *Birth of the Bomb*, p. 20; OHI Kowarski by Weiner, October 1969.

38. Halban to Weisskopf, May 9, 1939, copy in author's files.

39. F. Joliot to Bohr, Apr. 12, 1945, reel 21, Bohr Scientific Correspondence, AHQP. See also Halban, Affidavit, June 5, 1957, PCB, para. 18; KoT, pp. 438-439; OHI Kowarski by Weiner, October 1969; Clark, *Birth of the Bomb*, p. 20; Goldschmidt, *Rivalités atomiques*, pp. 27-28; F. Joliot, *Textes choisis*, p. 154. Joliot's opposition in principle to secrecy in 1939 was stressed by Biquard, interview, September 1976.

40. "Uranium Atoms, Split by Neutrons with Enormous Release of Atomic Energy, Also Give Off Other Particles Which May Help Perpetuate the Reaction," Science Service, London, Mar. 16, 1939, printed in *Science News Letter*, Apr. 1, 1939, p. 196; F. Joliot to Szilard, Apr. 19, 1939, F28,RI.

41. Bk.f.3, LS.

42. F28, RI; Bk.f.3, LS.

43. *Szilard: His Version*, p. 57; Szilard to Blackett, Apr. 14, 1939, Bk.f.3, LS; Anderson, E. Fermi, and Hanstein, *Phys. Rev.* 55 (1939): 797-798; Szilard and Zinn, *Phys. Rev.* 55 (1939): 799-800.

44. Halban, F. Joliot, and Kowarski, *Nature* 143 (1939): 680; Gowing, *Britain and Atomic Energy*, pp. 34-35; Thomson, "Anglo-U.S. Cooperation on Atomic Energy," p. 46; Irving, *Virus House*, pp. 32-34; Wilhelm Hanle, "A Half Century of Scientific Life" (Seminar at University of Missouri-Rolla, n.d.), BBA:Hanle, AIP; Bagge, Diebner, and Jay, *Von der Uranspaltung bis Calder Hall*, pp. 19-20; Golovin, *Kurchatov*, pp. 31-32.

6. A CONTRACT FOR NUCLEAR ENERGY

1. HalbT, p. 92; OHI F. Perrin, November 1973; "Francis Perrin," *Current Biography* (1951), pp. 483-484.

2. OHI Kowarski by Weiner, March 1969; OHI F. Perrin, November 1973.

3. F. Perrin, *Notice sur les travaux scientifiques*; OHI F. Perrin, November 1973.

4. F. Perrin, *CR* 208 (1939): 1394-1396.

5. Halban, F. Joliot, Kowarski, and F. Perrin, French Patent 971,324, applied for May 4, 1939, in Joliot-Curie, *Oeuvres*, pp. 687-691.

6. OHI F. Perrin, November 1973; OHI Kowarski by Weiner, October 1969; Clark, *Birth of the Bomb*, pp. 27-28.

7. F. Perrin, *CR* 209 (1939): 301-303.

8. OHI Kowarski by Weiner, October 1969; OHI F. Perrin, November 1973. Perrin had not read *The World Set Free* but "it was mentioned often by friends and I read many books of H. G. Wells." OHI F. Perrin, November 1977.

9. Halban, F. Joliot, Kowarski, and F. Perrin, French Patent 976,541, applied for May 1, 1939, in Joliot-Curie, *Oeuvres*, pp. 678-683; HalbT, p. 122.

10. F. Perrin, *CR* 208 (1939): 1394-1396.

11. Halban and Preiswerk, *Nature* 136 (1935): 1027; 137 (1936): 905-906. Adler and Halban, *Nature* 143 (1939): 793; Halban, F. Joliot, Kowarski, and F. Perrin, French Patent 976,542, applied for May 2, 1939, in Joliot-Curie, *Oeuvres*, pp. 684-686; OHI Kowarski, August 1973; CEA.51.C, RI.

12. French Patent 976,541, in Joliot-Curie, *Oeuvres*, pp. 678-683.

13. F. Joliot preface to Kowarski, *Avant-projets de distribution du gaz*, p. vi.

14. Appell, *Gabriel Lippmann* (P: Gaignault, 1923), p. 7; Forman, Heilbron, and Weart, *Physics circa 1900*, pp. 48-49; papers in 7A, 7B, Curie papers, Bibliothèque Nationale, Paris; Reid, *Marie Curie*, pp. 63, 69, 317; Barrès in *J. O., Débats, Chambre* (Dec. 10, 1921): 4876.

15. Palmer, *University Research and Patent Policies*, p. 8; Isaiah Bowman, "Summary Statement of the Work of the National Research Council—1934-1935," *Science* 82 (1935): 337-342. See also Kevles, *The Physicists*, p. 268; Noble, *America by Design*, ch. 6.

16. Frank Cameron, *Cottrell: Samaritan of Science* (Garden City, N.Y.: Doubleday, 1952), pp. 149-169, 280-318.

17. Szilard, *Collected Works*, pp. 605-651, 729-731; Szilard to Lindemann, June 3, 1935, Hist.-J, LS; Szilard to Rutherford, May 27, 1936, R-1, LS.

18. L. Fermi, *Atoms in the Family*, p. 101; Segrè, *Fermi*, pp. 83-84; Gian Carlo Wick to author, November 1976; Segrè to Gabrielle Giannini, Apr. 24, 1939, 9:10, E. Fermi Papers, University of Chicago Library; Lawrence to Howard A. Poillon, Apr. 27, 1939, 15:18, EOL.

19. U.S. Patent application 263,017, in Szilard, *Collected Works*, pp. 655-690; Szilard to Pegram, n.d. [October 1940?], 3bs.8, LS; Pegram to Albert C. Barrows, Mar. 16, 1935, 32:Patents, Pegram Collection, CUL.

20. French Patent 502,913, applied for May 29, 1916; Biquard, *Langevin*, p. 59;

Frederick V. Hunt, *Electroacoustics: The Analysis of Transduction, and Its Historical Background*, Harvard Monographs in Applied Science no. 5 (Cambridge: Harvard University Press, 1954), pp. 46-52; Clark, *Birth of the Bomb*, p. 28; OHI Kowarski by Weiner, March 1969; HalbT, pp. 92-93; OHI F. Perrin, November 1973; Goldschmidt, *Rivalités atomiques*, p. 50.

21. Halban, "L'Oeuvre scientifique de . . . Joliot," p. 44S.

22. Laugier, "Le Centre National de la Recherche Scientifique en France," p. 7; OHI Kowarski, July 1973. See also Auger, "Hommage à Henri Laugier"; "Henri Laugier," *Current Biography* (1948), pp. 371-372; Laugier, *Combat de l'exil*, p. 45; Laugier, "Reconstruction in France: Educational Equality."

23. Clark, *Birth of the Bomb*, p. 22. A sum of 50,000 francs was formally requested and granted in July. Chef du Service Central de la Recherche Scientifique to F. Joliot, July 6, 1939, CEA.10, RI. For additions, see 1939 budget, K.II, CF. For 1939 Joliot's laboratory in the Collège de France was originally granted 69,000 francs for overhead, 50,000 for matériel, and 30,000 for personnel. This did not include the Ivry Laboratory, whose projected budget for 1939 was 355,000 francs, of which 155,000 was for personnel. A few other people there received salaries directly from the CNRS. Chef du Service Central de la Recherche Scientifique to F. Joliot, Jan. 31, Feb. 14, 1939, F36, RI. Actual 1939 expenditures at Ivry were 369,520 francs plus 35,365 for construction. F. Joliot to M. l'Administrateur, CNRS, Jan. 9, 1941, F36, RI. The laboratory had received some 5 million francs over several years for new construction and equipment. Charles Jacob, "Rapport à M. le Ministre de l'Instruction Publique sur le Centre National de la Recherche Scientifique," December 1940, "Rapports préparatoires à la création du CNRS," CNRS, p. 31.

24. HalbT, pp. 125, 139, 141; OHI Kowarski by Weiner, October 1969.

25. Reid, *Marie Curie*, p. 93.

26. Ydewalle, *L'Union Minière du Haut-Katanga*, pp. 71-72, 79-80; H. E. Bishop, "The Present Situation in the Radium Industry," *Science* 57 (1923): 341-345; Eggleston, *Canada's Nuclear Story*, pp. 41-43; E. Curie, *Mme. Curie*, p. 200; Reid, *Marie Curie*, pp. 143-144. There is hearsay evidence that the Union Minière carried radium on its books at roughly $2500 a gram, the Canadians at twice this amount; the heavy costs of preparing, packaging, standardizing, and marketing minute amounts of radium compounds accounted for the rest. A. J. Pregel, "Memorandum Re: World Prices and Markets for Radium," in bound volumes of documents "Rex vs. Boris Pregel," assembled by Lowell Wadmond (counsel), Boris Pregel papers, New York, vol. 7; "Agreement, 30 April 1938, Between the Eldorado Gold Mines Ltd. and the Union Minière du Haut Katanga," copy in "Rex vs. Boris Pregel." According to Pregel memorandum, the peacetime consumption of radium ran about 25 g./year, of which about a tenth was sold for luminous compounds and the rest was rented, chiefly for medical use.

27. OHI Boris Pregel, January 1975; MC.16, dossier 309, RI; Lechien to F. Joliot, May 16, 1939, CEA.8, RI.

28. F. Joliot to L. G. Lecointe and J. van Stappen, May 4, 1939, and reply, May 5, F28, RI. By one account I. Curie went to one of these meetings and contributed to the discussion. Ydewalle, *Union Minière*, p. 80. See also Clark, *Birth of the Bomb*, p. 28; HalbT, p. 94; Strauss, *Men and Decisions*, p. 316.

29. Interview of B. Pregel, December 1974; Ydewalle, *Union Minière*, pp. 64-71, 78.

30. F. Joliot to Lecointe and Lecointe to F. Joliot, both May 10, 1939, CEA.13, RI.

31. Strauss, *Men and Decisions*, p. 316; HalbT, pp. 94-95; "Projet de convention," n.d., attached to note dated May 13, 1939, and probably preceding the initialed

Gentlemen's Agreement, CEA.12, RI. An early draft of the Agreement is in CEA.8, RI, and the text is in Gowing, *Dossier secret*, App. 2.

32. "Projet de lettre des inventeurs à l'Union Minière" (draft), n.d., CEA.12, RI; F. Joliot to Lechien, July 29, 1939, CEA.11, RI.

33. F. Joliot to Lechien, July 29, 1939, CEA.11, RI; CEA.16, RI. The Union Minière may have been reluctant to make a final agreement. Goldschmidt, *Rivalités atomiques*, p. 51. On Dec. 11, 1940, Halban wrote B. G. Dickens that at one point the team had planned to "limit our personal profit to a small amount (practically five per cent. to each of the four members forming the team of collaboration), and that the rest of the profit should be employed for salaries and apparatus for pure scientific workers in France." AB.1.19(62B), PRO. "Projet de Convention . . . entre MM. Halban, Joliot, Kowarski et Francis Perrin . . . et le Centre National de la Recherche Scientifique," May 1, 1940, "Brevets de base":6, CEA.

34. F. Joliot-Lechien correspondence, CEA.8, RI; Jonemann (s.a. de Transports) to F. Joliot, June 1, 1939, CEA.11, RI.

7. FISSION RESEARCH ON THE EVE OF WAR

1. Halban to Placzek, May 9, 1939; F. Perrin, *CR* 208 (1939): 1573-1575; Adler, *CR* 209 (1939): 301-303; Halban, F. Joliot, and Kowarski, *Nature* 143 (1939): 939; CEA.51.A(26), RI.

2. Interview of Auger, October 1973; Crowther, *Science in Liberated Europe*, p. 19; Auger to author, Mar. 20, 1978.

3. Goldschmidt, *Rivalités atomiques*, pp. 20-24; OHI Kowarski, July 1973; Haïssinsky, *Laboratoire Curie*, p. 15; I. Curie and Mme. Tsien San-Tsiang, *J. Phys.* 10 (1939): 495-496; Lab. dossiers 28-29, RI; L. Goldstein, A. Rogozinski, and R. J. Walen, *Nature* 144 (1939): 201-202; Goldstein, Rogozinski, and Walen, *J. Phys.* 10 (1939): 477-486; G. Haenny and A. Rosenberg, *CR* 208 (1939): 898-900; Chilowsky, French Patent 861,930, requested July 28, 1939, delivered Oct. 28, 1940.

4. My counts, by nationality of laboratory where work was done, from Turner, *Reviews of Modern Physics* 12 (1940): 1-29.

5. Solomon, *CR* 208 (1939): 570-572, 896-898; Bohr and Wheeler, *Phys. Rev.* 56 (1939): 426-450. See also Rosenfeld, "Nuclear Reminiscences" 1968, transcript at AIP, p. 15; Wheeler, "The Mechanism of Fission," p. 52; Wheeler, "Niels Bohr and Nuclear Physics," pp. 41-43.

6. Chadwick to E. Appleton, Dec. 5, 1939, AB.1.219, PRO.

7. Irving, *Virus House*, ch. 2; Golovin, *Kurchatov*, pp. 31-34; Kramish, *Atomic Energy in the Soviet Union*, ch. 2; Weart, "Secrecy, Simultaneous Discovery, Nuclear Reactors."

8. Thomson, *Nuclear Energy in Britain*, p. 4; "Summary of 'Report on Experiments on Uranium Oxide' submitted by Professor G. P. Thomson, April 1940," May 9, 1940, AB.1.219, PRO. See also J. L. Michels, G. Parry, and G. P. Thomson, *Nature* 143 (1939): 760; Gowing, *Britain and Atomic Energy*, pp. 34-40; Clark, *Birth of the Bomb*, pp. 34-37; interview of Thomson, December 1973.

9. Oliphant to Lawrence, Aug. 24, 1939, 14:6, EOL; Oliphant in *Meanjin Quarterly* (Melbourne), quoted in Snow, *Science and Government*, App., pp. 8-9.

10. Merle [Tuve] to Gregory [Breit], Apr. 12, 1939, copy furnished by Roberts from Dept. of Terrestrial Magnetism papers; Tuve to Szilard, Mar. 27, 1939, sbs.7, LS; Roberts to author, May 4, 1976; Tuve to Lawrence, June 17, 1939, 17:34, EOL.

11. Strauss, *Men and Decisions*, pp. 180, 436-437; Kenneth H. Kingdon and Herbert C. Pollack, "Pioneering in the Atomic Age," G[eneral] E[lectric] *Research Lab Bulletin*, Summer 1965, pp. 9-16; Fisk and Shockley, "A Study of Uranium as a

Source of Power," July 1940, BC:4, OSRD; OHI Fisk by L. Hoddeson, June 1976; Fisk to author, June 14, 1977.

12. Szilard to Wigner, May 21, 1939, O.0, LS; Anderson, "Early Days of the Chain Reaction," p. 9.

13. Anderson, E. Fermi, and Szilard, *Phys. Rev.* 56 (1939): 284-286. See also Amaldi, "George Placzek," *Ricerca Scientifica* 26 (1956): 2038-2042; OHI Frisch by Weiner, May 1967, p. 35; Rosenfeld, "Nuclear Reminiscences" 1968, transcript at AIP, pp. 13-14; Wheeler, "Mechanism of Fission," p. 52.

14. Szilard to Fermi, July 8, 1939, in Szilard, *Collected Works*, pp. 195-196; E. Fermi to Pegram, July 11, 1939, PCB. See also Bk.f.4, LS; Anderson in E. Fermi, *Collected Papers*, II, 11, 15, 18; Alexander Sachs to Lyman J. Briggs, June 3, 1940, 29: Originals Doc. 18, OSRD.

15. Bk.f.2, LS; *Szilard: His Version*, pp. 86-90, 102, 106-107.

16. Halban to Placzek, May 9, 1939, and Placzek [to Halban], "Optimale konzentration für der kettenreaktion," [May? 1939], copies in author's files; OHI Kowarski by Weiner, October 1969.

17. OHI Kowarski by Weiner, October 1969; Szilard note, "History. France. Kowarski" [September 1946?], Bk.f, LS.

18. Even without a moderator. Karl K. Darrow to Donald Cooksey, July 23, 1939, 6:9, EOL.

19. F. Joliot to Lechien, July 3, 29, 1939, CEA.11,RI. See also CEA.51.A(17,19), RI.

20. CEA.51.A(17, 19, 23, 24, 25), RI.

21. Halban to "Dearest Daddy" and Halban to Placzek, both Aug. 7, 1939, copies in author's files.

22. F. Joliot to Halban, Aug. 8, 1939; Halban to Halban, Sr., Aug. 7, 1939; copies in author's files.

23. Halban memorandum [to F. Joliot and Kowarski? late August 1939], CEA. 51.C, RI.

24. Halban, F. Joliot, Kowarski, and F. Perrin, *J. Phys.* 10 (1939): 428-429. See also Halban [to F. Joliot?], Aug. 21, 1939, CEA.51.C, RI; OHI F. Perrin, November 1973.

25. *Szilard: His Version*, p. 115; Halban [to F. Joliot?], Aug. 23, 1939, II.B.1, LK. See also *Time* 35 (Feb. 12, 1940): 44.

26. Darrow to Cooksey, July 23, 1939, 6:9, EOL; KoT, p. 449; OHI Kowarski by Weiner, October 1969.

27. André J.-L. Breton, "Proposition de résolution," Sénat no. 536, 2e session extraordinaire, Nov. 30, 1939, in "Rapports préparatoires à la création du CNRS," CNRS, p. 246; "Comptes-rendus," September (?), October 1938, in "Affaires divers . . . personnel," F^{17}17,492, AN; Zay, *Souvenirs*, pp. 269-270.

28. J. Perrin, *L'Organisation de la recherche scientifique en France*; R. Legendre, "L'Office National"; "Historique antérieure à 1949," unsigned typescript in "Articles sur le CNRS 1940-1956," CNRS; F29, RI.

29. Charles Jacob, "Rapport à M. le Ministre de l'Instruction Publique sur le Centre National de la Recherche Scientifique," December 1940, "Rapports préparatoires à la création du CNRS," CNRS, p. 28; H.II.e.6,7,23,40, CF; Joliot's *Carnet de fonctionnaire*, C.XII(Joliot), CF.

30. "Procès-verbal de la réunion extraordinaire des directeurs de laboratoire, Collège de France," Sept. 14, 1938, H.II.e.6, CF; Halban to Kowarski, [c. August 1939], CEA.51.C, RI.

31. Angelo Tosca [A. Rossi], *Les Communistes français pendant la Drôle de Guerre* (P: Iles d'or, 1951), p. 31.

32. H. Longchambon for the CNRS, Sept. 23, 1939, MI, CF. See also "Schémas d'organisation des laboratoires du Centre à la date du l^e Juin 1940," in "Centre National de la Recherche Scientifique 1939-1945," CNRS; Faral to Ministre de l'Education Nationale, June 4, 1940, G.IV.K.52, CF; notebook of carbon copies of memos, "Service intérieur," Mar. 12, 1940, H.II.e.24, CF.

33. F29, RI.

34. Jacob, "Rapport . . . sur le CNRS," in "Rapports préparatoires," CNRS.

35. F. Joliot to Laporte, Apr. 30, 1940, F28, RI; F. Joliot, *Textes Choisis*, p. 126; Biquard, *Langevin*, pp. 65, 87-88; Laugier, "How Science Can Win the War," p. 58. See also statement of Laugier, U.S. Senate, 77th, Committee on Military Affairs, *Technological Mobilisation: Hearings* (Washington, D.C.: U.S. Government Printing Office, 1942), II, 628, 642-643; Canac, "L'Organisation de la recherche scientifique," pp. 363-364.

36. Feb. 18, 1940, CEA.7, RI.

37. Interview of Dodé, October 1973; F. Joliot to Ministre de l'Armament (?), memorandum beginning, "Les recherches sur la rupture explosive de l'uranium," [Spring 1940?], LK.

38. OHI F. Perrin, November 1973; F. Joliot's 1940 appointments book, RI.

8. CONDITIONS FOR A WORKING REACTOR

1. Halban, F. Joliot, Kowarski, and F. Perrin, *J. Phys.* 10 (1939): 428-429.

2. CEA.51.C, RI; III.A, LK; OHI Kowarski, August 1973; Weart, "Secrecy, Simultaneous Discovery, Nuclear Reactors," p. 1053.

3. Adler and Halban, *Nature* 143 (1939): 793; Halban to Weisskopf, May 9, 1939, copy in author's files.

4. The formula gave k as simply the product of three factors: the probability that a neutron would escape resonant capture, the probability that a neutron would be absorbed in uranium, and the number of neutrons produced in uranium per neutron absorbed: $k_\infty = (1-P)(1-r)\nu'$, where P is the probability of resonance capture, r the fraction of neutrons that escape absorption in uranium, and ν' the number of neutrons produced per neutron absorbed, equal to ν times the probability that a neutron, once absorbed, will cause fission. Originally they also had a term to allow for neutrons escaping through the surface, but they later assumed an infinitely large system.

5. *Pli cacheté* no. 11,620, deposited Oct. 30, 1939, opened Aug. 18, 1948, in Joliot-Curie, *Oeuvres*, pp. 673-677.

6. Weart, "Secrecy, Simultaneous Discovery, Nuclear Reactors."

7. "Divergent Chain Reaction in Systems Composed of Uranium and Carbon," Feb. 6, 1940, in Szilard, *Collected Works*, p. 208; *Szilard: His Version*, pp. 81-82.

8. D. R. Pye to Appleton, Feb. 25, 1940. See also "Summary of 'Report on Experiments on Uranium Oxide' submitted by Prof. G. P. Thomson, April 1940," May 9, 1940, both AB.1.219, PRO; Heisenberg, "Bericht über die Möglichkeit technischer Energiegewinnung aus der Uranspaltung (II)," Feb. 29, 1940, G-40, pp. 19-21; Heisenberg, G-39 and G-93.

9. A unit central neutron source is assumed. Weart, "Secrecy, Simultaneous Discovery, Nuclear Reactors," p. 1053.

10. CEA.51.A(P.14), RI; Halban, F. Joliot, and Kowarski, *CR* 229 (1949): 909-914.

11. A. J. E. Welch, "Separation of Isotopes by Thermal Diffusion," *Annual Reports on the Progress of Chemistry for 1940*, 37 (London: The Chemical Society, 1941), pp. 153-164; F. Joliot to Martelly, Jan. 2, 1940, F29: Divers, RI; F. Joliot to Administrateur, CNRS, Jan. 9, 1941, "Travaux scientifiques en cours," F36, RI; OHI

Kowarski by Weiner, October 1969. See also Halban, F. Joliot, and Kowarski, *CR* 229 (1949): 909-914; Joliot-Curie, *Oeuvres*, p. 692.

12. Turner, *Phys. Rev.* 57 (1940): 334; Kowarski, "Solutions U et Am," June 1940, CEA.51.B, RI; KoT, pp. 493-494.

13. OHI Kowarski, August 1973; Halban, F. Joliot, and Kowarski, *CR* 229 (1949): 909-914; Irving, *Virus House*, pp. 159, 171-172; Szilard, *Collected Works*, pp. 216-256; E. Fermi, *Collected Papers*, II, 119-127; Eugene Wigner, "Nuclear Chain Reactions," in Goodman, ed., *The Science and Engineering of Nuclear Power*, pp. 99-109.

14. CEA.51.C, RI.

15. P. Langevin, *Annales de Physique* 17 (1942): 303-317. Langevin sent an offprint to Joliot on Feb. 17, 1943, with the note, "A mon cher Fred Joliot En souvenir d'une question posée par lui il y a trois ans." Folder of Joliot's lectures at Collège de France, 1947/1948, in possession of Dr. H. Langevin. See also F. Joliot, "Paul Langevin," *Obituary Notices of Fellows of the Royal Society (London)* 7 (1951): 413.

16. CEA.51.A(P.22,23), RI; notebook of Kowarski, CEA.51.B, RI; Halban, F. Joliot, and Kowarski, "Production et absorption des neutrons thermiques dans un milieu hétérogène contenant de l'uranium et de l'hydrogène," *pli cacheté* no. 11,700, deposited in Kowarski's name Apr. 22, 1940, opened Jan. 28, 1957, Archives of the Académie des Sciences, Paris.

17. Ibid.; CEA.51.A(P.19,21), RI.

18. OHI Kowarski, August 1973.

9. MODERATORS AND MINISTERS

1. Maurois, "Raoul Dautry," p. 28. See also *Hommage à Raoul Dautry;* "Raoul Dautry"; Silva, "In Memoriam Raoul Dautry."

2. Dautry testimony in France, Commission Chargée d'Enquêter, *Les Evènements survenus en France de 1933 à 1945*, Annexes, VII, 1951-1952, 1960, 2010-2011.

3. I. Joliot, "Discours prononcé à l'occasion de la remise d'une épée d'honneur à R. Dautry," June 15, 1948, Dautry Dossier, CEA. Dautry was president of the Comité Exécutif, Institut des Recherches Scientifiques Appliquées à la Défense Nationale.

4. Clark, *Birth of the Bomb*, p. 67. Halban was perhaps exaggerating. For a typical, more accurate if still spectacular, popularization, see *Les Echos*, no. 1094, June 15, 1939. See also Blondel, "Le Ravitaillement de la France," p. 436; Bouchérie, "La Forêt française au secours de la défense nationale."

5. Interview of J. Serruys, October 1973.

6. Szilard to Jean Cattier, Oct. 13, 1939, LS; Sachs to Briggs, June 3, and Briggs to Sachs, June 5, 1940, 29: Originals Doc. 18, OSRD; Clark, *Birth of the Bomb*, pp. 23, 190; Clark, *Tizard*, p. 220; HalbT, p. 116; Hewlett and Anderson, *The New World*, pp. 26, 65; Groves, *Now It Can Be Told*, p. 33; Groueff, *Manhattan Project*, pp. 50-51.

7. Frisch, Halban, and Koch, *Kgl. Danske Videnskabernes Selskab, Mathematisk-fysiske Meddelelser* 15, no. 10 (1938).

8. CEA.51.A(P.12), RI.

9. Clark, *Birth of the Bomb*, pp. 67-68; CEA.51.A(P.13), RI. There was indeed serious boron contamination in this Acheson Co. graphite, according to the Germans who looked into the problem. W. Hanle, G-85 (1941).

10. OHI Kowarski by Weiner, October 1969; Halban and Kowarski, Report BR-94.

11. Frisch, Halban, and Koch, *Kgl. Danske Videnskab. Selsk. Math.-fys.* 15, no. 10 (1938); Halban, *CR* 206 (1938): 1170-1171; Szilard to Fermi, July 3, and Fermi to Szilard, July 9, 1939, in Szilard, *Collected Works*, pp. 193-197.

12. F. Joliot to Wallich, Nov. 24, and to Haut-Commissaire à l'Economie Nationale, Dec. 8, 1939, F29:Economie Nationale, RI; Direction des Poudres to F. Joliot, Jan. 27, 1940, JA; F. Joliot to Ingénieur en Chef, Chef du Service D, Direction des Poudres, Feb. 2, 1940, F29:Poudres, RI.

13. F. Joliot, Halban, and Kowarski, Report "Sur les produits chimiques nécessaires pour continuer les expériences sur la libération de l'énergie atomique de l'uranium," with cover letter, Joliot to Ministre de l'Armament, Feb. 13, 1940, Joliot Dossier, CEA. A draft of this report, apparently not delivered, is dated November 1939. See also copy in F29:Armament, RI, and Joliot's 1940 appointments book, RI.

14. Cf. Szilard, "Conversation with Halban 2-3-44," 3b.s.-15A, LS, p. 2.

15. Letters between Soc. Norv. Fredrik Sejersted and I. G. Farbenindustrie, Jan. 19, 24, Feb. 7, 1940, JA; "Une Déclaration de M. Dautry: Le Rôle de la France dans les recherches sur l'énergie atomique," Aug. 11, 1945, *Le Monde*, Aug. 15, 1945, p. 3; Allier, "Affaire de l'eau lourde," report for Dautry, February 1945, JA; Allier, report to ministers, Mar. 16, 1940, JA.

16. Allier, "Affaire de l'eau lourde," February 1945, JA. See also interview of Allier, November 1973; Allier, "Les Premières piles atomiques et l'effort français"; Clark, *Birth of the Bomb*, ch. 4; Eisenmann, "Les Savants français et l'énergie atomique."

17. Szilard, "Conversation with Halban 2-3-44," 3b.s.-15A, LS; F. Joliot to Ministre de l'Armament (?), memorandum beginning, "Les recherches sur la rupture" [Spring 1940?], LK.

18. OHI Kowarski by Weiner, October 1969; Szilard, "Conversation with Halban," LS.

19. Allier to author, September 1976; Gentlemen's Agreement, Mar. 9, 1940 (signed Mar. 11) and other documents, JA.

20. Cf. Fernand Mossé to Jean Morin, June 23, 1947, copy in CEA.19, RI. Those aiding Allier were Fernand Mossé, a Sorbonne professor, Jean Muller, an officer, and Jehan Knall-Demars, a reserve officer in the counterespionage service. Commandant André Pérruche watched over the group from France. See also Dautry, *Le Monde*, Aug. 15, 1945.

21. Interview of Allier, November 1973; Paul Reynaud, *Mémoires*, vol. 2, *Envers et contre tous* (P: Flammarion, 1963), pp. 335-336.

22. Wallis to Jean Serruys, Oct. 29, 1973, kindly shown me by Serruys. See also Clark, *Birth of the Bomb*, p. 26.

23. Dautry, *Le Monde*, Aug. 15, 1945. It is possible that the 1940 French military conception of a uranium bomb is echoed in Lt.-Col. Sabatier, "Note sur la bombe-uranium," extract from *Bulletin d'information technique et scientifique* 9/G, July 1945 (Mâcon: Perroux, 1945). Sabatier supposed a mass of uranium-238 whose heat would create a firestorm.

10. FROM RESEARCH TO WAR PROJECT

1. Allier to Dautry, Mar. 27, 1940, JA.

2. Directeur Générale de la Production, Ministère de l'Armament to F. Joliot, May 1, 1940, F29:Armament, RI; Compagnie Industrielle Savoie-Acheson to F. Joliot, May 4, 1940, CEA.10, RI; HalbT, p. 97; Halban, Affidavit, June 5, 1957, PCB, para. 25; Lechien to F. Joliot, Mar. 5, 1940, CEA.11, RI.

3. F. Joliot to Ministre de l'Armament (?), memorandum beginning, "Les recherches sur la rupture" [Spring 1940?], LK; Collège de France, "Répartition des crédits de laboratoires pour 1940," and Faral to CNRS, Mar. 14, 1940, IV.3, CF.

4. Goldschmidt, *Rivalités atomiques*, p. 29; interview of Goldschmidt, November 1973.

5. Roberts et al., *Phys. Rev.* 55 (1939): 510, 664; CEA.51.A(P.21), CEA.51.B, and notebook, January-February 1940, RI; Halban, F. Joliot, and Kowarski, *pli cacheté* no. 11,699, deposited in Joliot's name on Apr. 22, 1940, Archives of the Académie des Sciences, Paris; Halban and Kowarski, Report BR-94 of the British Maud Committee, June 1940, LK.

6. Halban, F. Joliot, and Kowarski, French patent 971,384, applied for Apr. 30, 1940, in Joliot-Curie, *Oeuvres*, p. 692.

7. Halban and Kowarski, Report BR-94, LK; *Plis cachetés* nos. 11,699 and 11,700 and Halban, F. Joliot, and Kowarski, *pli cacheté* no. 11,703, deposited in Joliot's name on Apr. 29, 1940, opened Jan. 28, 1957. These notes are each signed by all three scientists; they were assigned individual names only to make it easier to have them opened later. OHI Kowarski, August 1973. See also Halban, F. Joliot, and Kowarski, French patents 971,384, applied for Apr. 30, 1940, and 971,386, applied for May 1, 1940, in Joliot-Curie, *Oeuvres*, pp. 692-695; CNRS "Instruction générale no. 4," Dec. 8, 1939, Brevets, CEA; Laugier, "National Center of Scientific Research," p. 9.

8. F. Joliot to Courthial, Direction des Fabrications, May 24, 1940, F29:Armament, RI.

9. Copy of envelope in III.A, LK; KoT, pp. 466, 469; F. Joliot, "Rapport sur l'activité du Laboratoire de Synthèse Atomique en cours de l'anée 1940," with covering letter to Administrateur, CNRS, Jan. 9, 1941, F36, RI.

10. Bernal, "In Memory Paul Langevin," p. 18.

11. Gowing, *Britain and Atomic Energy*, p. 36; Oliphant to Lawrence, Aug. 24, 1939, 14:6, EOL; OHI Chadwick by Weiner, April 1969; Karl K. Darrow to Donald Cooksey, July 23, 1939, 6:9, EOL; Frisch to Meitner, Oct. 8, 1939, 5:21, Lise Meitner papers, Churchill College, Cambridge, Eng.; OHI Frisch by Weiner, May 1967; interview of Frisch, December 1973; Frisch to author, Nov. 1, 1975.

12. Alfred O. Nier, E. T. Booth, J. R. Dunning, E. V. Grosse, *Phys. Rev.* 57 (1940): 546, 748.

13. OHI Grosse by Weiner, April 1974; Booth and Dunning in *Beginnings of the Nuclear Age*, pp. 8, 20-21. The earliest mention I have seen of isotope separation is Szilard to Lewis L. Strauss, Feb. 13, 1939, Bk.f.3, LS. See also Bohr to Frisch, draft [c. March 1939], Bohr Scientific Correspondence, AHQP; Hewlett and Anderson, *The New World*, pp. 13-14, 22-24, 32.

14. Irving, *Virus House*, pp. 48-51, 68-71.

15. Bagge and Diebner in *Von der Uranspaltung bis Calder Hall*, pp. 25-26; Irving, *Virus House*, ch. 3; P. Harteck et al., G-36 (1940).

16. Golovin, *Kurchatov*, chs. 9-10; *Bulletin de l'Académie des Sciences de l'URSS, Série physique* 5 (1941): 555-587, trans. Eugene Rabinowitch, Report CP-3021, June 5, 1945, available from National Technical Information Service, Oak Ridge, Tenn.

17. *Szilard: His Version*, p. 115. See also Wigner, historical notes, Apr. 16, 1941, bbs.41, LS; Briggs, "Memorandum Report on Proposed Experiments with Uranium," Aug. 14, 1940, S-1 Historical File A, OSRD(BC).

18. Szilard to Compton, Dec. 29, 1942, bbs.1, LS. See also Szilard to Alexander Sachs, Apr. 14, 22, 1940, 13:Sachs, OSRD; Sachs in U.S. Senate, 79th, Special Com-

mittee on Atomic Energy, *Hearings*, Nov. 27, 1945.

19. Szilard memorandum for Dr. Alexander Sachs, Apr. 22, 1940, in *Szilard: His Version*, pp. 123-125.

20. Segrè, *Fermi*, p. 103; Briggs to Edwin M. Watson, Feb. 20, 1940, PCB.

21. The first warning to watch purity came from Tuve. Szilard to Tuve, Mar. 22, 1939, Bk.f.2, LS. See also C. P. Brower to Szilard, July 6, 1939, B.5, LS, and s.b.s.9, LS.

22. Anderson and E. Fermi, in E. Fermi, *Collected Papers*, II, 32-40; Anderson, "Fermi, Szilard, and Trinity"; Hewlett and Anderson, *The New World*, ch. 2.

23. Szilard, "Memorandum for Professor Urey," May 30, 1940, 3b.s.6, LS.

24. Szilard, *Collected Works*, pp. 207-215.

25. *Szilard: His Version*, p. 116.

26. Szilard to Turner, May 30, 1940, in *Szilard: His Version*, pp. 127-129; folder "Szilard," XC:5, Barton Collection, AIP.

27. McMillan to Briggs, Aug. 31, 1940, 12:30, EOL; E. McMillan and P. H. Abelson, *Phys. Rev.* 57 (1940): 1185; OHI McMillan by C. Weiner, October 1972, pp. 152-155, 159-168, 172.

28. Childs, *An American Genius*, p. 206. See also Raymond T. Birge, "History of the Physics Department, University of California, Berkeley," vol. 4, "The Decade 1932-1942," copy at AIP; various OHIs at AIP and at the Bancroft Library, University of California, Berkeley; EOL; Robert Seidel, "Physics Research in California," Ph.D. diss., University of California, Berkeley, 1978.

29. Breit to J. McKeen Cattell, June 17, 1940, copy in 28:10, EOL; interview of Breit, December 1975; Breit to Tate, May 31, and reply, June 3, 1940, XC:6, Barton Collection, AIP. Many of the letters on secrecy received by Tate were handed over to Breit, who gave them for safekeeping to Henry Barton; these are now in Carton XC, Barton Collection, AIP. Other materials remained in the *Physical Review* files until those files were weeded by Samuel Goudsmit, who removed the secrecy materials for preservation; they are in General Correspondence: 1940-1941, Goudsmit Papers, Library of Congress, Washington, D.C. Materials on matters handled by the Breit committee but not judged secret are in Administration: Executive Board 1940: Committee on Scientific Publications: Adv. Reference Com.: Nuclear Physics, Archives of the National Academy of Sciences, Washington, D.C.

30. Thomson, *Nuclear Energy in Britain*, pp. 4-5; Chadwick to Appleton, Dec. 5, 1939, AB.1.219, PRO; Clark, *Tizard*, pp. 213-214; Gowing, *Britain and Atomic Energy*, pp. 40, 43; Frisch to author, Nov. 1, 1975.

31. OHI Peierls by Weiner, August 1969; Clark, *Birth of the Bomb*, p. 43; Peierls, *Cambridge Philosophical Society Proceedings* 35 (1939): 610-615.

32. Frisch to Meitner, Feb. 24, 1940, 5:23, Meitner Papers, Churchill College, Cambridge, Eng.

33. At least one other person had already followed this line of thought—Chadwick. But unlike Frisch and Peierls, Chadwick noted a measurement done by Tuve's group around May or June and not published by them but quoted in Bohr and Wheeler, *Phys. Rev.* 55 (1939): 443-444. For natural uranium, the fission cross-section for 0.6 MeV neutrons was reported to be $0.003 \times 10^{-24} \text{cm}^2$. Bohr and Wheeler gave good reason to believe that uranium-238 cannot be fissioned by neutrons with energies below about 0.7 MeV; from this Chadwick reasoned correctly that Tuve's cross-section chiefly represented fission in the uranium-235 in the sample. Multiplying by 139, the ratio of all uranium atoms to uranium-235 atoms in natural uranium, gave an upper limit of 0.4×10^{-24} cm^2 for the uranium-235 fission cross-section. Chadwick asked Morris Pryce to put this into Peierls' formula and learned that this cross-

section implies a critical mass of many kilograms of uranium-235—more than Chadwick thought could be separated. He therefore dropped the subject. Fermi's group at Columbia was apparently led astray at just this same point. But over Christmas 1939, Chadwick began to worry that the derived cross-section was several times smaller than the geometrical cross-section; he started to doubt Tuve's experiment, which in fact gave a value three times too low. Chadwick was planning to make his own measurement of the cross-section and other quantities when he heard of Frisch and Peierls' work. OHI Chadwick by Weiner, April 1969; Clark, *Birth of the Bomb*, pp. 45-48; E. Fermi to Compton, Sept. 1, 1943, MUC-EF-12, MUC Memos, ANL; R. Ladenburg et al., *Phys. Rev.* 56 (1939): 168-170.

34. OHI Peierls by Weiner, August 1969; AB.1.210, PRO. The first estimate simply rounded up the geometrical cross-section to 1.0×10^{-23} cm^2 and hence gave a very small estimate for critical mass. OHI Frisch by Weiner, May 1967; Clark, *Birth of the Bomb*, pp. 49-53; Clark, *Tizard*, pp. 217-218; Gowing, *Britain and Atomic Energy*, pp. 40-41, 390.

35. Clark, *Tizard*, p. 219; Allier, "Souvenirs sur la collaboration franco-britannique"; Allier, "Note confidentielle," April 1940, JA; "Meeting of Sub-Committee on U. Bomb of CSSAW . . . April 10th, 1940," AB.1.8, PRO; Interview of Allier, November 1973; Allier, "Affaire de l'eau lourde," February 1945, IV.A, LK; Cockcroft correspondence, April-May 1940, AB.1.210, PRO.

36. Tizard to William Elliot, Apr. 4, 1940, AB.1.37, PRO; Gowing, *Britain and Atomic Energy*, ch. 1.

37. Weart, "Secrecy, Simultaneous Discovery, Nuclear Reactors."

38. Blackett, "Jean Frédéric Joliot, 1900-1958," p. 96; Szilard, "Book" ms. [1950s?], History-G, LS. If the French had relied only on Norway for heavy water, it would have taken them until 1943 to get enough for a reactor.

11. IN OCCUPIED PARIS

1. Dautry in France, Commission Chargée d'enquêter, *Les Evènements survenus en France de 1933 à 1945*, p. 1966; Jules Jeanneney, *Journal politique, Septembre 1939-Juillet 1942*, ed. Jean-Noël Jeanneney (P: Armand Colin, 1972), p. 51. See also Shirer, *The Collapse of the Third Republic*, ch. 30.

2. I. Curie to Meloney, May 14, 1940, Meloney Collection, CUL (with minor spelling errors corrected).

3. Moureu, "Un Episode peu connu de la bataille de l'eau lourde," November 1961, Archives of Académie des Sciences, Paris, and LK; Biquard, *Joliot-Curie*, pp. 65-69; Clark, *Birth of the Bomb*, pp. 95-96; OHI Kowarski by Weiner, October 1969; Eisenmann, "Savants français et l'énergie atomique," p. 19; G.IV.K.51-52, CF.

4. Dautry in France, Commission Chargée d'Enquêter, *Les Evènements survenus en France de 1933 à 1945*, pp. 1970-1973.

5. OHI Kowarski by Weiner, October 1969; Halban, Affidavit, June 5, 1957, PCB, para. 26; Goldsmith, *Joliot-Curie*, p. 92; interviews of Allier, Biquard, and Moureu.

6. Kowarski, "Broompark," 1945, "K's G," LK.

7. Goldschmidt, *Rivalités atomiques*, pp. 55-56; Biquard, "Mon ami Frédéric Joliot-Curie," p. 76.

8. Faral to Th. Rosset, Directeur de l'Enseignement Supérieur, May 23, 1940, H.II.e.33, CF; Maurain to F. Joliot [and I. Curie?], July 17, 1940, F28, RI; F. Joliot to Faral (from Clairvivre, Dordogne), July 20, 1940, H.II.e.40, CF.

9. Biquard, *Joliot-Curie*, p. 71.

10. OHI Gentner by Weiner, November 1971; Mitchell Wilson, *Passion To Know: The World's Scientists* (Garden City, N.Y.: Doubleday, 1972), p. 171.

11. Pierre Laval, *The Diary of Pierre Laval* (N.Y.: Scribner's, 1948), pp. 175-176; Allier to author, October 1976. See also Allier, report on 1940 heavy water discussions at Vichy, 1945, JA; France, *La Délégation Française auprès de la Commission Allemande d'Armistice: Recueil de documents publiés par le gouvernement français*, vol. 1, *29 Juin 1940-29 Septembre 1940* (P: Imprimerie Nationale, 1947), pp. 209-215 and passim.

12. F. Joliot, "L'Energie atomique en France," p. 188; Savel, "Vingt-sept années de collaboration," p. 81; Alsos Mission, "Interview with Professor F. Joliot, London, September 5th and 7th, 1944," 26: S-1 Intelligence Reports, MED (Harrison-Bundy); OHI Guéron, April 1978; F. Joliot, "Résumé de l'activité du laboratoire de synthèse atomique depuis Octobre 1940," Mar. 28, 1944, F36, RI; London *Times*, Nov. 3, 1945; Leprince-Ringuet, *Des atomes et des hommes*, p. 119.

13. Gentner to author, Sept. 3, 1976; Alsos Mission, "Interview with Professor F. Joliot"; Alsos Mission [S. Goudsmit?], "Interview with F. J. [Joliot], Paris, Tuesday, 28 August 1944," Aug. 31, 1944, 26: S-1 Intelligence Reports, MED (Harrison-Bundy).

14. OHI Gentner by Weiner, November 1971; Crowther, *Science in Liberated Europe*, p. 57; Irving, *Virus House*, pp. 71-72, 78, 89, 231; Goudsmit, *Alsos*; Groves, *Now It Can Be Told*, pp. 212-215; Bothe, "Report on the Work Done During the War," in Great Britain, Combined Intelligence Objectives Subcommittee [Field Intelligence Agencies Technical], *Cyclotron Investigation: Heidelberg*, Items 21 and 24, File XXIX-47 (London: H.M. Stationery Office, 1945), pp. 1-12; Bothe, "Die Forschungsmittel der Kernphysik," G-205, in *Probleme der Kernphysik* (Berlin: Deutsche Akademie der Luftfahrtforschung, 1943), G-323, p. 51; Bagge, Diebner and Jay, *Von der Uranspaltung bis Calder Hall*, p. 39.

15. Groves, *Now It Can Be Told*, p. 233. Gentner's opinion that the chain reaction was "nicht explosiv" leaked to the United States via a family friend. Gerhard Dessauer to Szilard, July 6, 1942, 3b.s.13, LS. For reaction, see Szilard to Compton, Sept. 7, 1942, LS.

16. Alsos Mission, "Interview with 'F.J.,' " 1944, 26: S-1 Intelligence Reports, MED (Harrison-Bundy).

17. Feb. 15, 1941, p. 1, copy in newspaper clippings, RI.

18. Letters in F26, RI.

19. France, *La Délégation Française auprès de la Commission Allemande d'Armistice*, p. 134.

20. Bichelonne, *Conférence*, p. 25 and passim; Bichelonne, *L'Etat actuel de l'organisation économique française*. See also Paxton, *Vichy France*, pp. 138, 198, 212, 221, 231, 264n., 322; "Bichelonne," c. November 1942, no. 24,743, "Advice of Bichelonne to Industrialists," May 31, 1944, no. 73,898, Office of Strategic Services files, RG 226, NA.

21. Biquard, *Joliot-Curie*, pp. 77-78. I have found no documents relating to SEDARS.

22. Bichelonne to F. Joliot, Aug. 25, and reply, Aug. 27, 1943, trans. in 26: S-1 Intelligence Reports, MED (Harrison-Bundy). See also Allier, "Conversation du Vendredi 1er Décembre 1943 avec M. Joliot-Curie," JA.

23. F. de la Rozière to Joliot, Mar. 9, 1944, F26, RI; Biquard, *Joliot-Curie*, pp. 79-80; interview of Allier.

24. Noguères et al., *Histoire de la Résistance en France de 1940 à 1945*, I, 118; Paxton, *Vichy France*, pp. 11, 38-40.

25. Paxton, *Vichy France*, pp. 174-175, 181; A. Langevin, *Langevin*, pp. 171-180; Gentner to author, Sept. 3, 1976.

26. Biquard, *Joliot-Curie*, p. 74; Noguères et al., *Histoire de la Résistance*, I, 310, III, 127. See also Maublanc, "L'Université française et la Résistance"; Voguet, "Souvenirs de l'année universitaire 1940-1941."

27. *L'Humanité* special no., May 25, 1941, in microfilm edition of André Marty papers (Cambridge: Harvard University Library, 1961), E.XIV, copy at Hoover Library, Stanford University, Cal.; F. Joliot, "Paul Langevin," p. 408.

28. Tiersky, *French Communism*, pp. 101-110; Paxton, *Vichy France*, pp. 226, 292.

29. Bataillon, "Une Matinée avec Joliot"; Rouzé, *Joliot-Curie*, p. 42; OHI Gentner by Weiner, November 1971; Gentner to author, Sept. 3, 1976; Goudsmit, *Alsos*, p. 35.

30. Thoma, "Le Martyre de Fernand Holweck"; Maublanc, "L'Université française et la Résistance," p. 56. According to Crowther, *Science in Liberated Europe*, p. 64, I. Curie was involved in concealing fugitives; perhaps this refers to Holweck's group.

31. Thoma, "Le Martyre de Fernand Holweck," p. 51; Duclos et al., eds., *Le Parti Communiste Français dans la Résistance*, pp. 125, 178-179; *Comité à la Mémoire des savants français victimes de la barbarie allemande* (P: Comité à la Mémoire, [1949?]), copy in CEA.31, RI.

32. Rouzé, *Joliot-Curie*, p. 43; Caute, *Communism and the French Intellectuals*, pp. 147-161; Biquard, *Joliot-Curie*, pp. 75-76.

33. "Interview with Pierre Villon on the FFI," Oct. 31, 1944, no. XL2131, Office of Strategic Services files, RG 226, NA; Duclos, *Mémoires*, III, 134; Hostache, *Le Conseil National de la Résistance*, pp. 69-76 and passim; Noguères, *Histoire de la Résistance*, III, 127; Barrabé, "Quelques souvenirs," p. 58; Michel, *Histoire de la Résistance en France*, pp. 30-31.

34. Villon to author, Feb. 10, 1978.

35. Biquard, *Joliot-Curie*, pp. 74-82; interview of Biquard, September 1976; Joliot, note adjoining Vagnet to Joliot-Curie, Aug. 18, 1946, F12, RI; Crowther, *Science in Liberated Europe*, p. 54; Bidault, *Resistance*, p. 24. For most events in France during the Occupation the full truth will never be known. Of course there are no documents, although a physicist told me of finding, long after, a clandestine map hidden in the Collège de France laboratory.

36. Goudsmit, *Alsos*, p. 35; Biquard, *Joliot-Curie*, p. 72; Groves, *Now It Can Be Told*, p. 233; René Zazzo, "Quand Joliot-Curie s'appelait Adrien," *La Rue* (St.-Etienne), Apr. 13, 1945, in clippings, RI.

12. EXILES AND INDUSTRIALISTS

1. "Ordre de Mission de MM. Halban et Kowarski," June 16, 1940, in Gowing, *Dossier secret*, App. 3, p. 234; Clark, *Birth of the Bomb*, pp. 101-102; Halban to Dickins, Dec. 11, 1940, AB.1.19(62A), PRO; KoT, pp. 474, 520; Halban and Kowarski, Report BR-94 of Maud Committee, [end of June 1940], LK.

2. "Record of a Meeting Held with Dr. von Halban and M. Kowarski at Savoy Hill House on June 24th, 1940," AB.1.223, PRO. See also "Minutes of M.A.U.D. Sub-Committee . . . 10th July, 1940," AB.1.8, PRO; OHI Kowarski by Weiner, October 1969.

3. Gowing, *Britain and Atomic Energy*, pp. 45-54; OHI Chadwick by Weiner, April 1969, p. 105.

4. "Halban and Kowarski, Cavendish Laboratory, Cambridge," diary in Kowarski's hand, LK; Clark, *Birth of the Bomb*, pp. 120-121; H. Halban [Sr.] to Joliot, Oct. 31, 1940, F70, RI.

5. Halban and Kowarski, "Report 1. Interaction of Slow Neutrons with Heavy Water and with U_3O_8-D_2O mixtures," Dec. 19, 1940, PCB; Halban and Kowarski, "Evidence for a Potentially Divergent Nuclear Reaction Chain," Report BR-4, [c. January 1941], copy in 29, OSRD; original AB.4.4, PRO.

6. W. Heisenberg and K. Wirtz, "Grossversuche zur Vorbereitung der Konstruktion eines Uranbrenners," in Bothe and Flügge, eds., *Nuclear Physics and Cosmic Rays*, II, 149. See also Döpel, Döpel, and Heisenberg, G-23; E. Fermi, Report C-207, July 25, 1942, in E. Fermi, *Collected Works*, II, 203-205.

7. A neutron plus uranium-238 makes neptunium, which quickly changes to plutonium. McMillan and Abelson, *Phys. Rev.* 57 (1940): 1185.

8. *Szilard: His Version*, p. 117; Hewlett and Anderson, *The New World*, pp. 33-34, 39; Irving, *Virus House*, p. 68. Von Weizsäcker considered the fission not of plutonium but of neptunium.

9. Halban and Kowarski, "Report 1," PCB; "Minutes of Maud Committee, Technical Sub-Committee, 3rd Meeting, April 9, 1941," 1: S-1 Historical File A, OSRD(BC); interview of Peierls, May 1977.

10. "Minutes of Maud Committee, Technical Sub-committee, 2nd Meeting, January 8, 1941," IV.A, LK; Clark, *Birth of the Bomb*, p. 125; Gowing, *Britain and Atomic Energy*, pp. 59-60, 70-71, 75, 79-80; OHI Chadwick by Weiner, April 1969.

11. Halban to Dickins, Mar. 3, 1941, AB.1.19(99A), PRO. See also Carroll L. Wilson, "Work on Uranium Problem by British," in Wilson to Briggs, Apr. 25, 1941, 27, OSRD.

12. Szilard to Slade, Oct. 2, 1936, B.3, LS; Reader, *Imperial Chemical Industries*, II, 82, 91-93, 353; testimony of M. Perrin, Feb. 21, 1958, PCB, pp. 258-259; OHI Kowarski by Weiner, October 1969; "Minutes of Maud Technical Sub-committee, 2nd Meeting," IV.A, LK; Halban to Dickins, Feb. 28, 1941, AB.1.19(130A), PRO.

13. Reader, *Imperial Chemical Industries*, I, 380, II, 135-136, 235; McGowan in ICI, *Annual Reports*.

14. OHI Kowarski by Weiner, October 1969.

15. "Nuclear Energy," ICI to British government, Oct. 1, 1941, IV.A, LK. See also draft ICI reports, c. July 1941, AB.1.220(370C), PRO; Reader, *Imperial Chemical Industries*, II, 290; Gowing, *Britain and Atomic Energy*, App. 2.

16. Gowing, *Britain and Atomic Energy*, App. 2.

17. Preliminary draft of Maud Committee Report, [c. July 1941], 3:Historical, OSRD(BC); McGowan in ICI, *Annual Report*, 16th, 1942, 17th, 1943; ICI, *Proceedings of the Twelfth Annual General Meeting, May 1939*; McGowan, *Speech to the Glasgow Chamber of Commerce*; Reader, *Imperial Chemical Industries*, II, 281-282.

18. Gowing, *Britain and Atomic Energy*, p. 105.

19. Clark, *Birth of the Bomb*, p. 195. See also Gowing, *Britain and Atomic Energy*, ch. 3; Reader, *Imperial Chemical Industries*, II, 291-292, 503-504; Wheeler-Bennett, *John Anderson*, pp. 180-181, 193-194, 354.

20. Oliphant to Edward Appleton, Oct. 27, 1941, AB.1.157, PRO.

21. Oliphant to Cockcroft, Nov. 3, 1941, AB.1.157, PRO.

22. Cockcroft to Ralph H. Fowler, Dec. 28, 1940, AB.1.157, PRO. See also Fowler to Lawrence, Jan. 28, 1941, 29:Originals Doc. 18 CEA, OSRD; Statement by A. Donald Messenheimer, May 23, 1967, PCB, p. 561; Carroll L. Wilson to Conant, Apr. 24, 1942, 1:S-1 British Relations, OSRD(BC).

23. E. Fermi to Compton, Sept. 1, 1943, MUC-EF-12, MUC Memos, ANL. The

importance of British prodding is widely attested, e.g. by Groves, *Now It Can Be Told*, p. 408.

24. Urey to Conant, Mar. 30, 1942, 3:Historical File—Special, OSRD(BC); Tizard to Rivett, Sept. 25, 1943, in Clark, *Tizard*, p. 355; Sherwin, *A World Destroyed*, pp. 72, 77; Bush, *Pieces of the Action*, p. 283.

25. Compton to Halban (c/o Urey), May 11, 1942, 3:Historical File—Special, OSRD(BC).

26. Conant to E. V. Murphree, May 15, 1942, 3:Historical File—Special, OSRD (BC); Halban, draft of report on trip to America, IV.A, LK; Gowing, *Britain and Atomic Energy*, ch. 4.

27. Kowarski to author, June 16, 1975; OHI Guéron, April 1978; letters on Halban-Kowarski difficulties in II.A.2, LK.

28. Goldschmidt, "Hans Halban," p. 4; interview of Peierls, May 1977; OHI Guéron, April 1978.

29. "Hans H. von Halban & Another and the Department of Scientific and Industrial Research. Agreement," Sept. 22, 1942, IV.A, LK; Gowing, *Britain and Atomic Energy*, pp. 212-213.

13. A FEW TONS OF HEAVY WATER

1. Gowing, *Britain and Atomic Energy*, pp. 105, 229; British Information Service Statement, "Britain and the Atomic Bomb," Aug. 12, 1945, in Smyth, *Atomic Energy for Military Purposes*, App. 7.

2. Urey to Briggs, Feb. 4, Briggs to Urey, Mar. 18, 1941, 29:Doc. 18 Pegram-Urey, OSRD.

3. Urey to Briggs, Mar. 18, 1941, 29: Doc. 18 Pegram-Urey, OSRD.

4. Chadwick, "Report on the Experiments of Halban and Kowarski," May 6, 1941, in W. L. Webster to Briggs, May 31, 1941, 29:23, OSRD.

5. Minutes of Advisory Committee of the National Academy of Sciences on Uranium Disintegration, Apr. 30, 1941, 27:1, EOL, p. 4.

6. Report of National Academy of Sciences Committee on Atomic Fission, Compton to F. B. Jewett, May 17, 1941, 27:1, EOL.

7. Briggs to Conant, July 8, 1941, 29, OSRD.

8. Conant to Urey, Apr. 7, 1943, 27:25, EOL.

9. Conant memorandum to Bush, Apr. 1, 1942, 1: S-1 Historical File B, OSRD (BC); Executive Committee of S-1 of the OSRD, minutes of 15th Meeting, Sept. 10-11, 1943, 27:28, EOL.

10. Compton, "Final Reports on Metallurgical Projects of OSRD," Apr. 30, 1943, 4: Reports—Compton, OSRD(BC); 6: Reports—Urey, and 10: Urey's Reports, OSRD(BC).

11. Urey to Conant, Sept. 10, 1942, and E. V. Murphree to Conant, Apr. 20, 1943, 6: Urey, OSRD(BC).

12. Executive Committee of S-1 of the OSRD, minutes of 8th Meeting, Nov. 14, 1942, 27:28, EOL.

13. E. I. DuPont de Nemours & Co., "Study Report: Production of Heavy Water by Direct Method," Dec. 4, 1942, Report E-45, 27:26, EOL; *Manhattan District History*, bk. III, *The P-9 Project*, [c. 1945], DOE, p. 2.12; N. Hilberry, "Time Schedules and Cost Estimates," in Compton, "Report on the Feasibility of the '49' Project," Nov. 26, 1942, 1: S-1 Technical Reports, OSRD(BC); Compton, "Final Reports," Apr. 30, 1943, 4: Final Reports—Compton, OSRD(BC); Hewlett and Anderson, *The New World*, chs. 6, 9. The design power was 250 megawatts each, but the actual power, showing plutonium production, is still secret.

14. Seaborg, *History of Met Lab*, I, 94, 254, 437-438; Anderson, "Organization of the Metallurgical Project," Sept. 17, 1942, bbs.40, LS; Wigner, "Short History of Project," Nov. 20, 1942, 3bs.2, LS; Compton, *Atomic Quest*, pp. 109-110; Hewlett and Anderson, *The New World*, pp. 90-91, 105-106, 190-191; Met Lab Technical Council Minutes, Sept. 18, 1942, CS-274, and Sept. 29, 1942, CS-281, copies provided by Seaborg.

15. H. D. Smyth, E. P. Wigner, and H. C. Vernon, "Memorandum on P-9 for Laboratory Council Meeting, July 28, 1943," MUC-HDS-10, MUC memos, ANL; *Szilard: His Version*, pp. 179, 192-196; Seaborg, *History of Met Lab*, I, 86; James Franck to Compton, Dec. 7, 1943, and E. Fermi to Compton, Dec. 9, 1943, MUC-EF-20, MUC memos, ANL; E. D. Eastman to Compton, Dec. 15, 1943, Hilberry file, ANL.

16. Sherwin, *A World Destroyed*, ch. 2; Conant, *My Several Lives*, pp. 295-296; Szilard, "Proposed Conversation with Bush," 1944, in *Szilard: His Version*, pp. 178-179.

17. James Franck to J. C. Stearns, "Memorandum on Organization and Morale in the Metallurgical Laboratories," Jan. 25, 1945, MUC-JF-132, MUC memos, ANL; Roger Williams, memorandum to file, Aug. 21, 1943, NDN-55276, Hilberry file, ANL. See also Seaborg, *History of Met Lab*, I, 299, 437-438, 742; Hewlett and Anderson, *The New World*, pp. 190-191, 198; Smith, *A Peril and a Hope*, pp. 15-16; Compton, *Atomic Quest*, p. 169; Wigner, "Brief History of Planning for W," Jan. 7, 1944, MUC-EPW-45, Hilberry file, ANL.

18. Szilard to Marshall MacDuffie, June 13, 1956, MacDuffie Collection, CUL; Noble, *America by Design*, ch. 6.

19. Szilard, "Proposed Conversation with Bush," Feb. 28, 1944, Part II, C-1, LS; *Szilard: His Version*, p. 172.

20. Minutes of Executive Committee of S-1 of the OSRD, e.g. 12th Meeting, Feb. 10-11, 1943, 27:28, EOL.

21. Urey, "Memorandum of Conference Between Prof. E. Fermi and Prof. H. C. Urey on March 6, 7, and 8, 1943," Report A-554, II.B.4, LK; Urey to Conant, Mar. 11, 1943, 1: S-1 British Relations 2, OSRD(BC). See also "Report of [Lewis] Committee on Heavy Water Work," Aug. 19, 1943, 19: P-9 Committee, OSRD.

22. Smyth, Wigner, and Vernon, "Memorandum on P-9," MUC-HDS-10, MUC memos, ANL; Williams, memorandum to file, "Visit of P-9 Committee August 9 and 10, 1943," Aug. 11, 1943, NDN-55271, Hilberry file, ANL; Hewlett and Anderson, *The New World*, pp. 202-203.

23. Kramish, *Atomic Energy in the Soviet Union*, pp. 27-30; Kurchatov, "Fission of Heavy Nuclei," in *Bulletin de l'Académie des Sciences de l'URSS, Série Physique*, 5 (1941), trans. Rabinowitch, CP-3021, National Technical Information Service, pp. 41-44, 72.

24. The German decision was based on an experiment by Bothe and P. Jensen, G-71 (1941). They found a small diffusion length for neutrons in graphite, indicating strong absorption. To test for impurities, they reduced a sample of their graphite to ash and measured the ash's absorption of thermal neutrons. Finding this absorption far too small to account for all the absorption in the unreduced graphite, they concluded that the carbon itself absorbed neutrons and was useless as a moderator. There are various ways in which error could have entered this experiment, including an elementary error: heating the graphite in order to reduce it to ash could have driven off volatile impurities. Bothe and Jensen considered this possibility and rejected it, but the Germans never seem to have fully studied the question. The Americans were more thorough, as they were throughout fission work with their greater urgency and larger teams and facilities. For example, at about the same time Bothe and Jensen were circu-

lating their results, Szilard and Wigner were pointing out that if graphite is reduced to ash in an oxygen atmosphere, boron, one of the most neutron-absorbent impurities, may be driven off as an oxide. Szilard to G. E. F. Lundell, Jan. 17, 1941, 3: Szilard, OSRD; Pegram to Briggs, Feb. 14, 15, 1941, Reports S-35, S-36, 10: Pegram's Reports, OSRD. It is remarkable, and unexplained, that the Germans accepted this experiment, which contradicted an earlier, correct measurement by Bothe himself, G-12 (1940). The different routes chosen by the Americans and the Germans were separated by many things beyond one flawed experiment. At any rate, had the Germans pursued graphite, they might well have failed, for they did not have the unusually boron-free petroleum deposits which the Americans used as the basis for their graphite. Bagge and Diebner in *Von der Uranspaltung bis Calder Hall*, p. 39; Irving, *Virus House*, ch. 4; Alan Beyerchen, *Scientists under Hitler: Politics and the Physics Community in the Third Reich* (New Haven: Yale University Press, 1977), ch. 9. No German calculation of the critical mass of plutonium or uranium-235—the crucial calculation—has yet been found.

25. Urey to Murphree, Sept. 3, 1943, copy in 27:25, EOL; *Manhattan District History*, DOE, III, 5.5.1, 5.17-5.19, S15; Smyth, *Atomic Energy for Military Purposes*, ch. 9; Hewlett and Anderson, *The New World*, pp. 203-204.

26. Hewlett and Anderson, *The New World*, ch. 10 and passim; Smith, *A Peril and a Hope*; Sherwin, *A World Destroyed*, ch. 8.

27. Compton to K. D. Nichols, June 4, 1945, MUC-AC-1307, ANL.

14. FAILURES OF SECRECY

1. C. J. MacKenzie, Oct. 8, 1942, in Eggleston, *Canada's Nuclear Story*, p. 57.

2. OHI Kowarski by Weiner, October 1969; "Akers no. 11," Montreal to London, Feb. 18, 1943, II.A.2, LK. See also papers in AB.1, PRO.

3. "Laugier," *Current Biography*, 1948, pp. 371-372; OHI Perrin, November 1977; Henri de Kerillis, *De Gaulle dictateur: Une Grande mystification de l'histoire* (Montreal: Beauchemin, 1945), pp. 325-326; Biquard, *Joliot-Curie*, p. 82.

4. Crowther, "M. Louis Rapkine." See also "Provisional Report of Meeting in Memory of Louis Rapkine," RI; Crowther, *Science in Liberated Europe*, pp. 65-68; "Louis Rapkine," *Atomes* no. 35 (1949): 66. I call de Gaulle's organization Free French throughout, although in July 1942 its name changed to France Combattante, or Fighting French.

5. [Kowarski], note to Free French, 1944, II.A.2., LK.

6. Laugier, statement (trans. Rapkine) in U.S. Senate, 77th, Committee on Military Affairs, *Technological Mobilization. Hearings* (Washington, D.C.: U.S. Government Printing Office, 1942), II, 630-652; "Provisional Report of Meeting in Memory of Louis Rapkine," RI.

7. Crowther, *Science in Liberated Europe*, pp. 19, 38; Eggleston, *Canada's Nuclear Story*, pp. 58-59; Auger to author, Mar. 20, 1978.

8. Goldschmidt, *Rivalités atomiques*, p. 15.

9. Claude Lévi-Strauss, *Tristes tropiques* (P: Plon, 1955), p. 23; Goldschmidt, *Rivalités atomiques*, pp. 30-33; OHI Goldschmidt, September 1976; "Curriculum Vitae of Bertrand Goldschmidt," [c. 1941], bbs.11, LS.

10. Interview of Seaborg by Arthur Norberg with Weart, October 1974, Bancroft Library, University of California, Berkeley. See also Seaborg, *History of Met Lab*, I, 139, 152, 221-222, 251, 310-311, 317; C. D. Coryell in J. Franck to Compton, "Historical Résumé," Apr. 9, 1943, MUC-JF-41, MUC Memos, ANL.

11. Goldschmidt, *Rivalités atomiques*, pp. 25-42; interview of Goldschmidt, November 1973.

12. Gowing, *Britain and Atomic Energy*, p. 215; OHI Guéron, April 1978.

13. OHI Goldschmidt, September 1976.

14. Eggleston, *Canada's Nuclear Story*, p. 108.

15. Eggleston, *Canada's Nuclear Story*, pp. 37-38; Gowing, *Britain and Atomic Energy*, ch. 4.

16. Eggleston, *Canada's Nuclear Story*, pp. 59-60, 64-65, 71-72, 79; Conant memorandum to Bush, Mar. 25, 1943, 1: S-1 British Relations 2, OSRD(BC).

17. Bush, "Proposed Future Relations with the British and Canadians," extract from Report to the President, Dec. 16, 1942; F. D. Roosevelt to Bush, Dec. 28, 1942, 1: S-1 British Relations 2, OSRD(BC); Eggleston, *Canada's Nuclear Story*, p. 73

18. Goldschmidt, *Rivalités atomiques*, pp. 44-45. See also Seaborg, *History of Met Lab*, I, 499; Eggleston, *Canada's Nuclear Story*, p. 73.

19. Halban, Report on Chicago team [1942].

20. Cf. Edwin T. Layton, Jr., "American Ideologies of Science and Engineering," *Technology and Culture* 17 (1976): 696; Noble, *America by Design*, p. 34.

21. Gowing, *Britain and Atomic Energy*, pp. 194-196; Eggleston, *Canada's Nuclear Story*, pp. 111-115.

22. Eggleston, *Canada's Nuclear Story*, pp. 121-124; OHI Guéron, April 1978.

23. Halban to "Edward" (Sir Edward Appleton), Aug. 6, 1943.

24. Goldschmidt, *Rivalités atomiques*, p. 74.

25. Gowing, *Britain and Atomic Energy*, pp. 168n, 214, 439; Wheeler-Bennett, *John Anderson*, pp. 291-295, 330-331.

26. Anderson to Halban, Apr. 21, 1944; Gowing, *Britain and Atomic Energy*, pp. 272-275; Eggleston, *Canada's Nuclear Story*, pp. 97-102, 121.

27. OHI Guéron, April 1978; Goldschmidt, *Rivalités atomiques*, pp. 88-89; de Gaulle, *Mémoires de guerre*, vol. 2, *L'Unité, 1942-1944* (P: Plon, 1956), p. 242.

28. Larry Collins and Dominique Lapierre, *Is Paris Burning?* (New York: Simon & Schuster, 1965), pp. 109, 120; Goldsmith, *Joliot-Curie*, pp. 122-123.

29. Cogniot, "A la mémoire d'Henri Wallon," *La Pensée* n.s. no. 112 (1963): 28-37; *La Pensée* n.s. no. 85 (1959): 4; Maublanc, "L'Université française et la Résistance," p. 64; Biquard, *Joliot-Curie*, pp. 83-84.

30. F. Joliot, "L'Energie atomique en France," p. 188; Crowther, *Science in Liberated Europe*, p. 67.

31. Groves, *Now It Can Be Told*, pp. 142-143; Sherwin, *A World Destroyed*, pp. 62-63.

32. Groves, *Now It Can Be Told*, pp. 234, 240-241. See also John J. McCloy in U.S. Atomic Energy Commission, *In the Matter of J. Robert Oppenhelmer: Transcript of Hearings Before Personnel Security Board, 1954* (Cambridge: Massachusetts Institute of Technology Press, 1971), p. 732; Goudsmit, *Alsos*, pp. 96ff.; Pash, *The Alsos Mission*, chs. 13, 14, 22; John Lansdale, report on "Operation Harborage," [c. April 1945], "Top Secret" files: 7A, MED.

33. OHI Goldschmidt, September 1976; Gowing, *Britain and Atomic Energy*, p. 293.

34. "The substance of recommendations which will be made by Major General Groves," Jan. 20, 1945, attached to "Memorandum on Security Problems," n.d., 12, MED(Groves). See also Gowing, *Britain and Atomic Energy*, pp. 214-215, 289-296, 342-346; Gowing with Arnold, *Independence and Deterrence*, I, 9-10; Sherwin, *A World Destroyed*, pp. 108, 132, 290-291; Hewlett and Anderson, *The New World*, pp. 331-336; Goldschmidt, *Rivalités atomiques*, pp. 89-92; Groves, *Now It Can Be Told*, ch. 16; folder 12, MED(Groves); "Top Secret" files: 26A-26L, MED.

35. Interviews of Auger, Guéron, Goldschmidt, Kowarski, and Perrin.

36. OHI Kowarski by Weiner, October 1969.

37. OHI Kowarski by Weiner, October 1969.

38. Goldschmidt, *Rivalités atomiques*, pp. 82-83; OHI Goldschmidt, September 1976; Gowing, *Britain and Atomic Energy*, pp. 285-286.

39. OHI Kowarski by Weiner, October 1969.

40. Wells, *The World Set Free*, pp. 101-102, 145-146, 118.

41. Wells, *The World Set Free*, p. 241; OHI Guéron, April 1978.

42. Eggleston, *Canada's Nuclear Story*.

43. Golovin, *Kurchatov*, pp. 38-55; V. S. Fursov, "Work of the U.S.S.R. Academy of Sciences on an Uranium-Graphite Reactor," in *Conference of the Academy of Sciences of the USSR on the Peaceful Uses of Atomic Energy* (Washington, D.C.: U.S. Atomic Energy Commission, 1956), pp. 1-13; Kramish, *Atomic Energy in the Soviet Union*, pp. 112-113.

15. ORGANIZATION FOR NUCLEAR ENERGY

1. OHI Guéron, April 1978.

2. See Werth, *France, 1940-1955*.

3. Gilpin, *France in the Age of the Scientific State*, p. 152. See also Weisskopf, "Physics in France."

4. "Exposé des motifs," Ordonnance of Nov. 2, 1945, *J.O.*, Nov. 3, 1945, pp. 7192-7193.

5. F. Joliot, "Le Centre National de la Recherche Scientifique," p. 3; *Le Soir*, Sept. 5, 1944. See also F. Joliot, "Note sur la science soviétique"; L. Kopelmanars, "Le Centre National de la Recherche Scientifique," p. 126; Lumière, "A propos de la recherche scientifique libre ou dirigée"; Lévy-Bruhl, "Pas de science dirigée"; Alfaric, "Pour la science dirigée."

6. Williams, *Crisis and Compromise*, p. 13; Crowther, *Science in Liberated Europe*, p. 60; *L'Humanité*, Feb. 14, 1946; OHI Kowarski by Weiner, May 1970; Biquard, *Joliot-Curie*, p. 85; Pleven to F. Joliot, Jan. 17, 1945, F70, RI.

7. Crowther, *Science in Liberated Europe*, p. 22. See also Documentation Française, "Le Centre National de la Recherche Scientifique"; Kopelmanars, "Le Centre National de la Recherche Scientifique."

8. Crowther in *Manchester Guardian*, Jan. 18, 1947.

9. Weisskopf, "Physics in France," p. 8; Goldsmith, *Joliot-Curie*, ch. 7; Jean Combisson, "Physics in France," *Physics Today* 20, no. 11 (November 1967): 55-61; Robert G. Gilpin, Jr., "Science, Technology, and French Independence," in Long and Wright, eds., *Science Policies of Industrial Nations*, pp. 110-132.

10. Crowther, *Science in Liberated Europe*, pp. 26-33; F. Joliot, "Le Centre National de la Recherche Scientifique."

11. Allier memoranda, "Conversation du Vendredi 1er Decembre 1944 avec M. Joliot-Curie" and "Conversation du Jeudi 29 Mars 1945 avec M. Joliot-Curie," JA; Groves to Secretary of War [H. L. Stimson], May 13, 1945, 36, MED(Harrison-Bundy).

12. Dautry in France, Commission Chargée d'Enquêter, *Les Evènements survenus en France de 1933 à 1945*, VII, 1973, 1980; Allier to F. Joliot, Feb. 9, Mar. 14, 1945, F21, RI; F. Joliot to Villon, Mar. 28, 1945, F32, RI; Scheinman, *Atomic Energy Policy in France*, pp. 5-6; Allier, "Les Premières Piles," pp. 340-341; Documentation Française, "Le Développement nucléaire français," p. 6; Goldschmidt, *Rivalités atomiques*, pp. 177-178; Auger in *Echos du CEA* special no. (October 1965): 29.

13. De Gaulle, *The War Memoirs of General Charles de Gaulle*, vol. 3, *Salvation*

1944-1946, trans. Richard Howard (New York: Simon & Schuster, 1960), pp. 109, 269-270; Toutée, "Naissance du Commissariat"; Allier, "Les Premières piles," p. 342n.

14. *New York Times*, Apr. 8, 1954; A. Peyrefitte, *Le Mal français* (P: Plon, 1976), p. 83; OHI F. Perrin, November 1977.

15. Herriot, *Créer*, I, ch. 5; Richard F. Kuisel, *Ernest Mercier: French Technocrat* (Berkeley: University of California Press, 1967), pp. 114-119; Vilan, *La Politique de l'énergie en France*; Peyret, *La Bataille de l'énergie*; Flanner, *Paris Journal*, p. 47; Fabre, "La France manque de Charbon." A thorough history of energy waits to be written.

16. F. Joliot to Villon, Mar. 28, 1945, F32, RI.

17. Jacques Lucius and Allier to author, Feb. 14, 1978.

18. [Jean Toutée], "Projet d'ordonnance," typescript [October 1945], ms. additions by Allier and Dautry, JA; interview of Allier, September 1976.

19. Scheinman, *Atomic Energy Policy in France*, ch. 1.

20. Auger in *Echos du CEA* special no. (October 1965): 29.

21. Paraphrase, in Frances Henderson to Don Bermingham, Mar. 8, 1946, bbs.49, LS.

22. The engineer estimated that with a full effort the Russians could do it in four to five years. Transcript of telephone call of May 21, 1945, 12, MED(Groves); F. Joliot, "La Recherche scientifique est-elle manacée?" p. 17. See also Orcel, "Frédéric Joliot-Curie," p. 55; Biquard, *Joliot-Curie*, pp. 104-105.

23. OHI Kowarski, August 1976; interviews of Goldschmidt, November 1973, and Auger, October 1973; Goldschmidt, *Rivalités atomiques*, pp. 92-93; Gowing, *Britain and Atomic Energy*, p. 296.

24. OHI Kowarski by Weiner, October 1969; notes in II.B.2, LK.

25. Goldschmidt, "Les Premiers pas," p. 11; OHI Kowarski by Weiner, May 1970

26. F. Joliot to Thomson, Mar. 13, to Bohr, May 4, to J. F. Allen, June 18, and to Chadwick, Oct. 24, 1946, CEA.23, RI; [F. Joliot?], "Rapport sur l'activité du Laboratoire de Synthèse Atomique pendant l'année 1947," Jan. 15, 1948, CEA.32, RI; F. Joliot to Walter Ill, Oct. 10, 1946, CEA.23, RI. Orders to destroy the French cyclotron were given the German scientists but resisted by Prof. W. Riezler. Gentner to author, Sept. 3, 1976.

27 Pierre Audiat, "Au Collège de France," pp. 99-102; letters of Oct. 5, 18, 1945, F11, RI; F5, F8, F11, F19, F21, F33, CEA.18-CEA.26, CEA.40, RI.

28. F. Joliot in World Federation of Scientific Workers, *World Federation of Scientific Workers* (London: WFSW, 1947), p. 4. See also Goldsmith, *Joliot-Curie*, pp. 174-177.

29. "Bidault," *Current Biography*, 1945, p. 51; Irving, *Christian Democracy in France*.

30. Goldschmidt, "Les Premiers pas," p. 9; Lydia Cassin and Julien Vergne in *Echos du CEA* special no. (October 1965): 30-31, 38.

31. Gowing, *Britain and Atomic Energy*, ch. 11.

32. Hélène Emmanuel in *Echos du CEA* special no. (October 1965): 31; Goldschmidt, *Rivalités atomiques*, pp. 181-182; Pash, *Alsos Mission*, pp. 85-86.

33. Allier, "Les Premières piles," pp. 340-342; Dautry, "Note pour M. le Président," May 26, 1945; Allier to Dautry, Aug. 8, Oct. 1, 1945; Allier, "Note au sujet de la fourniture de 2 tonnes," June 11, 1947, JA; Commissariat à l'Energie Atomique, Comité Scientifique, Minutes, no. 90, Dec. 9, 1947, CEA.49, RI; U.S. Central Intelligence Agency, *National Intelligence Survey: France*, (1952), CIA Archives, Washing-

ton, D.C. (released under Freedom of Information Act), p. 73-19.

34. Goldschmidt, "Les Premiers pas," p. 11; Comité Scientifique, Minutes, no. 1, Mar. 5, 1946, CEA.49, RI.

35. Goldschmidt, "Les Premiers pas," pp. 11-12; F. Joliot, "Le Programme du Commissariat," Mar. 19, 1946, printed in *Echos du CEA* special no. (October 1965): 15-16.

36. OHI Kowarski by Weiner, May 1970.

16. THE COMMISSARIAT A L'ENERGIE ATOMIQUE

1. OHI Kowarski by Weiner, May 1970.

2. Orcel, "Irène Joliot-Curie," p. 59.

3. Commissariat à l'Energie Atomique, Comité Scientifique, Minutes, no. 130, Jan. 11, 1949, CEA.50, RI, p. 8.

4. Crowther, *Science in Liberated Europe*, pp. 13, 24; Biquard, *Joliot-Curie*, pp. 77-79; Kowarski, "Psychology and Structure of Large-Scale Physical Research."

5. OHI Kowarski, August 1976. See also OHI Kowarski by Weiner, May 1970; Biquard, *Joliot-Curie*, pp. 111, 208.

6. Hunebelle, "Une Epopée de l'après-guerre," pp. 110-111.

7. F. Joliot, "La Recherche scientifique est-elle menacée?" p. 17; Crowther, *Science in Liberated Europe*, p. 14; Kowarski, "Atomic Energy Developments in France," p. 60; [Kowarski], "Atomic Energy Developments in France During 1946-1950"; Scheinman, *Atomic Energy Policy in France*, p. 26; Etienne Gibert, "Le Commissariat à l'Energie Atomique à cinq ans," *Le Monde*, Jan. 20-21, 1951, p. 3; OHI Guéron, April 1978; France, Présidence du Conseil, CEA, *Rapport d'activité, 1946-1950*, pp. 60-61; Puget, "Les Problèmes administratifs," pp. 18, 34. The 1610 CEA employees included 1210 scientific and technical personnel, 149 administrators, and 251 general services people; there were 200 at the Paris headquarters, 664 at Châtillon, 116 at Le Bouchet, and the rest elsewhere in prospecting and mining work.

8. Anatole Abragam and Jules Horowitz in *Echos du CEA* special no. (October 1965): 29, 34.

9. Kowarski, "Atomic Energy Developments in France," p. 60; F. Joliot, "La Recherche scientifique est-elle menacée?" p. 16; Scheinman, *Atomic Energy Policy in France*; Puget, "Problèmes administratifs," p. 34.

10. Emmanuel and Guéron in *Echos du CEA* special no. (October 1965): 31, 33; Echard's Curriculum Vitae, [c. February 1942], F26, RI; OHI Kowarski by Weiner, May 1970; Goldschmidt, *Rivalités atomiques*, p. 183.

11. F. Joliot in Comité Scientifique, Minutes, no. 94, Dec. 23, 1947, CEA.49, RI; Surdin, "L'Electronique au Commissariat à l'Energie Atomique"; Guéron, "Chimie analytique et pureté nucléaire," p. 56; [Kowarski], "Atomic Energy Developments in France During 1946-1950"; Scheinman, *Atomic Energy Policy in France*, p. 28.

12. OHI Goldschmidt, September 1976.

13. Biquard, *Joliot-Curie*, pp. 96-97; Scheinman, *Atomic Energy Policy in France*, pp. 32-33; OHI Kowarski by Weiner, May 1970; OHI Kowarski, August 1976; Goldschmidt, *Rivalités atomiques*, p. 183; Guéron in *Echos du CEA* special no. (October 1965): 33.

14. Julien Vergne in *Echos du CEA* special no. (October 1965): 38; Scheinman, *Atomic Energy Policy in France*, pp. 30, 36-37.

15. Dautry in France, Présidence du Conseil, CEA, *Rapport d'activité, 1946-1950*, pp. 66, 89. See also Dautry, *L'Organisation de la vie sociale*; Dupouy, "Raoul Dautry."

16. Joliot, "Le Programme du Commissariat," Mar. 19, 1946, *Echos du CEA* special no. (October 1965): 15-16.

17. Gowing, *Independence and Deterrence*, I, 155-157 and passim; Goldschmidt, *Rivalités atomiques*, pp. 46, 180-181; OHI Kowarski by Weiner, May 1970.

18. F. Joliot to Cockcroft, Apr. 5, 1946, CEA.23, RI; Goldschmidt, *Les Perspectives actuelles*, pp. 8, 12; Dexter Masters and Katherine Way, eds., *One World or None* (New York: McGraw-Hill, 1946); Comité Scientifique, Minutes, no. 10, Apr. 1, 1946, no. 18, May 3, 1946, CEA.49, RI. See also Joliot, "L'Energie atomique en France;" pp. 190-191; Guéron, "Energie nucléaire et politique générale," p. 5; Dautry, "L'Energie atomique."

19. France, Présidence du Conseil, CEA, *Rapport d'activité, 1946-1950*, p. 8; Comité Scientifique, Minutes, no. 11, Apr. 5, 1946, CEA.49, RI; Biquard, *Joliot-Curie*, pp. 102-103; OHI Kowarski by Weiner, May 1970.

20. In a national survey held at the end of 1945, 56 percent felt France should build an atomic bomb and 12 percent were undecided. But the question was presented right after one asking whether the bomb would be used "in the next war," which could have biased the results. Institut Française d'Opinion Publique, *Sondages 8*, no. 2 (Jan. 16, 1946), p. 18. A poll of the readership of *Atomes* found 78 percent of respondents opposed a nation's having any nuclear secrets. *Atomes* no. 6 (August-September 1946): 2. See also Mendl, *Deterrence and Persuasion*.

21. Kowarski, "Atomic Energy Developments in France," p. 139; Scheinman, *Atomic Energy Policy in France*, pp. 20-22, 36, and passim.

22. Comité Scientifique, Minutes, 1946-1950, CEA.49, CEA.50, RI; Ailleret, *L'Aventure atomique française*, p. 35 and passim.

23. Kevles, *The Physicists*, chs. 22-23; various OHIs at AIP.

24. Scheinman, *Atomic Energy Policy in France*, pp. 31-32; "Procès Verbaux du Comité de l'Energie Atomique" and "Comité Scientifique (Principales Décisions Prises)," IV.B, LK; Comité Scientifique, Minutes, no. 60, CEA.49, RI.

25. Jacques Stohr in "French Atomic Scientists Report," p. 302. See also Labeyrie, "Au Fort de Châtillon"; Louis Vautrey in *Echos du CEA* special no. (October 1965): 37; France, Présidence du Conseil, Commissariat à l'Energie Atomique, *Commissariat à l'Energie Atomique 1945-1956* (P: [CEA, 1956]), p. 15.

26. Auger in *Echos du CEA* special no. (October 1965): 29; F. Perrin, "Vingt ans après," p. 46. See also Kowarski in "French Atomic Scientists Report," p. 301; OHI Kowarski by Weiner, May 1970.

27. Gilpin, *France in the Age of the Scientific State*, p. 169.

28. "Situation au 20 février 1949, des personnes recevant du CEA un salaire," unsigned typescript, CEA.50, RI; Comité Scientifique, Minutes, no. 99, Feb. 10, 1948, no. 132, Feb. 22, 1949, no. 134, Apr. 5, 1949, no. 139, May 1, 1949, CEA.49-50, RI; France, Présidence du Conseil, CEA, *Rapport d'activité, 1946-1950*, p. 72.

29. Kowarski, "Atomic Energy Developments in France," p. 64; F. Joliot, "L'Energie atomique en France," p. 190; Biquard, *Joliot-Curie*, pp. 100-101.

30. F. Perrin, "Joliot"; "La Bataille de l'eau lourde" (screenplay by Jean Marin), 1948.

31. Biquard, *Joliot-Curie*, p. 111. See also Scheinman, *Atomic Energy Policy in France*, pp. 37-40.

32. Bellow, "A Strictly Personal Syllabus," *New Yorker*, July 12, 1976, p. 82; Charles A. Micaud, "The Bases of Communist Strength in France," *Western Political Quarterly* 8 (1955): 365; Duclos to "Cher camarade," July 10, 1947, F19, RI.

33. World Federation of Scientific Workers, *WFSW*, p. 3. See also F. Joliot, *Textes Choisis*.

34. Gowing, *Independence and Deterrence*, I, 243-244, 339-340.

35. *New York Times*, Mar. 19, 1948, p. 1, Mar. 20, 1948, p. 1, Mar. 23, 1948, p. 26, Apr. 1, 1948, p. 15; *Time* 51, no. 13 (Mar. 29, 1948): 28.

36. *J.O., Débats, Conseil de la République*, no. 26, Mar. 18, 1948, pp. 807-812, 861; Scheinman, *Atomic Energy Policy in France*, pp. 38-39.

17. THE FIRST FRENCH REACTOR

1. "Procès Verbaux du Comité de l'Energie Atomique," IV.B, LK; Comité Scientifique, Minutes, no. 74, July 8, 1947, CEA.49, RI.

2. F. Joliot, "L'Energie atomique en France," p. 189; [Kowarski], "Atomic Energy Developments in France During 1946-1950"; OHI Kowarski by Weiner, May 1970; Orcel, "Irène Joliot-Curie," p. 56; Barrabé, "Quelques souvenirs," p. 60; Marcel Roubault, "La Découverte de l'uranium français"; Roubault, "La Course à l'uranium," *Le Monde*, Nov. 16-18, 1950; Jacques Geffroy in *Echos du CEA* special no. (October 1965): 33.

3. Roubault, "La Recherche et l'exploitation des minerais d'uranium," p. 130.

4. Rousseau, "L'Avenir de l'énergie atomique," p. 126; France, Présidence du Conseil, CEA, *Rapport d'activité, 1946-1950*, pp. 31, 41-42; Roubault with Georges Jurain, *Géologie de l'uranium* (P: Masson, 1958).

5. *Pechiney, compagnie de produits chimiques et électrométallurgiques* (P: Draeger, 1951); Claude J. Gignoux, *Histoire d'une entreprise française* (P: Hachette, 1955); Lucius and Allier to author, Feb. 14, 1978; OHI Guéron, April 1978. "Pechiney" was the traditional appellation, but until 1950 the official company name was Produits Chimiques et Electrométallurgique Alais Froges et Camargue.

6. Guéron, "L'Eau lourde et le graphite"; Guéron in *Echos du CEA* special no. (October 1965): 33-34; P. Legendre, L. Mondet, Ph. Arrongon, P. Cornuault, J. Guéron, and H. Hering, "The Production of Nuclear Graphite in France," in International Conference on the Peaceful Uses of Atomic Energy, Geneva, 1955, *Proceedings*, vol. 8, United Nations Doct. A/CONF. 8/1 (New York: United Nations, 1955), pp. 474-477.

7. Guéron, "Chimie analytique et pureté nucléaire."

8. Goldschmidt, "La Purification de l'uranium."

9. Comité Scientifique, Minutes, no. 14, Apr. 12, 1946, nos. 81-83, October, 1947, CEA.49, RI; "Rapport sur l'activité des Services de Physique et Chimie . . . 1948," CEA.50, RI; Denivelle to F. Joliot, Feb. 26, 1948, F5, RI.

10. Goldschmidt, *Rivalités atomiques*, p. 182; Kowarski, "Zoé: Le Départ des piles françaises," p. 21.

11. Guéron in *Echos du CEA* special no. (October 1965): 33; OHI Kowarski by Weiner, May 1970; Comité Scientifique, Minutes, no. 74, July 8, 1947, CEA.49, RI; F. Joliot in "Procès-verbal de la 62ème réunion du Comité de l'Energie Atomique . . . 5 Novembre 1947," CEA.53, RI.

12. Stohr, "La Fabrication des comprimés d'oxyde d'uranium"; *Echos du CEA* special no. (October 1965): 34-35, 37-38; OHI Kowarski by Weiner, May 1970.

13. Le Meur, "La Mécanique de la pile"; OHI Kowarski by Weiner, May 1970; Biquard, *Joliot-Curie*, pp. 105-107.

14. *L'Aube*, Dec. 17, 1949. See also *Atomes* no. 35 (February 1949): 65.

15. *The Economist* 156 (Jan. 15, 1949): 96; [Stephen White], "France on the Atomic Trail," *The Economist* 156 (Jan. 15, 1949): 108-109; Biquard, *Joliot-Curie*, pp. 112-114.

16. *L'Humanité*, Jan. 6, 1949, p. 1, Jan. 22, 1949, p. 2; *France-Soir* as quoted in San Francisco *Chronicle*, Jan. 16, 1949.

17. *Histoire du Parti Communiste Français*, III, 63. See also Fauvet with Duhamel, *Histoire du Parti Communiste Français*, II, 224-231; Caute, *Communism and the French Intellectuals*, pp. 34, 184-189.

18. No. 546 was the only cell that sent Joliot birthday greetings (Mar. 22, 1947, F70, RI), and the Collège de France would have been regarded as his permanent workplace and hence the location of his cell.

19. Williams, *Crisis and Compromise*, p. 81. See also Brayance, *Anatomie du Parti Communiste Français*, pp. 28-29, 38; Henry W. Ehrmann, "The Decline of the Socialist Party," in Earle, ed., *Modern France*, pp. 181-199.

20. Goldschmidt, *Rivalités atomiques*, p. 186.

21. Casanova, *Responsabilités de l'intellectuel communiste*, pp. 13-15, 30. See also Cohen et al., *Science bourgeoise et science prolétarienne*; Loren Graham, *Science and Philosophy in the Soviet Union* (New York: Knopf, 1972).

22. "Décisions du Secrétariat [Comité Central, PCF] du 29/IX/48," in G.IV, microfilm edition of André Marty papers (Cambridge: Harvard University Library, 1961).

23. Biquard, *Joliot-Curie*, pp. 129-135; Goldsmith, *Joliot-Curie*, pp. 185-190; Caute, *Communism and the French Intellectuals*, pp. 188-190.

24. Audax Minor, "Letter from Paris," *New Yorker*, May 7, 1949, p. 84.

25. See e.g. "Une Famille de savants," *Annales Politiques et Littéraires* 106, no. 2544 (1935):609.

26. Opening discourse, Congrès Mondial des Partisans de la Paix, Paris, in *Journal du Congrès Mondial de la Paix*, no. 27, Apr. 21, 1949 (copy in RI).

27. *Le Monde*, Apr. 25, 26, 1949.

28. *L'Aurore*, Apr. 21, 22, Dec. 5, 1949.

29. Mar. 21, 1948, CEA.40, RI; Apr. 26, 1949, F5, RI; Brayance, *Anatomie du PCF*, p. 30; Williams, *Politics in Postwar France*, p. 48n10; Bertrand Poirot-Delpech, "Les Etudiants sont en quête de regroupements politiques," *Le Monde*, Nov. 17, 1956. One of the four cells at Châtillon was the cell "des Chalets," suggesting a residential-location cell among the non-CEA employees in the chalets built to house the workers.

30. OHI Goldschmidt, September 1976; Scheinman, *Atomic Energy Policy in France*, pp. 42-43; OHI Kowarski by Weiner, May 1970; OHI Kowarski, August 1976; Goldschmidt, *Rivalités atomiques*, p. 190; OHI Guéron, April 1978.

31. Comité Scientifique, Minutes, no. 130, Jan. 11, 1949, CEA.50, RI, pp. 12-13.

32. R. G. Arneson to Secretary of State, May 12, 1949, U. S. State Dept. Archives; Dautry to Premier, Apr. 5, 1949, CEA.40, RI; Vincent Auriol, *Journal du Septennat 1947-1954* (P: Colin, 1977), III, 169; F. Joliot in Comité Scientifique, Minutes, no. 150, Oct. 19, 1949, CEA.50, RI; *Le Monde*, Apr. 3, 1949.

33. Goldschmidt, *Rivalités atomiques*, p. 192; Hunebelle, "Une Epopée de l'après-guerre," p. 112.

34. France, Présidence du Conseil, CEA, *Rapport d'activité, 1946-1950*. Figures do not include the effects of a swift inflation, higher during the early years. About a third was spent on building the center at Saclay and a quarter on prospecting and mining. Gibert, "Le CEA à 5 ans."

35. Comité Scientifique, Minutes, no. 155, Feb. 7, 1950, CEA.50, RI.

36. Sous-comité de la Pile de Châtillon, Minutes, January-June 1949, CEA.53, RI; Charles Fisher, "Statistical Data on the Production and Use of Artificial Radioisotopes in France," in International Conference on the Peaceful Uses of Atomic Energy, Geneva, 1955, *Proceedings*, XIV, 21-23; Fisher, "Comment on fabrique de la radioactivité," *Atomes* no. 69 (December 1951): 399-402.

37. Comité Scientifique, Minutes, no. 131, Jan. 28, 1949, CEA.50, RI.

38. Kowarski, "Programme du Commissariat à l'Energie Atomique pour la

période 1949-1953," [c. December 1948], IV.B, LK; Goldschmidt, "Les Perspectives actuelles," p. 16.

39. Kowarski, "Après Zoé la deuxième étape"; Kowarski, "La Pile de Saclay," La Documentation Française, *Cahiers Français d'Information* no. 251 (May 1, 1954): 13-17; Jacques Yvon, "The Saclay Reactor: Two Years of Experience in the Use of a Compressed Gas as a Heat Transfer Agent," in International Conference on the Peaceful Uses of Atomic Energy, Geneva, 1955, *Proceedings*, II, 337-350.

40. Charles Eichner, Goldschmidt, and Paul Vertès, "L'Elaboration de l'uranium métallique à l'usine du Bouchet," *Bulletin de la Société Chimique de France*, 5th ser., 18 (1951): 140-142; "Rapport sur l'activité des Services de Physique et Chimie . . . 1948," CEA.50, RI.

41. Pierre Regnaut in *Echos du CEA* special no. (October 1965): 35; "French Atomic Scientists Report," pp. 300-302.

42. Vergne in *Echos du CEA* special no. (October 1965): 38.

18. POLITICS AND PLUTONIUM

1. OHI Goldschmidt, September 1976. See also Goldschmidt, *Rivalités atomiques*, p. 187; OHI Guéron, April 1978.

2. OHI Kowarski by Weiner, May 1970.

3. Union Française Universitaire, *La Guerre froide de M. Yvon Delbos contre l'Université Française: L'Affaire Georges Teissier* (P: Union Française Universitaire, [1950]; "How Russia Got U.S. Secrets: 10,000 Spies in High Places," *US News and World Report* 28, no. 7 (Feb. 17, 1950):11-13; *L'Humanité*, Feb. 7, 1950. Because the French press often closely reflected the views of ministers, the French tended to read the U.S. press as if it functioned in the same way.

4. U.S. Central Intelligence Agency, *National Intelligence Survey: France* (1952), pp. 73-3, 73-9, CIA Archives; U.S. Dept. of State, *Foreign Relations of the United States: 1949*, I, 466, 488, 626.

5. See G.IV and G.IX, microfilm edition of André Marty papers; Pickles, *French Politics*, pp. 121-125.

6. Jacques Fauvet, *Le Monde*, Apr. 4, 5, 1950; *L'Humanité*, Apr. 6, 1950.

7. OHI Perrin, November 1977. See also OHI Kowarski by Weiner, May 1970.

8. Pickles, *French Politics*, p. 252; OHI Kowarski by Weiner, May 1970.

9. *Le Monde*, Apr. 6, 29, 1950. See also clippings, RI; Jean Legendre, Interpellation, *J.O., Débats, Assemblée*, Apr. 25, 1950, p. 2806, May 9, 1950, pp. 3378-3379.

10. OHI Goldschmidt, September 1976; Biquard, *Joliot-Curie*, pp. 109-110.

11. Orcel, "Frédéric Joliot-Curie," p. 58; *J.O., Débats, Assemblée*, May 9, 1950, pp. 3383, 3380; Scheinman, *Atomic Energy Policy in France*, pp. 45-46.

12. Goldschmidt, *Rivalités atomiques*, p. 189; Kowarski, "Pouvons-nous construire de grands accélérateurs?" Report to F. Perrin, December 1950, LK.

13. Copies of letters, Feb. 24, 1951, II.B.2, Mar. 6, 19, 1951, II.A.4, LK; Schienman, *Atomic Energy Policy in France*, pp. 46-53, 70-74; Jacques Fauvet, "L'Epuration administrative anticommuniste," *Le Monde*, Jan. 5, 1951.

14. OHI Kowarski by Weiner, May 1970. For the second reactor, see *Atomes* no. 85 (April 1953).

15. Elsasser, *Memoirs of a Physicist*, p. 165.

16. Scheinman, *Atomic Energy Policy in France*, pp. 49-57, 97; Goldschmidt, *Rivalités atomiques*, pp. 188-192; OHI F. Perrin, November 1977.

17. Lucius and Allier to author, Feb. 14, 1978.

18. "Comité Scientifique (Principales Décisions Prises)," IV.B, LK.

19. F. Perrin, "Vingt ans après," p. 42; OHI Kowarski, August 1976; Allier, "Les Premières piles," p. 345.

20. Mendl, *Deterrence and Persuasion*; Ailleret, *L'Aventure atomique française*, pp. 65, 103; Kohl, *French Nuclear Diplomacy*, ch. 1.

21. Perrin in France, Présidence du Conseil, CEA, *Rapport d'activité, 1946-1950*, p. 21; OHI Perrin, November 1977. See also Mendl, *Deterrence and Persuasion*, p. 137.

22. Scheinman, *Atomic Energy Policy in France*, pp. 63-68, 85-87, 95-96; Nicholas Vichney, "La Bombe française aura exigé dix ans d'études et d'efforts," *Le Monde*, Dec. 31, 1959; Goldschmidt, *Rivalités atomiques*, pp. 193-194; OHI Kowarski by Weiner, May 1970; OHI Kowarski, May 1974, August 1976; "Réunion d'études du 13 Septembre 1951 à Gif," LK; F. Perrin, "Eléments du programme du CEA pour la 2e. période quinquenniale (II)," Oct. 6, 1951, LK; OHI Guéron, April 1978.

23. J. O., *Débats, Assemblée*, July 4, 1952, p. 3461; Scheinman, *Atomic Energy Policy in France*, pp. 75-79; Puget, "Problèmes administratifs," p. 7.

24. U.S. CIA, *National Intelligence Survey: France* (1952), p. 73-20, CIA Archives.

25. P. Chambadal and M. Pascal, "Recovery of the Energy Produced in Air Cooled Graphite Reactor G1," in International Conference on the Peaceful Uses of Atomic Energy, Geneva, 1955, *Proceedings*, III, 81-83.

26. OHI Perrin, November 1977; Goldschmidt, *Rivalités atomiques*, p. 195; Yvon, "Les Piles à graphite," *Echos du CEA* special no. (October 1965): 25; Scheinman, *Atomic Energy Policy in France*, pp. 62-63, 101-102; Hunebelle, "Une Epopée de l'après-guerre," p. 157; Taranger, "Les Rapports de l'industrie avec le CEA," p. 1.

27. See Bupp and Derian, *Light Water*.

28. OHI Goldschmidt, September 1976; OHI Perrin, November 1977.

29. F. Perrin, "Vingt ans après," p. 45.

30. Eggleston, *Canada's Nuclear Story*. Efficiency: *Reviews of Modern Physics* 50 (Jan. 1978): S143; J.A.L. Robertson, "The CANDU Reactor System: An Appropriate Technology," *Science* 199 (1978): 657-664.

31. Gowing, *Independence and Deterrence*.

32. F. Perrin, "Vingt ans après," p. 46.

33. See Gilpin, "Science, Technology, and French Independence," in Long and Wright, eds., *Science Policies of Industrial Nations*, pp. 110-132.

34. F. Joliot in Biquard, *Joliot-Curie*, pp. 141-142; de Broglie, "Notice sur . . . Joliot," p. 248.

AFTERWORD

1. Ellul, *Technological Society*, p. 99, quoting Jacques Soustelle.

2. For scientists' psychology, see Anne Roe, *Psychological Monographs* 65, no. 14, 67, no. 2 (1953); Roe, *Genetic Psychology Monographs* 43 (1951): 121-239; Lewis M. Terman, "Scientists and Nonscientists in a Group of 800 Gifted Men," *Psychological Monographs* 68, no. 7 (1954); Roe, "The Psychology of Scientists," in Karl Hill, *The Management of Scientists* (Boston: Beacon, 1964), pp. 49-71; Holton, *The Scientific Imagination*, pp. 229-252. I implicitly use Jung's personality theory, or at least that part of it dealing with suppression of feeling. Carl Jung, *Collected Works*, vol. 6 (Princeton: Princeton University Press, 1971), ch. 10; Jung et al., *Man and His Symbols* (Garden City, N.Y.: Doubleday, 1964), pp. 58-61.

3. See e.g. Joseph Campbell, *The Masks of God*, vol. 1, *Primitive Mythology* (New York: Viking, 1970), p. 281 and passim; Weston LaBarre, *The Ghost Dance:*

Origins of Religion (New York: Dell, 1972), pp. 486-488 and passim; Giorgio de Santillana, *The Origins of Scientific Thought* (Chicago: University of Chicago Press, 1961), pp. 55-57; Mircea Eliade, *The Forge and the Crucible: The Origins and Structure of Alchemy*, trans. Stephen Corrin (New York: Harper & Row, 1962), 88n; Joseph Needham, *Science and Civilization in China*, vol. 2 with Wang Ling, *History of Scientific Thought* (Cambridge, Eng.: Cambridge University Press, 1956), ch. 10. For features of shamanism not generally shared by scientists, see Eliade, *Shamanism: Archaic Techniques of Ecstasy* (Princeton: Princeton University Press, 1964). See also Joseph Ben-David, *The Scientist's Role in Society: A Comparative Study* (Englewood Cliffs, N.J.: Prentice-Hall, 1971), pp. 180-185.

4. John K. Galbraith, *The New Industrial State*, rev. ed. (Boston: Houghton Mifflin, 1971).

5. See Theodore Roszak, *Where the Wasteland Ends: Politics and Transcendence in Postindustrial Society* (Garden City, N.Y.: Doubleday, 1972); Robert M. Pirsig, *Zen and the Art of Motorcycle Maintenance* (New York: William Morrow, 1974); Holton, *The Scientific Imagination*, pp. 84-110.

Index